CARBON NANOTUBES AND NANOPARTICLES

Current and Potential Applications

AAP Research Notes on Nanoscience & Nanotechnology

CARBON NANOTUBES AND NANOPARTICLES

Current and Potential Applications

Edited by

Alexander V. Vakhrushev, DSc

Vladimir I. Kodolov, DSc

A. K. Haghi, PhD

Suresh C. Ameta, PhD

Apple Academic Press Inc.	Apple Academic Press Inc.
3333 Mistwell Crescent	1265 Goldenrod Circle NE
Oakville, ON L6L 0A2	Palm Bay, Florida 32905
Canada USA	USA

First issued in paperback 2021

Exclusive worldwide distribution by CRC Press, a member of Taylor & Francis Group

ISBN 13: 978-1-77463-413-4 (pbk)
ISBN 13: 978-1-77188-734-2 (hbk)

CIP data on file with Canada Library and Archives

Library of Congress Cataloging-in-Publication Data

Names: Vakhrushev, Alexander V., editor. | Kodolov, Vladimir I. (Vladimir Ivanovich), editor. | Haghi, A. K., editor. | Ameta, Suresh C., editor.

Title: Carbon nanotubes and nanoparticles : current and potential applications / editors, Alexander V. Vakhrushev, Vladimir I. Kodolov, A.K. Haghi, Suresh C. Ameta.

Description: Toronto; New Jersey : Apple Academic Press, 2019. | Includes bibliographical references and index.

Identifiers: LCCN 2019006193 (print) | LCCN 2019006469 (ebook) | ISBN 9780429463877 (ebook) | ISBN 9781771887342 (hardcover : alk. paper)

Subjects: | MESH: Nanotubes, Carbon | Nanoparticles

Classification: LCC RS201.N35 (ebook) | LCC RS201.N35 (print) | NLM QT 36.5 | DDC 610.28--dc23

LC record available at https://lccn.loc.gov/2019006193

Apple Academic Press also publishes its books in a variety of electronic formats. Some content that appears in print may not be available in electronic format. For information about Apple Academic Press products, visit our website at **www.appleacademicpress.com** and the CRC Press website at **www.crcpress.com**

ABOUT THE EDITORS

Alexander V. Vakhrushev, DSc

Alexander V. Vakhrushev, DSc, is a Professor at the M. T. Kalashnikov Izhevsk State Technical University in Izhevsk, Russia, where he teaches theory, calculating, and design of nano- and microsystems. He is also the Chief Researcher of the Department of Information-Measuring Systems of the Institute of Mechanics of the Ural Branch of the Russian Academy of Sciences and Head of the Department of Nanotechnology and Microsystems of Kalashnikov Izhevsk State Technical University. He is a corresponding member of the Russian Engineering Academy. He has over 400 publications to his name, including monographs, articles, reports, reviews, and patents. He has received several awards, including an academician A. F. Sidorov Prize from the Ural Division of the Russian Academy of Sciences for significant contribution to the creation of the theoretical fundamentals of physical processes taking place in multilevel nanosystems and honorable scientist of the Udmurt Republic. He is currently a member of editorial board of several journals, including *Computational Continuum Mechanics, Chemical Physics and Mesoscopia,* and *Nanobuild.* His research interests include multiscale mathematical modeling of physical–chemical processes into the nanohetero systems at nano-, micro-, and macro-levels; static and dynamic interaction of nanoelements; and basic laws relating the structure and macro characteristics of nanohetero structures.

Vladimir I. Kodolov, DSc

Vladimir I. Kodolov is Professor and Head of the Department of Chemistry and Chemical Technology at M. T. Kalashnikov Izhevsk State Technical University in Izhevsk, Russia, as well as Chief of Basic Research at the High Educational Center of Chemical Physics and Mesoscopy at the Udmurt Scientific Center, Ural Division at the Russian Academy of Sciences. He is also the Scientific Head of Innovation Center at the Izhevsk Electromechanical Plant in Izhevsk, Russia.

He is Vice Editor-in-Chief of the Russian journal *Chemical Physics and Mesoscopy* and is a member of the editorial boards of several Russian journals. He is the Honorable Professor of the M. T. Kalashnikov Izhevsk State Technical University, Honored Scientist of the Udmurt Republic, Honored Scientific Worker of the Russian Federation, Honorary Worker of Russian Education, and also Honorable Academician of the International Academic Society.

A. K. Haghi, PhD

A. K. Haghi, PhD, is the author and editor of 165 books, as well as over 1000 published papers in various journals and conference proceedings. Dr. Haghi has received several grants, consulted for a number of major corporations, and is a frequent speaker to national and international audiences. Since 1983, he served as professor at several universities. He is currently Editor-in-Chief of the *International Journal of Chemoinformatics and Chemical Engineering* and *Polymers Research Journal* and on the editorial boards of many international journals. He is also a member of the Canadian Research and Development Center of Sciences and Cultures (CRDCSC), Montreal, Quebec, Canada. He holds a BSc in urban and environmental engineering from the University of North Carolina (USA), an MSc in mechanical engineering from North Carolina A&T State University (USA), a DEA in applied mechanics, acoustics and materials from the Université de Technologie de Compiègne (France), and a PhD in engineering sciences from Université de Franche-Comté (France).

Suresh C. Ameta, PhD

Suresh C. Ameta, PhD, is currently Dean, Faculty of Science at PAHER University, Udaipur, India. He has served as Professor and Head of the Department of Chemistry at North Gujarat University Patan and at M. L. Sukhadia University, Udaipur, and as Head of the Department of Polymer Science. He also served as Dean of Postgraduate Studies. Prof. Ameta has held the position of President of the Indian Chemical Society, Kolkata and is now a life-long Vice President. He was awarded a number of prestigious awards during his career such as national prizes twice for writing chemistry books in Hindi. He also received the Prof. M. N. Desai Award (2004), the Prof. W. U. Malik Award (2008), the National Teacher Award (2011), the Prof. G. V. Bakore Award (2007), a Life-time Achievement Award by the Indian Chemical Society (2011) as well as the Indian Council of Chemist (2015), etc. He has successfully guided 81 PhD students. Having more than 350 research publications to his credit in journals of national and international repute, he is also the author of many undergraduate- and postgraduate-level books. He has published three books with Apple Academic Press: *Chemical Applications of Symmetry and Group Theory*, *Microwave-Assisted Organic Synthesis*, and *Green Chemistry: Fundamentals and Applications* and two with Taylor and Francis: *Solar Energy Conversion and Storage and Photocatalysis*. He has also written chapters in books published by several other international publishers. Prof. Ameta has delivered lectures and chaired sessions at national conferences and is a reviewer of number of international journals. In addition, he has completed five major research projects from different funding agencies, such as DST, UGC, CSIR, and Ministry of Energy, Govt. of India.

ABOUT THE AAP RESEARCH NOTES ON NANOSCIENCE & NANOTECHNOLOGY BOOK SERIES

AAP Research Notes on Nanoscience & Nanotechnology reports on research development in the field of nanoscience and nanotechnology for academic institutes and industrial sectors interested in advanced research.

Books in the AAP Research Notes on Nanoscience & Nanotechnology Book Series:

Nanostructure, Nanosystems, and Nanostructured Materials: Theory, Production, and Development
Editors: P. M. Sivakumar, PhD, Vladimir I. Kodolov, DSc, Gennady E. Zaikov, DSc, A. K. Haghi, PhD

Nanostructures, Nanomaterials, and Nanotechnologies to Nanoindustry
Editors: Vladimir I. Kodolov, DSc, Gennady E. Zaikov, DSc, and A. K. Haghi, PhD

Foundations of Nanotechnology: Volume 1: Pore Size in Carbon-Based Nano-Adsorbents
A. K. Haghi, PhD, Sabu Thomas, PhD, and Moein MehdiPour MirMahaleh

Foundations of Nanotechnology: Volume 2: Nanoelements Formation and Interaction
Sabu Thomas, PhD, Saeedeh Rafiei, Shima Maghsoodlou, and Arezo Afzali

Foundations of Nanotechnology: Volume 3: Mechanics of Carbon Nanotubes
Saeedeh Rafiei

Engineered Carbon Nanotubes and Nanofibrous Material: Integrating Theory and Technique
Editors: A. K. Haghi, PhD, Praveen K. M., and Sabu Thomas, PhD

Carbon Nanotubes and Nanoparticles: Current and Potential Applications
Editors: Alexander V. Vakhrushev, DSc, Vladimir I. Kodolov, DSc, A. K. Haghi, PhD, and Suresh C. Ameta, PhD

Advances in Nanotechnology and the Environmental Sciences: Applications, Innovations, and Visions for the Future
Editors: Alexander V. Vakhrushev, DSc, Suresh C. Ameta, PhD, Heru Susanto, PhD, and A. K. Haghi, PhD

Chemical Nanoscience and Nanotechnology: New Materials and Modern Techniques
Editors: Francisco Torrens, PhD, A. K. Haghi, PhD, and Tanmoy Chakraborty, PhD

CONTENTS

CONTRIBUTORS

Rakshit Ameta
Department of Chemistry, J.R.N. Rajasthan Vidyapeeth, Udaipur 313003, Rajasthan, India

Suresh C. Ameta
Department of Chemistry, PAHER University, Udaipur 313003, Rajasthan, India

Jayesh Bhatt
Department of Chemistry, PAHER University, Udaipur 313003, Rajasthan, India

Tanmoy Chakraborty
Department of Chemistry, Manipal University Jaipur, Jaipur 303007, India

Shailesh S. Chalikwar
Department of Pharmaceutics, R.C. Patel Institute of Pharmaceutical Education and Research, Shirpur, Dhule, Maharashtra, India

R. Narayana Charyulu
Department of Pharmaceutics, NGSMIPS, Mangalore, Karnataka, India

I. A. Cherenkov
Department of Medical Biology, Izhevsk State Medical Academy, Izhevsk, Russia

N. N. Chuckova
Department of Medical Biology, Izhevsk State Medical Academy, Izhevsk, Russia

N. V. Cormilina
Department of Medical Biology, Izhevsk State Medical Academy, Izhevsk, Russia

Pankaj V. Dangre
Department of Pharmaceutics, R.C. Patel Institute of Pharmaceutical Education and Research, Shirpur, Dhule, Maharashtra, India

Kakoli Dasgupta
Department of Chemistry, Birla Institute of Technology, Mesra, Ranchi 835215, Jharkhand, India

Barnali Dasgupta Ghosh
Department of Chemistry, Birla Institute of Technology, Mesra, Ranchi 835215, Jharkhand, India

Anujit Ghosal
Plant NanoBiotechnology Laboratory, School of Biotechnology, Jawaharlal Nehru University, New Delhi 110067, India
School of Life Science, Beijing Institute of Technology, Beijing, China

Sanjay Jain
Department of Pharmacognosy, Indore Institute of Pharmacy, Indore, Madhya Pradesh, India

Neha Kapoor
Department of Chemistry, PAHER University, Udaipur 313003, Rajasthan, India

Sonia Khanna
Department of Chemistry, School of Basic Science and Research, Sharda University, Greater Noida, India

Vladimir I. Kodolov
Basic Research—High Educational Center of Chemical Physics and Mesoscopy, Udmurt Scientific Centre, Ural Division, RAS, Izhevsk, Russia
Kalashnikov Izhevsk State Technical University, Izhevsk, Russia

A. A. Kopylova
Basic Research—High Educational Center of Chemical Physics and Mesoscopy, Udmurt Scientific Centre, Ural Division, RAS, Izhevsk, Russia
Kalashnikov Izhevsk State Technical University, Izhevsk, Russia

Anuradha Kumari
Department of Chemistry, Birla Institute of Technology, Mesra, Ranchi 835215, Jharkhand, India

A. A. Lapin
Kazan State Energy University, Kazan, Russia

T. M. Makhneva
Basic Research—High Educational Center of Chemical Physics and Mesoscopy, Udmurt Scientific Centre, Ural Division, RAS, Izhevsk, Russia
Institute of Mechanics, Udmurt Scientific Centre, Ural Division, RAS, Izhevsk, Russia

V. M. Merzljakova
Izhevsk State Agricultural Academy, Izhevsk, Russia

R. V. Mustakimov
Basic Research—High Educational Center of Chemical Physics and Mesoscopy, Udmurt Scientific Centre, Ural Division, RAS, Izhevsk, Russia
Scientific Innovation Centre, Izhevsk Electromechanical Plant—"KUPOL," Izhevsk, Russia

Sukanchan Palit
Department of Chemical Engineering, University of Petroleum and Energy Studies, Energy Acres, Post Office Bidholi via Premnagar, Dehradun 248007, Uttarakhand, India
43, Judges Bagan, Post Office Haridevpur, Kolkata 700082, India

P. K. Panda
Materials Science Division, CSIR-National Aerospace Laboratory, Bengaluru 560017, India

Bhavya Pathak
Department of Chemistry, PAHER University, Udaipur 313003, Rajasthan, India

S. A. Pigalev
Basic Research—High Educational Center of Chemical Physics and Mesoscopy, Udmurt Scientific Centre, Ural Division, RAS, Izhevsk, Russia
Kalashnikov Izhevsk State Technical University, Izhevsk, Russia

Prabhat Ranjan
Department of Mechatronics Engineering, Manipal University Jaipur, Jaipur 303007, India

Neha Kanwar Rawat
Materials Science Division, CSIR-National Aerospace Laboratory, Bengaluru 560017, India

I. N. Shabanova
Basic Research—High Educational Center of Chemical Physics and Mesoscopy, Udmurt Scientific Centre, Ural Division, RAS, Izhevsk, Russia
Kalashnikov Izhevsk State Technical University, Izhevsk, Russia
Physicotechnical Institute, Udmurt Scientific Centre, Ural Division, RAS, Izhevsk, Russia

Shalini
Department of Chemistry, Alankar P.G. Girls College, Jaipur 302012, India
Department of Chemistry, Manipal University Jaipur, Jaipur 303007, India

Saumya Shalu
Department of Chemistry, Birla Institute of Technology, Mesra, Ranchi 835215, Jharkhand, India

I. A. Shestakov
Department of Mechanics of Nanostructures, Institute of Mechanics, Udmurt Federal Research Center, Ural Division, Russian Academy of Sciences, Izhevsk, Russia
Department of Nanotechnology and Microsystems, Kalashnikov Izhevsk State Technical University, Izhevsk, Russia

C. K. Sudhakar
Department of Pharmaceutics, School of Pharmaceutical Sciences, Lovely Institute of Technology (Pharmacy), Lovely Professional University, Jalandhar-Delhi G.T. Road, Phagwara 144411, Punjab, India

Vandana Suhag
Department of Chemistry, Manipal University Jaipur, Jaipur 303007, India

Hiteshi Tandon
Department of Chemistry, Manipal University Jaipur, Jaipur 303007, India
Department of Chemistry, Alankar P.G. Girls College, Jaipur 302012, India

N. S. Terebova
Basic Research—High Educational Center of Chemical Physics and Mesoscopy, Udmurt Scientific Centre, Ural Division, RAS, Izhevsk, Russia
Kalashnikov Izhevsk State Technical University, Izhevsk, Russia
Physicotechnical Institute, Udmurt Scientific Centre, Ural Division, RAS, Izhevsk, Russia

V. V. Trineeva
Basic Research—High Educational Center of Chemical Physics and Mesoscopy, Udmurt Scientific Centre, Ural Division, RAS, Izhevsk, Russia
Institute of Mechanics, Udmurt Scientific Centre, Ural Division, RAS, Izhevsk, Russia

Nitish Upadhyay
Department of Quality Assurance, Cipla Pharmaceutical Ltd., Indore, Madhya Pradesh, India

A. A. Vakhrushev
Department of Simulation and Modeling of Metallurgical Processes, Montan University, Leoben, Austria

Alexander V. Vakhrushev
Department of Mechanics of Nanostructures, Institute of Mechanics, Udmurt Federal Research Center, Ural Division, Russian Academy of Sciences, Izhevsk, Russia
Department of Nanotechnology and Microsystems, Kalashnikov Izhevsk State Technical University, Izhevsk, Russia

S. S. Vidrina
Department of Nanotechnology and Microsystems, Kalashnikov Izhevsk State Technical University, Izhevsk, Russia

ABBREVIATIONS

AC	activated carbon
AD	Alzheimer's disease
AD	arc discharge
AFM	atomic force microscope
AOP	advanced oxidation processes
AOT	aerosol OT
APP	ammonium polyphosphate
BCE	Biopharmaceutical Classification System
BRB	berberine
CNH	carbon nanohorn
CNT	carbon nanotubes
COD	chemical oxygen demand
CPT	camptothecin
CQAs	critical quality attributes
CVD	chemical vapor deposition
DC	direct current
DEP	diethyl phthalate
DETA	diethylenetriamine
DFT	density functional theory
DNP	2,4-dinitrophenol
DNPC	2, 6-dinitro-p-cresol
DOX	doxorubicin
ECoG	electrocorticography
ECPs	electrically conducting polymers
EMI	electromagnetic interference
ENPs	engineered nanomaterial
EPO	erythropoietin
EPR	electron paramagnetic resonance
FA	folic acid
f-CNTs	functionalized carbon nanotubes
FI	fluorescein isothiocyanate
FMWCNT	functionalized multiwalled carbon nanotubes
GIT	gastrointestinal tract

GNR	graphene nanoribbon
HiPco	high-pressure carbon monoxide
HLB	hydrophilic and lipophilic balance
LDA	local density approximation
LFCS	Lipid Formulation Classification System
MD	molecular dynamic
MS	mild steel
MWCNTols	hydroxylated multiwall carbon nanotube
MWCNTs	multiwalled carbon nanotubes
MWNT	multiwall carbon nanotubes
NaDDBS	sodium dodecyl benzene sulfonate
NC	nanocomposite
NCE	new chemical entities
NE	nanoelements
NOM	natural organic matter
NSS	nephron-sparing surgery
NT	nanotechnology
NTs	nanotubes
OLED	organic light-emitting diode
o/w	oil-in-water
PAMAM	polyamidoamine
PANI	polyaniline
PANI-ES	polyaniline in its emeraldine salt form
PDE	photocatalytic decomposition efficiency
PEDOT	poly(3,4-ethylenedioxythiophene)
PES	polyether sulfone
PPy	polypyrrole
PTh	polythiophene
PVA	poly(vinyl acetate)
PVP	polyvinyl pyrrolidone
PW	plane wave
PWSD	poorly water-soluble drug
RhB	rhodamine B
RR2	reactive red 2
SDBS	sodium dodecylbenzenesulfonate
SDS	sodium dodecyl sulfate
SE	self-emulsifying
SMEDDS	self-microemulsifying drug delivery system
SNEDDS	self-nanoemulsifying drug delivery system

SWCNT	single-walled carbon nanotubes
STM	scanning tunneling microscope
SWNT	single-walled carbon nanotubes
TDDS	transdermal drug delivery system
TEM	transition electron microscopy
TiV	V-doped titania
TNB	titanium n-butoxide
TNT	titanate nanotubes
TOC	total organic carbon
TPU	thermoplastic polyurethane
VLD	visible-light-driven
XPS	X-ray photoelectron spectroscopy
XRD	X-ray diffraction

PREFACE

Carbon nanotubes (CNTs) and nanoparticles have gained attention in recent times due to their extraordinary properties of strength, flexibility, sensors, conduction, etc. They have been observed in various forms, each having its characteristic behavior.

CNTs and nanoparticles have been extensively studied making way for basic understanding and potential for various applications. This volume has discussed several applications of CNTs and nanoparticles.

In the first chapter, it is shown that CNTs make unique skeleton formulation that attributes to the amalgamation of superlative mechanical, thermal, and electronic properties. It is also shown that CNTs are relishing increasing popularity as building blocks for novel drug delivery systems as well as for bioimaging and biosensing.

In the second chapter, a detailed analysis of CNTs based on DFT techniques has been presented. The chapter has been divided into four sections; in the first section, emergence of CNT, whereas in the second section, structure and their properties have been discussed. In third section, detailed investigation of CNTs by using density functional theory (DFT) method have been elaborated, whereas in fourth section, recent works based on CNTs and their applications in numerous fields have been reported.

CNT composites as photocatalytic materials are discussed in detail in Chapter 3.

The aim of Chapter 4 is devoted to the study of methods for extracting cholesterol from its solutions using sorbents based on CNTs. Numerical modeling of processes was carried out with the purpose of revealing mechanisms of interaction of nanotubes with cholesterol, as well as with high- and low-density lipoproteins. Numerical modeling was carried out using molecular dynamics. An experiment was performed to adsorb cholesterol from its solution using powdered nanotubes. It qualitatively confirmed the fact of adsorption of cholesterol by CNTs from its solutions.

A concise review on the synthesis techniques, properties, and applications of CNTs is presented in Chapter 5.

The objective of Chapter 6 is to show that conducting polymers/CNT nanocomposites have emerged as one of the most fascinating polymers in the field of material science and engineering. Their marvelous resourceful properties make them useful in every field of research. They find application in varying arenas such as supercapacitors, sensors, electronic materials, drug release, biosensors, tissue engineering, as well as stimuli responsive and biomimetic polymeric materials. The chapter discusses role of conducting polymers and CNTs with their vast applications as nanocomposites and their significant contribution in materials science and engineering as well as their prospects.

In Chapter 7, CNTs and their applications in chemical engineering are discussed in detail.

Chapter 8 deals with the safe design of the hydrogen storage tank, which includes a set of cylindrical metal containers made up of stainless steel filled with CNTs; a cooling jacket made of liquid nitrogen in the likeness of a Dewar vessel into which metal containers with CNTs (SWNT) are dipped and filled with gaseous hydrogen. The analyses of the scheme of the experiment are presented and the results of the study of hydrogen storage at a liquid nitrogen temperature are shown. It is confirmed that hydrogen at the temperature of liquid nitrogen is retained by the SWNT in the accumulator tank, and at normal temperature, it is released.

In Chapter 9, a critical overview on CNTs and material science is developed.

The properties and applications of nanotubes are discussed in Chapter 10.

CNTs are one of the most sought topics after in the field of nanotechnology. CNTs are the cylindrical carbon molecules with novel properties of extraordinary strength, flexibility, sensing, conduction, etc. They have the potential to be used in electronics due to their unequalled electronic properties, and in medical and pharmacological fields due to their unique surface area and stiffness. They have added advantage of being a potential device for drug delivery.

The aim of Chapter 11 is to study the change in the electronic structure and magnetic properties of modified copper/carbon nanocomposites.

In Chapter 12, for the first time, the metal/carbon nanocomposites, which contain silicon, phosphorus, or sulphur, are produced within active media by means of mechanochemical modification. In this case, the change of element oxidation states as well as the increasing of metal (copper, nickel, and iron) atomic magnetic moment takes place. At the same time, above elements and functional groups with them appear in carbon shell of nanocomposites. These facts open new era for further investigations and development of metal/carbon nanocomposites application fields.

In Chapter 13, self-nanoemulsifying drug delivery system (SNEDDS): formulation development and quality attributes are discussed in detail.

The investigation of copper/carbon nanocomposite aqueous sols for application at the cultivation of lilies is presented in Chapter 14.

Metal/carbon nanocomposites and modified analogous: possible participation in vital processes are discussed in Chapter 15.

CHAPTER 1

CARBON NANOTUBES AS VERSATILE CARRIERS IN DRUG DELIVERY

C. K. SUDHAKAR[1*], NITISH UPADHYAY[2], SANJAY JAIN[3], and R. NARAYANA CHARYULU[4]

[1]*Department of Pharmaceutics, School of Pharmaceutical Sciences, Lovely Institute of Technology (Pharmacy), Lovely Professional University, Jalandhar-Delhi G.T. Road, Phagwara 144411, Punjab, India*

[2]*Department of Quality Assurance, Cipla Pharmaceutical Ltd., Indore, Madhya Pradesh, India*

[3]*Department of Pharmacognosy, Indore Institute of Pharmacy, Indore, Madhya Pradesh, India*

[4]*Department of Pharmaceutics, NGSMIPS, Mangalore, Karnataka, India*

**Corresponding author. E-mail: ckbhaipharma@gmail.com*

ABSTRACT

Carbon nanotubes (CNTs) are the graphite sheet rolled nanocargo delivery system used to deliver many drugs to specific area of body depending upon the targeted system and route of administration. Chemistry of CNTs depends on different method of preparation of the carbon nanotubes. CNTs have novel distinct properties that make them potentially useful in a wide variety of solicitations in nanotechnology in relation to medicine field. CNTs act as drug cargo or Trojan horse, where it can entrap the potent to toxic drug by slight modification in the structure of CNTs and it can deliver to various tissue of the body without harming the normal cells and tissue and targeting the perpetrator tissue. In this chapter, the current state of knowledge of CNTs in the field of medicine with their biomedical applications–in particular for different disease is discussed.

1.1 INTRODUCTION

Carbon nanotubes (CNTs) make unique skeleton formulation, which attributes to the amalgamation of superlative mechanical, thermal, and electronic properties. There are many nanotechnology formulations available such as dendrimers, liposomes, polymersomes, etc., but CNTs make it a way different from these nanotechnology formulations due to high loading capacity, cell penetration, and different functional groups for chemically drug loading. CNTs are relishing increasing popularity as building blocks for novel drug delivery systems as well as for bioimaging and biosensing.

1.2 CHEMISTRY OF CNTs

CNTs can be produced by different methods such as arc discharge method; laser ablation, plasma torch; catalyst-assisted chemical vapor deposition (CVD); ball milling; natural, incidental, and controlled flame environments; diffusion flame synthesis; electrolysis; use of solar energy; heat treatment of a polymer; low-temperature solid pyrolysis, etc.[68] The CNTs have diverse properties, one of the properties is specific surface escorted by adsorption site on the CNTs. CNTs' adsorption sites can be classified into (1) pore adsorption sites, (2) surface adsorption site, (3) groove adsorption sites, and (4) interstitial sites. CNTs produced by different methods can impact the adsorption sites of CNTs. CNTs developed by CVD method using catalyst only can produce surface adsorption site, groove adsorption sites, and interstitial adsorption sites. CNTs prepared by CVD method are not able to produce pore adsorption site using catalysis. CNTs with cap ends restrict the formation of pore adsorption inside the CNTs. Removing the cap of CNTs using mild or thermal treatment, can lead to pore or internal adsorption site. The drug molecule or a gaseous molecule adsorbed in pores of CNTs is bound much stronger than all other adsorption sites. Large-scale productions of CNTs can lead to bundles of CNTs which can lead different adsorption sites such as surface adsorption site, groove adsorption sites, and interstitial sites (Fig. 1.1). The adsorption site has different binding capacity such as π–π interaction, covalent bonding, and noncovalent bonding. In π–π interaction binding, aromatic group molecules interact with the π–π stacking of CNTs.[12] The covalent functionalization of CNT is narrow since it causes sp^3 hybridization, destruction of π-bond interaction; and loss of electrical conductivity properties on the carbon sites of CNTs. Covalent functionalization of CNT form stable chemical bond which can be done by oxidation, halogenation, amidation, thiolation, hydrogenation, etc.[36] The noncovalent functionalization of CNTs, done by van der Waal bonds, retains the structural integrity and network, causes no loss of electronic properties, and helps in wrapping of polymers on the surface of the CNTs.[36,61,77] Noncovalent functionalization of CNTs is done by adsorption of polymer, surfactant, nanoparticles, dendrimers, biomolecules, etc. Covalent functionalization is more preferred than noncovalent functionalization as they block the π-bond interaction, which allows the sustained release of therapeutic molecules.[36]

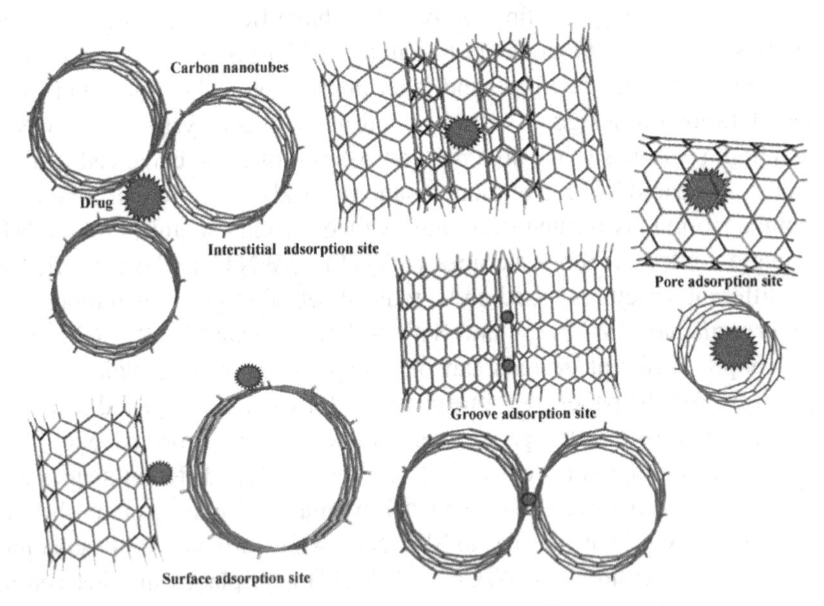

FIGURE 1.1 Different adsorption sites for gaseous and solid materials, which may be adsorption sites for drug moiety.

The groove sites are narrow troughs formed outside the bundles where two CNTs abut. The interstitial sites are channels between individual CNTs inside the bundle. Simulations showed that CNT interior has the highest binding energy for adsorbing molecules. The presence of the surrounding walls maximizes the attractive van der Waals interactions with the adsorbed molecule. On the other hand, the nanotube's external wall curves away from the adsorbed molecules, which implies that the adsorption energy must be smaller than that for the interior or for flat graphene. The adsorption energy for groove sites lies between the adsorption energies of external and internal sites.[7] Adsorption of molecules in the interstitial channels is a topic still under debate. Molecular simulations shed light on the ongoing controversy.

1.3 PHYSICOCHEMICAL PROPERTIES OF CNTs

CNT is a unique architecture, rolled-up or curling of two-dimensional honeycomb lattice structure of graphene sheet. CNTs can be classified into zigzag and armchair geometric structure.[20,70] CNTs are a hexagonal lattice of carbon sheet curled into cylindrical shape. CNTs can be single-layer sheet or multiple layers arranged cylindrically. CNTs have flexible physicochemical properties that allow the covalent and noncovalent bonding of drugs molecules, genes, proteins, peptides, and other pharmaceutical entities, and show coherent rationale design of novel nanodrug delivery system.[10,17,18] CNTs have a high specific surface area per unit weight, offering enhanced loading capacity compared to conventional nanomaterials of spherical shape.[62,90] Nanotubes form in two categories: multiwalled carbon nanotubes (MWCNT) and single-walled carbon nanotubes (SWNT).[1] SWNT can be classified in three different structures based on way the sheet of graphene is molded into cylinder: armchair SWCNT, chiral SWCNT, and zigzag SWCNT (Fig. 1.2). For example, different designs can be visualized by rolling plain sheet of paper; if we roll the paper from one corner, it forms one design and if we roll the paper from edge of the paper, then another design is formed. Similarly, if a sheet of graphite is rolled from its corner or edge, different shapes and designs of CNTs are developed. MWCNT are made of same graphite sheet of SWCNTs. MWCNTs are similar to SWCNTs, which are placed in each other to form multiple cylindrical SWCNTs; SWCNTs are placed in other central hollow SWCNTs in concentric cylinders way. The MWCNTs are classified into two different structures depending upon the arrangement of the rolling of graphite sheet, that is, "Russian doll"-like structure and parchment-like model. Russian doll-like structure is more preferred than parchment-like model; Russian doll-like structure is concentric cylinder of SWCNTs, whereas the parchment-like model is single sheet of graphite which is rolled around it forming parchment-like structure.[15] SWCNTS have low tensile strength than MWCNTs but are more stable nanostructures than MWCNTs as they have better defined structure of wall when compared to MWCNTs which has structure faultiness.[37] SWCNTs which are skinny and stiff have shown more penetrability in spherical bacteria than MWCNTs.[8] CNTs are superlative nanomaterials which have been used as polymer material in all the fields of bioscience, as electronic material in the field of engineering, and as nanocarrier in the pharmaceutical field. Their versatile physical properties such as high tensile strength, electronic properties, conductivity and surface area, and many more favorable properties make them unique from other materials.

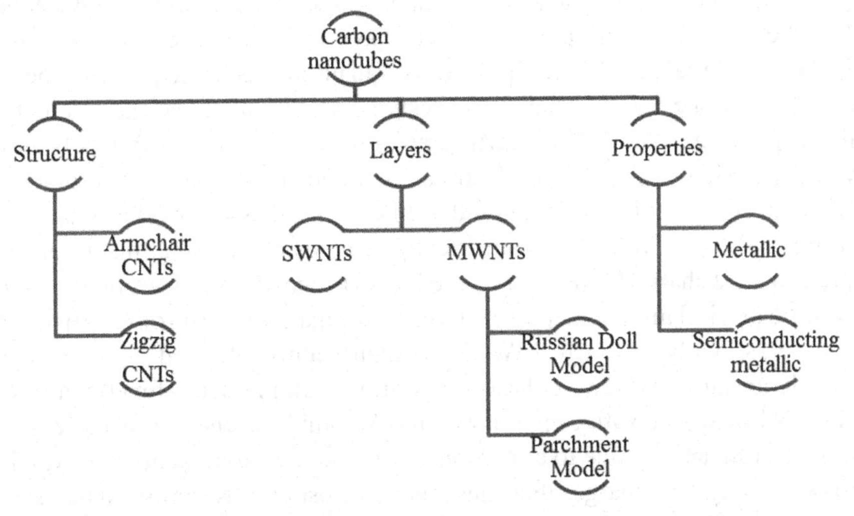

FIGURE 1.2 Classification of the CNTs based on structure, layers, and properties.

1.4 ELECTRICAL AND STRUCTURAL PROPERTIES OF CNTs

Electronic properties of CNTs are well known and they are important in different fields of sciences, which depend on the radius and rolling direction of graphite sheet.[44,48,52] CNTs have very good electronic properties; SWCNTs has shown well-defined system in footings of electronic properties than MWCNTs.[1] Depending on their sheet direction and diameter, CNTs may be classified into metallic and semiconducting categories. CNTs can be metallic or semiconducting reliant upon the structure. SWNTs with armchair-shaped structure sheet have metallic properties and with zigzag structure have both metallic and semiconducting properties. CNT vary their electrical conductance based on their location and environment as they are very sensitive to their surrounding and vary in their conductance if there is any change in electrostatic charges and covalent or noncovalent bonding of various molecules to CNTs molecules.[11,13,35] Kazemi-Beydokhti et al.[33] produced SWNTs–cisplatin complexes by three different methods: encapsulation, electrostatic interactions, and covalent binding of cisplatin and phospholipid PEG derivatives. DSPE–PEG2000 can improve the dispersibility of SWCNTs and the electrostatic interaction between SWCNTS platform and cisplatin molecules can be aided by DSPG phospholipid. The negative charge imparted by DSPG on SWCNTs is reason behind the cisplatin attraction and could retain more quantities of drugs on the external area of SWCNTs than the encapsulations method. CNTs have inimitable electrical properties that materialize as consequences of quantum mechanical occurrences at the nanoscale range. Their tube design, cylindrical shape, and helicity also explain the electrical properties, and different shapes of SWCNTs and MWCNTs show variation in electrical conductance of CNTs. The unique electrical conductivity of CNT could be hitched to advance novel mechanism of drug release.[63] MWCNT's electrical properties could affect macrophage integrity, and charged MWCNTs were less cytotoxic to the macrophages but own greater inflammatory prospective, as compared to control noncharged MWCNTs. Only charged MWCNTs significantly affected macrophage mitochondrial membrane polarity.[97,74] Cationic amino acids-functionalized MWCNTs impart positive charge which is favorable, as they show increased antibacterial activity due to attraction toward the bacterial membrane which possesses negative charges that cause the electrostatic adsorption on bacteria membrane. Functionalization with cationic amino acids with MWCNTs than SWCNTs, as they are lesser cytotoxicity than SWCNTs and could be a constructive methodology in the sterilization and disinfection industry.[89]

1.5 SPECIFIC SURFACE OF CNTs

External specific surface area of SWCNTs and MWCNTs is important in pharmaceutical field, and is calculated as a function of their features such as diameter of CNTs, number of concentric shells in MWCNTs, and number of nanotubes in a bundle.[55] The factors which impact the specific surface area of CNTs were number of concentric shells, diameter of the CNTs, impurities, and hydroxyl and carboxyl groups functionalization on the surface of CNTs.[4] The specific surface area of CNTs is a macroscopic factor which can be supportive to adjust the production conditions of CNTs.[55] Functionalized SWCNTs can offer higher ability of drug loading compared with the traditional liposomes and dendrimer drug carriers due to their extremely high specific surface area.[40] Diameter of CNTs is very important feature for specific surface area of CNTs; diameter of SWCNTs is in the range of 0.4–2 nm.[34] It was observed that during synthesis of the SWCNTs if the temperature is very high then the diameter of CNTs will be larger in size. The diameter of MWCNTs is larger than SWCNTs, and inner circles of CNTs have diameter of 1–3 nm and outer diameter of 2–100 nm. In contrast, MWCNTs are usually made from several cylindrical carbon layers with diameters in the range of 1–3 nm for the inner CNTs and 2–100 nm for the outer surface of CNTs.[3,96] Benefit of CNTs embraces of high drug-loading capacity and inherent firmness robust structure; all these features extend circulation time of CNTs in body which indirectly increases the bioavailability of drug moieties that are incorporated in CNTs.[37,100,102,105,106] These CNTs function with a larger inner volume to be used as the drug container, large aspect ratios for numerous functionalization attachments, and the ability to be readily taken up by the cell.[78]

1.6 SOLUBILITY OF CNTs

CNTs are versatile materials but they have some issues related to solubility. Solubility is an important factor in pharmaceutical field. Many researchers have tried to solve the solubility issue of CNTs; they have given many approaches such as wrapping of surfactant, functionalization of CNTs, polymer interaction with CNTs, and other novel carrier linkage such as dendrimers which enhance the solubility of CNTs. Among these functionalization, CNTs were more explored for drug delivery, misapplication, and protein drug delivery system. Functionalized CNTs have high specific surface area and good drug cargo delivery for therapeutic moiety as they are more soluble then the parent CNTs. Functionalization of CNTs with organic molecules or dressing the CNTs with specific molecules renders the solubility, enhances the dispersibility of the CNTs in water, and increases the conjugation or binding of drug toward the CNTs as more functional groups are available for drug binding and these functionalized CNTs–drug complex will be able to target and release the drug in the tissue.[19,22] Shen and coworker[69] prepared hydroxylated multiwall carbon nanotubes (MWCNTols) which were enfolded by polyvinyl pyrrolidone (PVP) by solution blending method, and MWCNTols–PVP has shown enhanced water solubility due to the water solubility properties of PVP; this additional solubility will upsurge the stability and dispersibility of MWCNTols in water (Fig. 1.3). The solubility of PVP–MWCNTols in aqueous phase was up to 2.34 mg/mL.[69] Increasing the solubility of CNTs do not impair the cells or macrophages as it was studied by Deng et al.[16] that water-soluble MWCNT do not cause any harm to the phagocytic activity of macrophages, instead they enhance the phagocytic activity of macrophages which play a vital role of defense action against the invading microorganisms. Sahoo et al.[95] revealed that polymer wrapping approach such as PVP wrapping on the CNTs containing camptothecin (CPT) shown enhanced solubility and stability (Fig. 1.4).

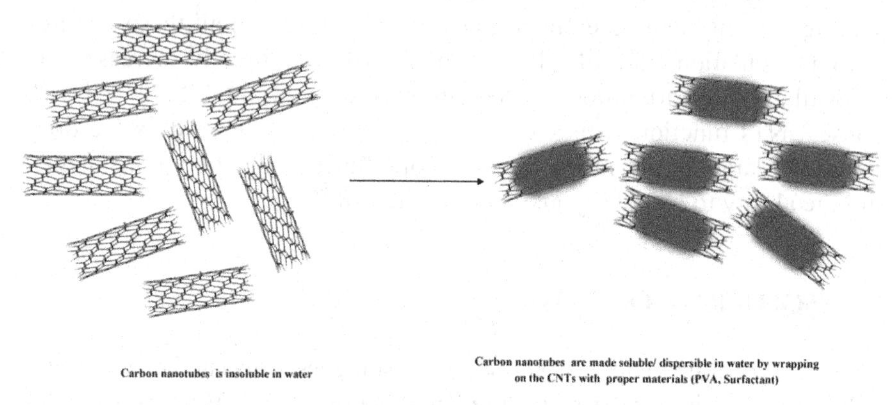

FIGURE 1.3 Enhancement of CNTs solubility by surfactant wrapping.

Polyvinylpyrrolidone (PVP)
Carbon nanotubes
PVP wrapping Carbon nanotubes

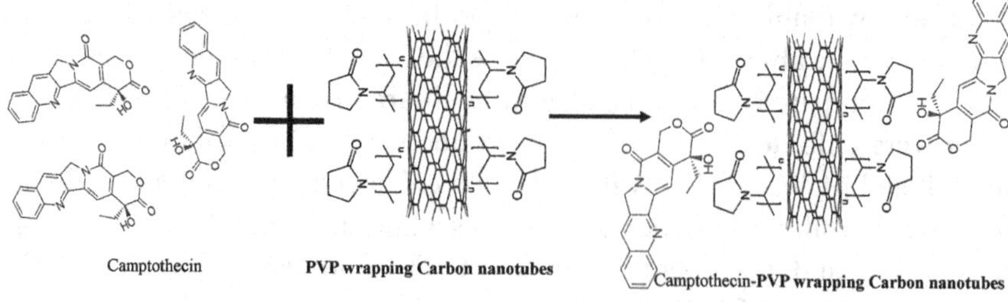

FIGURE 1.4 (See color insert.) Enhancement of solubility and stability of camptothecin by camptothecin–PVP wrapping CNT complexes.

1.7 CNTs ACTING AS HOMEOSTATIC

Nowacki and his coworker[94] studied that in breast-conserving therapy or nephron-sparing surgery (NSS), known as partial nephrectomy, two major problems persist: intraoperative hemorrhage and risk of local tumor relapse. CNT have numerous properties, one of them is homeostatic properties; in NSS of kidney cancer, there is always a chance of local tumor relapse due to excessive bleeding during the surgery. Combination of CNTs and cisplatin was innovative drug delivery method to prevent both local tumor relapse and bleeding in NSS surgery of the kidney cancer on the xenografted murine model. The CNTs will prevent the bleeding by acting as homeostatic and cisplatin as an oncostatic agent. Amplification of the procoagulant activity of MWCNTs was done by conjugation with carboxylation or amidation groups. Mechanistic studies demonstrate that MWCNTs enhance proliferation of the contact activation pathway (intrinsic) through a nonclassical mechanism strongly reliant on Christmas factor.[5] SWCNTs shortened the clotting time of platelet-poor plasma.[80] Meng and his coworker [46,47] studied that functionalization of CNTs with long carboxylated L-COOH and long aminated L-NH_2 groups persuaded a greater level of platelet activation than smaller carboxylated and small animated groups, correspondingly; the clot becomes soft and hard depending upon the functionalization group and the length of the group's chain. Long aminated CNTs amplified the clots' hard rigidity considerably, whereas long carboxylated and small animated CNTs make the clots softer and nonrigid in nature (Fig. 1.5).

1.8 ORNAMENTED CNTs WITH SURFACTANT

Lohan et al.[41] entail the systematic development of berberine (BRB)-loaded MWCNTs with polysorbate and phospholipid coatings for effective management of Alzheimer's disease (AD). They also demonstrated significant potential of polysorbate-/phospholipid-coated MWCNTs of BRB in holistic management of AD. Polysorbates offer more solubility to MWCNTs vis-à-vis the phospholipids.[42] CNTs are wrapped by the phospholipid surfactant along with mesoporous silica, the aperture or pore size can be varied by changing concentration ratio of phospholipids and surfactants. Sodium dodecylbenzenesulfonate (SDBS) is ionic surfactants wrapping around the CNTs and SDBS–CNTs has shown that dispersions are most stable as compared to other surfactants.[14,27,49,76] The ionic surfactant SDBS depends on two features, the geometry of CNTs such as outer surface area, inner diameter of CNTs, and governing the morphology of the surfactant masses which will cause the stabilization of CNTs in water.[14,29,49]

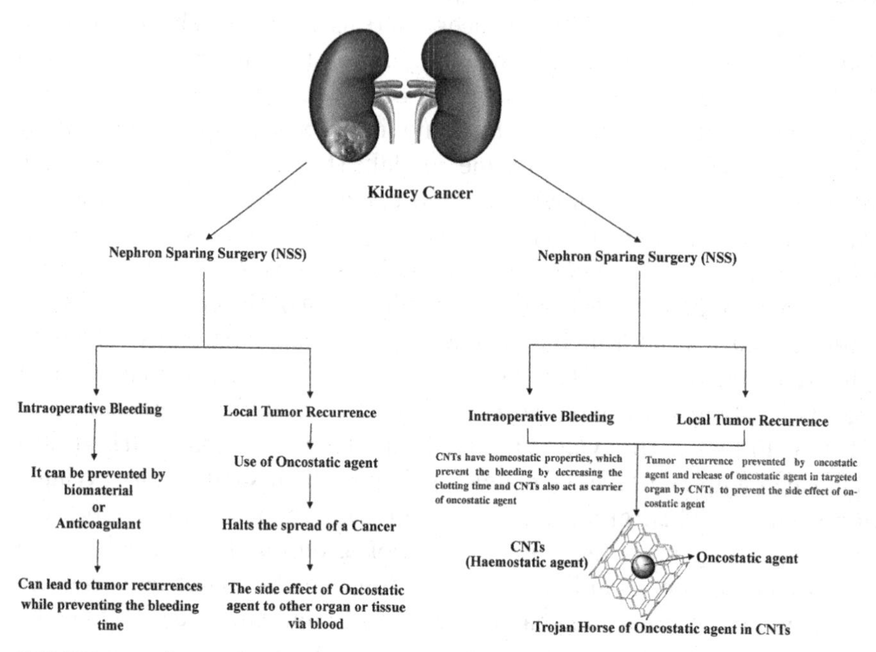

FIGURE 1.5 (See color insert.) Carbon nanotubes act as Trojan horse which carries oncostatic drug and CNTs act as homeostatic carbon nanotubes.

1.9 CNTs AS NANOCARGO DELIVERY SYSTEM IN CANCER

CNTs have been explored as a nanocargo delivery system[73] for chemotherapeutic agents[39] such as doxorubicin (DOX),[47] oxaliplatin,[104] mitoxantrone,[99] quercetin,[98] methotrexate,[101] cisplatin,[24] gemcitabine,[59] 10-hydroxycamptothecin,[85] paclitaxel,[25] docetaxel,[58] ruthenium complexes,[82] tamoxifen,[8] and si-RNA.[81] CNTs are used for lymphatic targeting in tumors, as the lymph and its capillaries are the major transport system for tumor to spread into different organs.[92] Xie et al.[86] validated in animals that nanocarrier such as liposomes drug delivery and polymer nanocarrier are good targeting cargo to the lymph, and release and hoards high amount of drug to lymph and surrounding lymphatics compared to traditional drug delivery. Hydrophilic MWCNTs ornamented with magnetite nanoparticles possesses all features characteristics which are required to target the lymphatic delivery system. Hydrophilic MWCNTs–magnetite have high affinity for lymph and lymphatic system and are able to release the majority amount of anticancer drug in lymph. Fabricated magnetite MWCNTs bearing gemcitabine deliver high amount of drug to lymph with aid of external magnetic field.[28,86] Anticancer agents were amalgamated into the apertures of *f*-MCNTs bearing a layer of magnetite nanoparticles on the internal hollow surface of the CNTs. In order to advance this system for more accuracy targeting, *f*-MCNTs–magnetite are noncovalently linked with folic acid (FA) which give more guidance to nanocarrier to target the tumor.[87,88,103]

Moreover, the CNTs are of cylindrical, elongated shape; the geometry plays an important role in lymph node targeting and retaining in the nodes. The nanovesicles such as liposomes which are mostly oval to spherical shape are not retained for longer time in lymph nodes, whereas the CNTs which are cylindrical in shape are retained for longer time than liposomes or spherical-shaped carriers.[71,87,88] CNTs–magnetite are not only used for the targeting the drug delivery to tumor but are also used as imaging process for diagnosis of cancer.[84] "Longboat delivery system" was developed by Dhar and coworkers, where SWCNTs are linked with anticancer drug (cisplatin) via FA derivative; long boat delivery sail to pool of body fluids and blood to reach its destination tumor cells, and via endocytosis, it enters into the cell and interacts with the nucleus of the cell. Zhang et al. engineered the SWCNTs with functionalizing with carboxylate group and wrapping or coating with polysaccharides material for efficient loading of chemotherapeutic agent DOX at neutral pH, and targeting and releasing drug to tumor site at lower pH.[91] G-5 dendrimers–MWCNTs which act as modified carrier of CNTs bearing DOX can be used as targeted and pH-responsive delivery of anticancer drug in tumor cell were explored.[83] G-5 dendrimers linked with fluorescein isothiocyanate (FI) as diagnostic agent and FA as targeting the cancer cell were linked to MWCNTs; drug is entrapped in the dendrimers. DOX-aminated dendrimer–MWCNTs–FI–FA complexes has shown effective anticancer efficiency, which is comparable to that of free drug and were able to target and impede the progress of the cancer cells.[83] Lay et al.[38] revealed that paclitaxel–PEG-g–CNTs are encouraging delivery system for cancer therapeutics. The paclitaxel-loaded PEG-g–SWCNTs have less loading efficiency then paclitaxel-loaded PEG-g–MWCNTs. Both paclitaxel-loaded PEG-g–SWCNTs and paclitaxel-loaded PEG-g–MWCNTs have shown better profile in terms of targeting and killing the cancer cells when compared to free paclitaxel (Fig. 1.6). Researcher prepared OX-SWCNTs and *f*-MWCNTs drug delivery for CPT drug. Ox-SWCNTs are oxidized SWCNTs by acid treatment to have carboxyl group and CPT moiety linked on the Ox-SWCNTs, as a nanocargo for lipophilic antitumor agent,[79] whereas MWCNTs were firstly layered with the poloxamer P-123 to show high water solubility. The poloxamer-coated MWCNTs can efficiently form noncovalent supramolecular complexes with anticancer drug CPT.

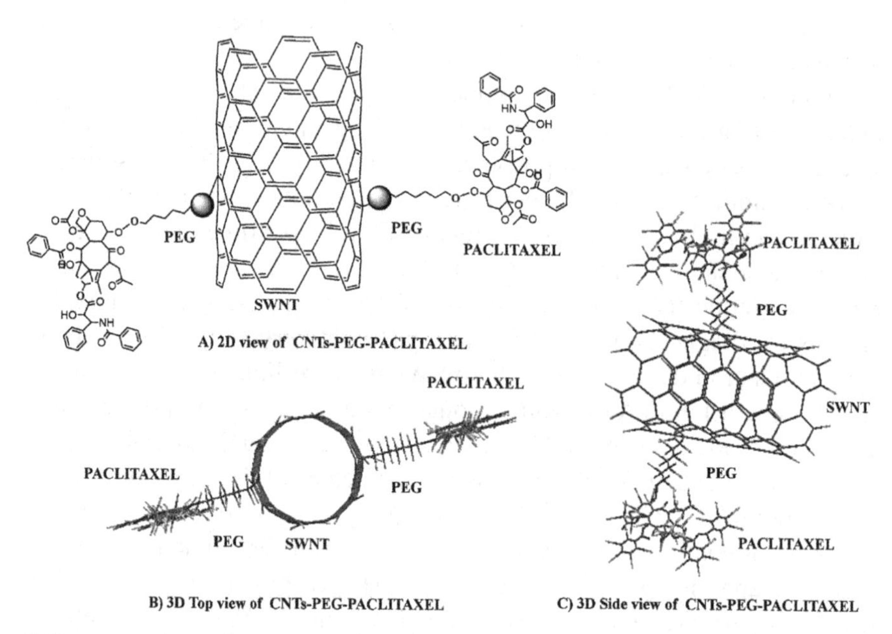

FIGURE 1.6 (See color insert.) 2D and 3D views of the CNT–PEG–paclitaxel as drug carrier.

These complex MWCNTs can assist intracellular delivery of antitumor drug and progress drug action.[56] SWCNTs–DOX complexes were formed by π–π interactions between DOX and the surface of SWCNTs, whereas MWCNT-block polymer–DOX complexes were formed by π–π stacking interaction between the DOX, block polymer, and MWCNT.[2,47] MWCNT-block polymer–DOX complexes acted as double targeting to the cancer cell by as targeting and pH sensitive. SWCNTs–DOX acted as cancer targeting drugs and have therapeutic properties (Fig. 1.7).

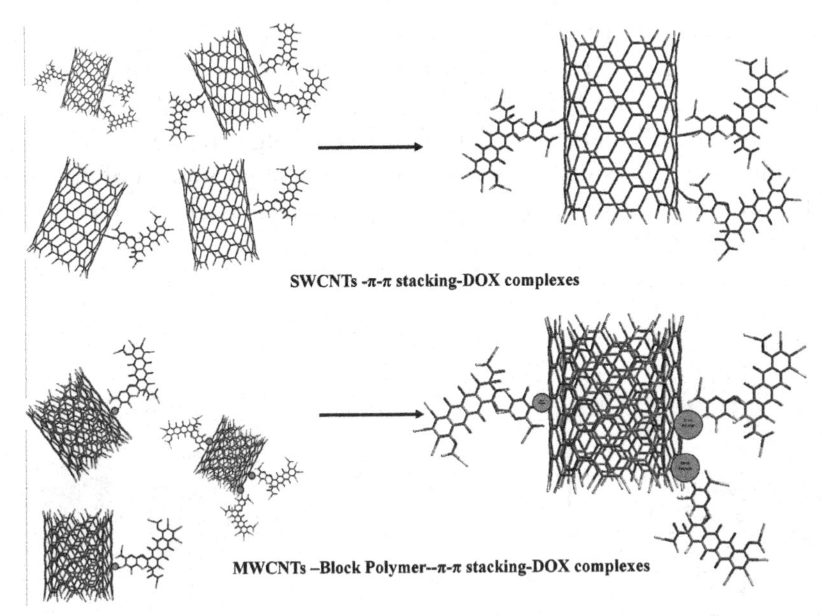

FIGURE 1.7 (See color insert.) SWCNT and MWCNT as carriers for the doxorubicin through π–π stacking.

1.10 DUAL COMBINATION OF CNTs–LIPOSOMES AS A DRUG DELIVERY TO TARGET INFECTED CELLS

CNTs are well known as drug delivery system and well explored in cancer and other diseases as delivery system. Functionalized CNTs are well linked to different moieties and their structure has ability to enter the cell membrane without any deterioration of the cell activity and its integrity. Such ability of CNTs makes them more preferable drug delivery system for different moieties. Every coin has two sides; similarly, CNT has some robust properties and also some limitations such as low dose loading; if the drug level is to be increased in the body then more CNTs are to be used to maintain the concentration of drug in the body. But increasing the CNTs concentration in cell may increase the toxicity in the cell.[60] In order to overcome these problems, CNTs can link or be incorporated to nanovesicles liposomes which act as carrier of drug in nanotrain drug delivery system.[31] Zhu et al.[93] studied about the nanotrain drug delivery of DOX–SWCNTs–liposome; it was formulated to have double effect of CNTs and liposomes to target and increase the DOX content in the tumor site (Fig. 1.8). In vitro MCF-7 cells showed that DOX–SWCNTs–liposome release the drug in slow controlled manner when compared to DOX solution but nanotrain DOX–SWCNTs–liposomes were able to penetrate faster into the cell membrane than free DOX drug. The benefit of these nanotrain drug–CNTs–liposomes looms is that a large amount of drug can be delivered into cells without having any deleterious effect on cell with large amount of CNTs.

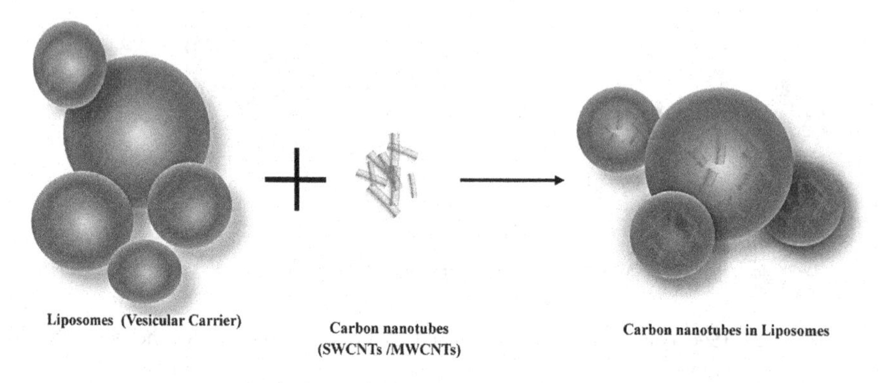

FIGURE 1.8 Nanotrain CNTs–liposomes to carry the drug.

1.11 CNTs USED IN TRANSDERMAL DELIVERY SYSTEM

Stinchcomb and Hinds have fabricated a CNT membrane-based transdermal drug delivery system (TDDS) that can dynamically pump drug through the skin with low power. The CNTs work on the principle of electroosmosis and electrophoresis to eject the drug molecule from reservoir to skin.[48,67] Im and coworkers[26] prepared an electrosensitive CNTs–TDDS by the electrospinning method to govern the drug release. The electrosensitive CNTs–TDDS consists of MWCNTs (additive electrical conductivity), polyethylene oxide, benzil α,α-dimethyl acetal (photoinitiator), and pentaerythritol triacrylate (copolymer), all these fabricated containing the ketoprofen drug. Due to outstanding electrical conductivity of the CNTs, on applying the gradient electric voltage, they will have gradient drug release. Increasing the electrical voltage enhances the drug release.[26] Novel transdermal bandage with concept of SWCNTs wrapped by iodine and PVP is used in wound healing; SWCNTs provide here slow and localized release of iodine which acts as antiseptic and the hollow cylindrical SWCNTs allow the oxygen toward the wound and help in the tissue growth.[65,66,72] Additionally, the intrinsic antimicrobial activity, hemostatic,[5,94] electrical conductivity,[1,52] light-sensitive protection, and controlled release properties of CNTs[53] would contribute to the improvement of transdermal patches with more distinct ability as novel dosage form. CNTs have various intrinsic properties such as antimicrobial activity which is needed in nonsterile pharmaceutical products such as transdermal patches.[53]

1.12 MACROMOLECULES PROTEIN-LINKED FUNCTIONALIZED CNTs

The combined antimicrobial cargo delivery such as silver wrapped on SWCNTs and linked with antimicrobial proteins or peptides can be used to overcome the antibiotics resistances as these combined antimicrobial cargo is effective against wide varieties of microbes and pathogen.[7] The length of MWCNTs plays vital role in permeation of cell membrane; longer dimension in length MWCNTs shows low penetration into the cell membrane when compared to the shorter length MWCNTs, as shorter MWCNTs permeate more effectively into the cell membrane by endocytosis.[32,50] Oxidation approach toward functionalization of CNTs is best loom to link the protein on CNTs, as acid treatment-functionalized CNTs are noncytotoxic. After functionalization, the CNTs surface becomes slightly jagged and corrugated.[51] The jagged and grooved oxidized CNTs impulsively engross various proteins and nonspecifically bind them onto the outer surface of the CNTs.[30,51] The protein-functionalized CNTs are able to retain the biological function of protein and enter cells comparatively much easier than parent proteins.[44] CNTs have been functionalized using various proteins, including bovine serum albumin, streptavidin[43] DNA, and other proteins.[51] Various proteins adsorb spontaneously on the sidewalls of acid-oxidized SWCNTs.[30] Salvador-Morales et al.[64] showed that pristine CNT activate the complement following both the classical and the alternative way by selective adsorption of some of its proteins.[54]

1.13 TWO VERSATILE CARGO COMBINATION: DENDRIMER AND MWCNTs

Using divergent technology, the dendrimers are connected to SWCNTs, where the core of the dendrimers is attached to CNTs and further new

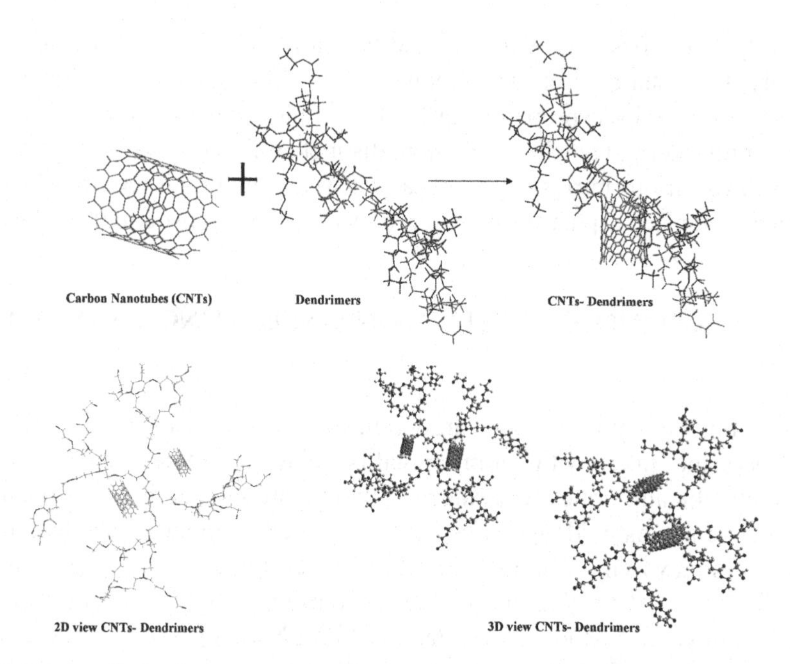

FIGURE 1.9 (See color insert.) Using convergent method to link the CNTs with dendrimers.

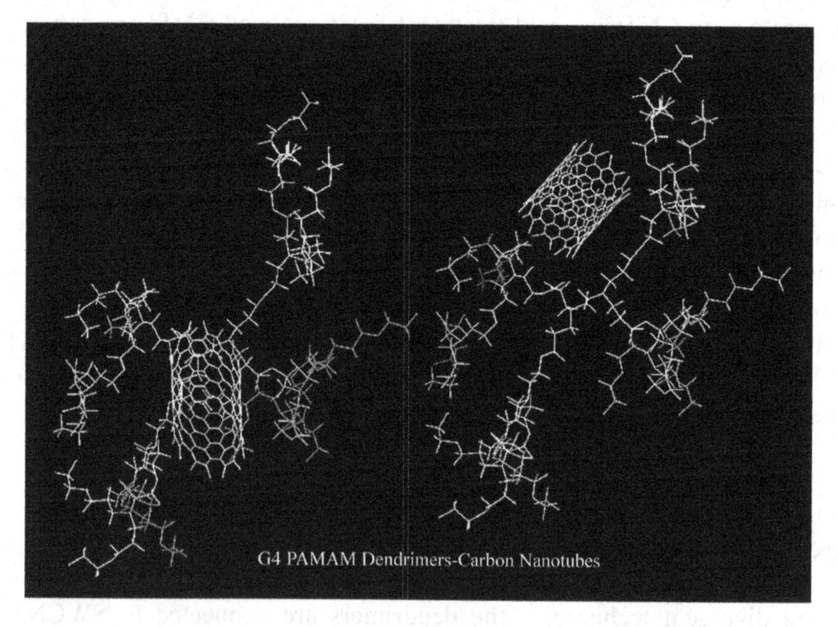

FIGURE 1.10 (See color insert.) G4 PAMAM dendrimers are enrolled to carbon nanotubes to form a versatile carrier.

periphery of the molecule is activated for reactions with more monomers. Such technology do not aggravate any damage to π-bond interaction of the SWCNTs.[6] Using convergent technology (Fig. 1.9), pristine SWCNTs and polyamidoamine (PAMAM) dendrimers hybrids (SWCNT–PAMAM) are prepared that lead to highly dispersed nanoparticles consistently.[23] Surface alteration of CNTs with polymers is an effective technique to resolve solubility. Linear polymer and hyperbranched polymers were used to enhance the solubility. Hyperbranched polymer such as dendrimers are more preferred than linear polymer to overcome the solubility of CNTs; dendrimers are three-dimensional highly branched polymers with high-density functional groups, which also help to link many drug moiety with these functional group.[75] When CNT and dendrimers are complexed together, their dispersion capability is analyzed by potential of mean forces. Nonprotonated dendrimer complexing with CNT do not show good dispersion capability when compared to protonated dendrimers linked with CNTs.[21,57] Mehdipoor et al.[45] prepared a novel drug delivery combination of two versatile carriers for anticancer drug. They initially prepared ferric oxide nanoparticles and linked them to CNTs (MWCNTs) to form iron oxide nanoparticles–MWCNTs complexes. Linear dendrimers of polyethylene glycol and poly(citric acid) are linked with cisplatin. After preparing iron oxide nanoparticles–CNT and cisplatin–PCA–PEG–PCA~PEG–PCA, dendrimers are complexed together to form hybrid nanocarrier cisplatin/PCA–PEG–PCA/CNT/γ–Fe_2O_3NP with significant features such as magnetic-based targeting to tumor, high loading capacity, water solubility, etc.[45] Protonated dendrimers have the capacity to solubilize the SWCNT, G3, and G4 protonated PAMAM dendrimer linked to SWCNT (Fig. 1.10); the complexes become more soluble which can be used as effective carrier for single-stranded DNA in soluble dendrimer–CNT complexes.

1.14 CONCLUSION

CNTs act as versatile carriers or Trojan horse and smuggle the toxic drug through modified dressed CNTs and target the infected cells. The physicochemical properties of CNTs are vital in pharmaceutical field and any changes in properties affect their carrier ability. Solubility is important concern in any formulation of drug and many approaches, such as wrapping of polymers, surfactant and phospholipid interaction, dendrimers linkage, functionalization of CNTs, etc., are applied to enhance the solubility of

CNTs. Inflection of the drug release profiles can be achieved by CNTs modification by covalent and noncovalent binding with other materials. CNTs are well explored as carrier or cargo for different categories of drug. CNTs are deeply researched in cargo delivery of anticancer and antifungal drugs, and proteins. CNTs are such ingredients that show distinct features in different fields of sciences and in nanomedicine; CNTs have opened new doorstep for novel and innovative formulation to overcome the limitations of other carriers.

KEYWORDS

- **carbon nanotubes**
- **carrier**
- **drug delivery**
- **nanomedicine**
- **nanocargo**
- **cylindrical graphite**
- **biomedical applications**

REFERENCES

1. Ajayan, P. M. Nanotubes from Carbon. *Chem. Rev.* **1999,** *99*, 1787–1799.
2. Ali-Boucetta, H.; Al-Jamal, K. T.; McCarthy, D.; Prato, M.; Bianco, A.; Kostarelos, K. Multiwalled Carbon Nanotube-doxorubicin Supramolecular Complexes for Cancer Therapeutics. *Chem. Commun. (Camb).* **2008,** *4*, 459–461.
3. Bekyarova, E.; Ni, Y.; Malarkey, E. B.; et al. Applications of Carbon Nanotubes in Biotechnology and Biomedicine. *J. Biomed. Nanotechnol.* **2005,** *1*, 3–17.
4. Birch, M. E.; Ruda-Eberenz, T. A.; Chai, M.; Andrews, R.; Ranal, L.; Hatfield, R. L. Properties that Influence the Specific Surface Areas of Carbon Nanotubes and Nanofibers. *Ann. Occup. Hyg.* **2013,** *57* (9), 1148–1166.
5. Burke, A. R.; Singh, R. N.; Carroll, D. L.; Owen, J. D.; Kock, N. D.; D'Agostino, R. Jr.; Torti, F. M.; Torti, S. V. Determinants of the Thrombogenic Potential of Multiwalled Carbon Nanotubes. *Biomaterials* **2011,** *32* (26), 5970–5978.
6. Campidelli, S.; Sooambar, C.; Lozano Diz, E.; Ehli, C.; Guldi, D. M.; Prato, M. Dendrimer-functionalized Single-wall Carbon Nanotubes: Synthesis, Characterization, and Photoinduced Electron Transfer. *J. Am. Chem. Soc.* **2006,** *128* (38), 12544–12552.

7. Chaudhari, A. A.; Ashmore, D.; Nath, S. D.; Kate, K.; Dennis, V.; Singh, S. R.; Owen, D. R.; Palazzo, C.; Arnold, R. D.; Miller, M. E.; Pillai, S. R. A Novel Covalent Approach to Bio-conjugate Silver Coated Single Walled Carbon Nanotubes with Antimicrobial Peptide. *J. Nanobiotechnol.* **2016,** *14* (1), 58.
8. Chen, C.; Hou, L.; Zhang, H.; Zhu, L.; Zhang, H.; Zhang, C.; Shi, J.; Wang, L.; Jia, X.; Zhang, Z. Single-walled Carbon Nanotubes Mediated Targeted Tamoxifen Delivery System using Aspargine-glycine-arginine Peptide. *J. Drug Target.* **2013,** *21* (9), 809–821.
9. Chen, H.; Wang, B.; Gao, D.; Guan, M.; Zheng, L.; Ouyang, H.; Chai, Z.; Zhao, Y.; Feng, W. Broad-spectrum Antibacterial Activity of Carbon Nanotubes to Human Gut Bacteria. *Small* **2013,** *9* (16), 2735–2746.
10. Chen, J.; Chen, S.; Zhao, X.; Kuznetsova, L. V.; Wong, S. S.; Ojima, I. Functionalized Single-walled Carbon Nanotubes as Rationally Designed Vehicles for Tumor-targeted Drug Delivery. *J. Am. Chem. Soc.* **2008,** *130* (49), 16778–16785.
11. Chen, R. J.; Bangsaruntip, S.; Drouvalakis, K. A.; Kam, N. W.; Shim, M.; Li, Y.; Kim, W.; Utz, P. J.; Dai, H. Noncovalent Functionalization of Carbon Nanotubes for Highly Specific Electronic Biosensors. *Proc. Natl. Acad. Sci. USA* **2003,** *100* (9), 4984–4989.
12. Choudhary, V.; Gupta, A. Polymer/Carbon Nanotube Nanocomposites. In *Carbon Nanotubes—Polymer Nanocomposites*; Yellampalli, S., Ed.; InTech: Croatia, 2011; Vol. 4, pp 65–90.
13. Collins, P. G.; Bradley, K.; Ishigami, M.; Zettl, A. Extreme Oxygen Sensitivity of Electronic Properties of Carbon Nanotubes. *Science.* **2000,** *287* (5459), 1801–1804.
14. Dai, L.; Sun, J. Mechanical Properties of Carbon Nanotubes-polymer Composites. In *Carbon Nanotubes—Current Progress of Their Polymer Composites*; Mohamed, B., Hafez, I. H., Eds.; InTech: United Kingdom, 2016; Vol. 6, pp 156–194.
15. Danailov, D.; Keblinski, P.; Nayak, S.; Ajayan, P. M. Bending Properties of Carbon Nanotubes Encapsulating Solid Nanowires. *J. Nanosci. Nanotechnol.* **2002,** *2*, 503–507.
16. Deng, X.; Xiong, D.; Wang, Y.; Chen, W.; Luan, Q.; Zhang, H.; Jiao, Z.; Wu, M. Water Soluble Multi-walled Carbon Nanotubes Enhance Peritoneal Macrophage Activity In Vivo. *J. Nanosci. Nanotechnol.* **2010,** *10* (12), 8663–8669.
17. Detection and Measurement of Carbon Nanotubes (CNT). https://www.malvern.com/en/industry-applications/sample-type-form/carbon-nanotubes (accessed Sept 27, 2017).
18. Drabu, S.; Khatri, S.; Babu, S.; Sahu, R. K. Carbon Nanotubes in Pharmaceutical Nanotechnology: An introduction to Future Drug Delivery System. *J. Chem. Pharm. Res.* **2010,** *2* (1), 444–457.
19. Elhissi, A. M. A.; Ahmed, W.; Hassan, I. U.; Dhanak, V. R.; D'Emanuele, A. Carbon Nanotubes in Cancer Therapy and Drug Delivery. *J. Drug Deliv.* **2012,** *2012*, 837327.
20. Gao, X. L.; Li, K. Finite Deformation Continuum Model for Single-walled Carbon Nanotubes. *Int. J. Solids Struct.* **2003,** *40* (26), 7329–7337.
21. García, A.; Herrero, M. A.; Frein, S.; Deschenaux, R.; Muñoz, R.; Bustero, I.; Toma, F.; Prato, M. Synthesis of Dendrimer–Carbon Nanotube Conjugates. *Phys. Stat. Sol. (A)* **2008,** *205,* 1402–1407.
22. Geckeler, K. E.; Premkumar, T. Carbon Nanotubes: Are They Dispersed or Dissolved in Liquids? *Nanoscale Res. Lett.* **2011,** *6,* 136–139.
23. Giacalone, F.; Campisciano, V.; Calabrese, C.; La Parola, V.; Syrgiannis, Z.; Prato, M.; Gruttadauria, M. Single-walled Carbon Nanotube-polyamidoamine Dendrimer Hybrids for Heterogeneous Catalysis. *ACS Nano.* **2016,** *10* (4), 4627–4636.

24. Guven, A.; Villares, G. J.; Hilsenbeck, S. G.; Lewis, A.; Landua, J. D.; Dobrolecki, L. E.; Wilson, L. J.; Lewis, M. T. Carbon Nanotube Capsules Enhance the In Vivo Efficacy of Cisplatin. *Acta Biomater.* **2017,** *58*, 466–478.
25. Hashemzadeh, H.; Raissi, H. The Functionalization of Carbon Nanotubes to Enhance the Efficacy of the Anticancer Drug Paclitaxel: A Molecular Dynamics Simulation Study. *J. Mol. Model.* **2017,** *23* (8), 222.
26. Im, J. S.; Bai, B. C.; Lee, Y. S. The Effect of Carbon Nanotubes on Drug Delivery in an Electro-sensitive Transdermal Drug Delivery System. *Biomaterials* **2010,** *31*, 1414–1419.
27. Islam, M. F.; Rojas, E.; Bergey, D. M.; Johnson, A. T.; Yodh, A. G. High Weight Fraction Surfactant Solubilization of Single-wall Carbon Nanotubes in Water. *Nano Lett.* **2003,** *3* (2), 269–273.
28. Ji, S. R.; Liu, C.; Zhang, B.; Yang, F.; Xu, J.; Long, J.; Jin, C.; Fu, D. L.; Ni, Q. X.; Yu, X. J. Carbon Nanotubes in Cancer Diagnosis and Therapy. *Biochim. Biophys. Acta* **2010,** *1806* (1), 29–35.
29. Ju, L.; Zhang, W.; Wang, X.; Hu, J.; Zhang, Y. Aggregation Kinetics of SDBS-dispersed Carbon Nanotubes in Different Aqueous Suspensions. *Colloids Surf. A Physicochem. Eng. Asp.* **2012,** *409,* 159–166.
30. Kam, N. W. S.; Dai, H. Carbon Nanotubes as Intracellular Protein Transporters: Generality and Biological Functionality. *J. Am. Chem. Soc.* **2005,** *127* (16), 6021–6026.
31. Karchemski, F.; Zucker, D.; Barenholz, Y.; Regev, O. Carbon Nanotubes–Liposomes Conjugate as a Platform for Drug Delivery into Cells. *J. Control. Release* **2012,** *160* (2), 339–345.
32. Karimi, M.; Solati, N.; Ghasemi, A.; Estiar, M. A.; Hashemkhani, M.; Kiani, P.; Mohamed, E.; Saeidi, A.; Taheri, M.; Avci, P.; Aref, A. R.; Amiri, M.; Baniasadi, F.; Hamblin, M. R. Carbon Nanotubes Part II: A Remarkable Carrier for Drug and Gene Delivery. *Expert Opin. Drug Deliv.* **2015,** *12* (7), 1089–1105.
33. Kazemi-Beydokhti, A.; Heris, S. Z.; Jaafari, M. R. Investigation of Different Methods for Cisplatin Loading Using Single-walled Carbon Nanotube. *Chem. Eng. Res. Des.* **2016,** *112*, 56–63.
34. Klumpp, C.; Kostarelos, K.; Prato, M.; Bianco, A. Functionalized Carbon Nanotubes as Emerging Nanovectors for the Delivery of Therapeutics. *Biochim. Biophys. Acta* **2006,** *1758*, 404–412.
35. Kong, J.; Franklin, N. R.; Zhou, C.; Chapline, M. G.; Peng, S.; Cho, K.; Da, H. Nanotube Molecular Wires as Chemical Sensors. *Science* **2000,** *287*, 622–625.
36. Kumar, S.; Rani, R.; Dilbaghi, N.; Tankeshwar, K.; Kim, K. H. Carbon Nanotubes: A Novel Material for Multifaceted Applications in Human Healthcare. *Chem. Soc. Rev.* **2017,** *46* (1), 158–196.
37. Kushwaha, S. S.; Ghoshal, S.; Rai, A. K.; Singh, S. Carbon Nanotubes as a Novel Drug Delivery System for Anticancer Therapy: A Review. *Braz. J. Pharm. Sci.* **2013,** *49* (4), 629–643.
38. Lay, C. L.; Liu, H. Q.; Tan, H. R.; Liu, Y. Delivery of Paclitaxel by Physically Loading Onto Poly(Ethylene Glycol) (PEG)-graft-carbon Nanotubes for Potent Cancer Therapeutics. *Nanotechnology* **2010,** *21* (6), 065101.
39. Lay, C. L.; Liu, J.; Liu, Y. Functionalized Carbon Nanotubes for Anticancer Drug Delivery. *Expert Rev. Med. Devices* **2011,** *8* (5), 561–566.

40. Liu, Z.; Sun, X.; Nakayama-Ratchford, N.; Dai, H. Supramolecular Chemistry on Water-soluble Carbon Nanotubes for Drug Loading and Delivery. *ACS Nano* **2007,** *1* (1), 50–56.
41. Lohan, S.; Raza, K.; Mehta, S. K.; Bhatti, G. K.; Saini, S.; Singh, B. Anti-Alzheimer's Potential of Berberine Using Surface Decorated Multi-walled Carbon Nanotubes: A Preclinical Evidence. *Int. J. Pharm.* **2017,** *530* (1–2), 263–278.
42. Lohan, S.; Raza, K.; Singla, S.; Chhibber, S.; Wadhwa, S.; Katare, O. P.; Kumar, P.; Singh, B. Studies on Enhancement of Anti-microbial Activity of Pristine MWCNTs Against Pathogens. *AAPS PharmSciTech.* **2016,** *17* (5), 1042–1048.
43. Lui, Z.; Galli, F.; et al. Stable Single-walled Carbon Nanotube—Streptavidin Complex for Biorecognition. *J. Phys. Chem. C* **2010,** *114*, 4345.
44. Mallick, K.; Strydom, A. M. Biophilic Carbon Nanotubes. *Colloids Surf. B Biointerfaces* **2013,** *105*, 310–318.
45. Mehdipoor, E.; Adeli, M.; Bavadi, M.; Sasanpour, P.; Rashidian, B. A Possible Anticancer Drug Delivery System Based on Carbon Nanotube–Dendrimer Hybrid Nanomaterials. *J. Mater. Chem.* **2011,** *21*, 15456–15463.
46. Meng, J.; Cheng, X.; Liu, J.; Zhang, W.; Li, X.; Kong, H.; Xu, H. Effects of Long and Short Carboxylated or Aminated Multiwalled Carbon Nanotubes on Blood Coagulation. *PLoS One* **2012,** *7* (7), e38995.
47. Meng, L.; Zhang, Z.; Lu, Q.; Fei, Z.; Dyson, P. J. Single Walled Carbon Nanotubes as Drug Delivery Vehicles: Targeting Doxorubicin to Tumors. *Biomaterials* **2012,** *33* (6), 1689–1698.
48. Miyako, E.; Kono, K.; Yuba, E.; Hosokawa, C.; Nagai, H.; Hagihara, Y. Carbon Nanotube–Liposome Supramolecular Nanotrains for Intelligent Molecular-transport Systems. *Nat. Commun.* **2012,** *3*, 1226.
49. Moore, V. C.; Strano, M. S.; Haroz, E. H.; Hauge, R. H.; Smalley, R. E.; Schmidt, J. Individually Suspended Single-walled Carbon Nanotubes in Various Surfactants. *Nano Lett.* **2003,** *3* (10), 1379–1382.
50. Mu, Q.; Broughton, D. L.; Yan, B. Endosomal Leakage and Nuclear Translocation of Multiwalled Carbon Nanotubes: Developing a Model for Cell Uptake. *Nano Lett.* **2009,** *9* (12), 4370–4375.
51. Nagaraju, K.; Reddy, R.; Reddy, N. A review on Protein Functionalized Carbon Nanotubes. *J. Appl. Biomater. Funct. Mater.* **2015,** *13* (4), e301–e312.
52. Noel, Y.; Demichelis, R. Nanotube Systems. http://www.theochem.unito.it/crystal_tuto/mssc2008_cd/tutorials/nanotube/nanotube.html (accessed Sept 8, 2017).
53. Olivi, M.; Zanni, E.; Bellis, G. D.; Talora, C.; Sarto, M. S.; Palleschi, C.; Flahaut, E.; Monthioux, M.; Rapino, S.; Uccelletti, D.; Fiorito, S. Inhibition of Microbial Growth by Carbon Nanotube Networks. *Nanoscale* **2013,** *5*, 9023–9029.
54. Pantarotto, D.; Partidos, C. D.; Graff, R.; Hoebeke, J.; Briand, J. P.; Prato, M.; Bianco, A. Synthesis, Structural Characterization and Immunological Properties of Carbon Nanotube Functionalized with Peptides. *J. Am. Chem. Soc.* **2003,** *125*, 6160–6164.
55. Peigney, A.; Laurent, C. H.; Flahaut, E.; Bacsa, R. R.; Rousset, A. Specific Surface Area of Carbon Nanotubes and Bundles of Carbon Nanotubes. *Carbon* **2001,** *39* (4), 507–514.
56. Permanaa, B.; Ohbaa, T.; Itohb, T.; Kanoha, H. Systematic Sorption Studies of Camptothecin on Oxidized Single-walled Carbon Nanotubes. *Colloids Surf. A Physicochem. Eng. Asp.* **2016,** *490*, 121–132.

57. Pramanika, D.; Maiti, P. K. Dendrimer Assisted Dispersion of Carbon Nanotubes: A Molecular Dynamics Study. *Soft Matter*. **2016,** *12*, 8512–8520.
58. Raza, K.; Kumar, D.; Kiran, C.; Kumar, M.; Guru, S. K.; Kumar, P.; Arora, S.; Sharma, G.; Bhushan, S.; Katare, O. P. Conjugation of Docetaxel with Multiwalled Carbon Nanotubes and Codelivery with Piperine: Implications on Pharmacokinetic Profile and Anticancer Activity. *Mol Pharm*. **2016,** *13* (7), 2423–2432.
59. Razzazan, A.; Atyabi, F.; Kazemi, B.; Dinarvand, R. In Vivo Drug Delivery of Gemcitabine with PEGylated Single-walled Carbon Nanotubes. *Mater. Sci. Eng. C Mater. Biol. Appl.* **2016,** *62*, 614–625.
60. Regev, O.; Barenholz, Y.; Peretz, S.; Zucker, D.; Bavli-Felsen, Y. Can Carbon Nanotube-liposome Conjugates Address the Issues Associated with Carbon Nanotubes in Drug Delivery? *Future Med Chem*. **2013,** *5* (5), 503–505.
61. Robinson, J. T.; Hong, G.; Liang, Y.; Zhang, B.; Yaghi, O. K.; Dai, H. In Vivo Fluorescence Imaging in the Second Near-infrared Window with Long Circulating Carbon Nanotubes Capable of Ultrahigh Tumor Uptake. *J. Am. Chem. Soc.* **2012,** *134* (25), 10664–10669.
62. Roldo, M. Carbon Nanotubes in Drug Delivery: Just a Carrier? https://researchportal.port.ac.uk/portal/files/3418926/Carbon_nanotubes_in_drug_delivery_submitted.pdf (accessed Oct 14, 2017).
63. Rosen, Y.; Gurman, P. Carbon Nanotubes for Drug Delivery Applications. In *Nanotechnology and Drug Delivery*; Arias, J. L., Ed.; CRC Press, 2014; Vol. 1, pp 233–248.
64. Salvador-Morales, C.; Flahaut, E.; Sim, E.; Sloan, J.; Green, M. L.; Sim, R. B. Complement Activation and Protein Adsorption by Carbon Nanotubes. *Mol Immunol*. **2006,** *43* (3), 193–201.
65. Schwengber, A.; Prado, H. J.; Zilli, D. A.; Bonelli, P. R.; Cukierman, A. L. Carbon Nanotubes Buckypapers for Potential Transdermal Drug Delivery. *Mater. Sci. Eng. C Mater. Biol. Appl.* **2015,** *57*, 7–13.
66. Scoville, C.; Cole, R.; Hogg, J.; Farooque, O.; Russell, A. Carbon Nanotubes https://courses.cs.washington.edu/courses/csep590a/08sp/.../CarbonNanotubes.pdf (accessed Sept 26, 2017).
67. Sealy, C. Carbon Nanotubes Patch Up Drug Delivery. *Nano Today* **2010,** *5* (5), 376–377.
68. Shah, M.; Agrawal, Y. Carbon Nanotube: A Novel Carrier for Sustained Release Formulation. *Fuller. Nanotub. Carbon Nanostruct.* **2012,** *20* (8), 696–708.
69. Shen, Q. J.; Liu, X. B.; Jin, W. J. Solubility Increase of Multi-walled Carbon Nanotubes in Water. *Carbon* **2013,** *60*, 562–563.
70. Shen, H. S. Postbuckling Prediction of Double-walled Carbon Nanotubes Under Hydrostatic Pressure. *Int. J. Solids Struct.* **2004,** *41*, 2643–2657.
71. Shi Kam, N. W.; Jessop, T. C.; Wender, P. A.; Dai, H. Nanotube Molecular Transporters: Internalization of Carbon Nanotube-protein Conjugates into Mammalian Cells. *J. Am. Chem. Soc.* **2004,** *126* (22), 6850–6851.
72. Simmons, T. J.; Lee, S. H.; Park, T. J.; Hashim, D. P.; Ajayan, P. M.; Linhardt, R. J. Antiseptic Single Wall Carbon Nanotube Bandages. *Carbon* **2009,** *47*, 1561–1564.
73. Son, K. H.; Hong, J. H.; Lee, J. W. Carbon Nanotubes as Cancer Therapeutic Carriers and Mediators. *Int. J. Nanomed.* **2016,** *11*, 5163–5185.
74. Stone, V.; Kermanizadeh, A.; Johnston, H.; Boyles, M.; Varet, J. Carbon Nanotube–Cellular Interactions: Macrophages, Epithelial and Mesothelial Cells. In *The Toxicology of Carbon Nanotubes*; Donaldson, K., Poland, C., Duffin, R., Bonner, J., Eds.; Cambridge: Cambridge University Press, 2012; pp 174–209.

75. Sun, J.; Hong, C.; Pana, C. Surface Modification of Carbon Nanotubes with Dendrimers or Hyperbranched Polymers. *Polym. Chem.* **2011,** *2*, 998–1007.
76. Suttipong, M.; Tummala, N. R.; Kitiyanan, B.; Striolo, A. Role of Surfactant Molecular Structure on Self-assembly: Aqueous SDBS on Carbon Nanotubes. *J. Phys. Chem. C* **2011,** *115*, 17286–17296.
77. Talapatra, S.; Rawat, D. S.; Migone, A. D. Possible Existence of a Higher Coverage Quasi-one-dimensional Phase of Argon Adsorbed on Bundles of Single-walled Carbon Nanotubes. *J. Nanosci. Nanotechnol.* **2002,** *2* (5), 467–470.
78. Tamsyn, A. H.; James, M. H. Modeling the Loading and Unloading of Drugs into Nanotubes. *Small* **2009,** *5* (3), 300–308.
79. Tian, Z.; Yin, M.; Ma, H.; Zhu, L.; Shen, H.; Jia, N. Supramolecular Assembly and Antitumor Activity of Multiwalled Carbon Nanotube–Camptothecin Complexes. *J. Nanosci. Nanotechnol.* **2011,** *11* (2), 953–958.
80. Vakhrusheva, T. V.; Gusev, A. A.; Gusev, S. A.; Vlasova, I. I. Albumin Reduces Thrombogenic Potential of Single-walled Carbon Nanotubes. *Toxicol. Lett.* **2013,** *221* (2), 137–145.
81. Varkouhi, A. K.; Foillard, S.; Lammers, T.; Schiffelers, R. M.; Doris, E.; Hennink, W. E.; Storm, G. SiRNA Delivery with Functionalized Carbon Nanotubes. *Int. J. Pharm.* **2011,** *416* (2), 419–425.
82. Wang, N.; Feng, Y.; Zeng, L.; Zhao, Z.; Chen, T. Functionalized Multiwalled Carbon Nanotubes as Carriers of Ruthenium Complexes to Antagonize Cancer Multidrug Resistance and Radioresistance. *ACS Appl. Mater. Interfaces* **2015,** *7* (27), 14933–14945.
83. Wen, S.; Liu, H.; Cai, H.; Shen, M.; Shi, X. Targeted and pH-responsive Delivery of Doxorubicin to Cancer Cells Using Multifunctional Dendrimer-modified Multi-walled Carbon Nanotubes. *Adv. Healthc. Mater.* **2013,** *2* (9), 1267–1276.
84. Wu, H.; Liu, G.; Wang, X.; et al. Solvothermal Synthesis of Cobalt Ferrite Nanoparticles Loaded on Multiwalled Carbon Nanotubes for Magnetic Resonance Imaging and Drug Delivery. *Acta Biomater.* **2011,** *7* (9), 3496–3504.
85. Wu, W.; Li, R.; Bian, X.; Zhu, Z.; Ding, D.; Li, X.; Jia, Z.; Jiang, X.; Hu, Y. Covalently Combining Carbon Nanotubes with Anticancer Agent: Preparation and Antitumor Activity. *ACS Nano.* **2009,** *3* (9), 2740–2750.
86. Xie, Y.; Bagby, T. R.; Cohen, M.; Forrest, M. L. Drug Delivery to the Lymphatic System: Importance in Future Cancer Diagnosis and Therapies. *Expert Opin. Drug Deliv.* **2009,** *6* (8), 785–792.
87. Yang, F.; Fu, D. L.; Long, J.; Ni, Q. X. Magnetic Lymphatic Targeting Drug Delivery System Using Carbon Nanotubes. *Med. Hypotheses* **2008,** *70* (4), 765–767.
88. Yang, F.; Hu, J.; Yang, D.; Long, J.; Luo, G.; Jin, C.; Yu, X.; Xu, J.; Wang, C.; Ni, Q.; Fu, D. Pilot Study of Targeting Magnetic Carbon Nanotubes to Lymph Nodes. *Nanomedicine (Lond)* **2009,** *4* (3), 317–330.
89. Zardini, H. Z.; Amiri, A.; Shanbedi, M.; Maghrebi, M.; Baniadam, M. Enhanced Antibacterial Activity of Amino Acids-functionalized Multi Walled Carbon Nanotubes by a Simple Method. *Colloids Surf. B Biointerfaces* **2012,** *92*, 196–202.
90. Zhang, W.; Zhang, Z.; Zhang, Y. The Application of Carbon Nanotubes in Target Drug Delivery Systems for Cancer Therapies. *Nanoscale Res. Lett.* **2011,** *6* (1), 1–22.
91. Zhang, X.; Meng, L.; Lu, Q.; Fei, Z.; Dyson, P. J. Targeted Delivery and Controlled Release of Doxorubicin to Cancer Cells Using Modified Single Wall Carbon Nanotubes. *Biomaterials* **2009,** *30* (30), 6041–6047.

92. Zhang, X. Y.; Lu, W. Y. Recent Advances in Lymphatic Targeted Drug Delivery System for Tumor Metastasis. *Cancer Biol. Med.* **2014,** *11* (4), 247–254.
93. Zhu, X.; Huang, H.; Zhang, Y.; Xie, Y.; Hou, L.; Zhang, H.; Zhang, Z. Investigation on Single-walled Carbon Nanotubes–Liposomes Conjugate to Treatment Tumor with Dual-mechanism. *Curr. Pharm. Biotechnol.* **2015,** *16* (10), 927–936.
94. Nowacki, M.; Wiśniewski, M.; Werengowska-Ciećwierz K.; Terzyk, A. P.; Kloskowski T.; Marszałek, A.; Bodnar, M.; Pokrywczyńska, M.; Nazarewski, Ł.; Pietkun, K.; Jundziłł, A.; Drewa, T. New Application of Carbon Nanotubes in Haemostatic Dressing Filled with Anticancer Substance. *Biomed Pharmacother* **2015,** *69*, 349–354.
95. Sahoo, N. G.; Bao, H.; Pan, Y.; Pal, M.; Kakran, M.; Cheng, H. K.; Li, L.; Tan, L. P. Functionalized Carbon Nanomaterials as Nanocarriers for Loading and Delivery of a Poorly Water-soluble Anticancer Drug: A Comparative Study. *Chem. Commun.* **2011,** *47*, 5235–5237.
96. Madani, S. Y.; Naderi, N.; Dissanayake, O.; Tan, A.; Seifalian, A. M. A New Era of Cancer Treatment: Carbon Nanotubes as Drug Delivery Tools. *Int. J. Nanomed.* **2011,** *6*, 2963–2979.
97. Fiorito, S.; Monthioux, M.; Psaila, R.; Pierimarchi P.; Zonfrillo, M.; D'Emilia, E.; Grimaldi, S.; Lisi, A.; Béguin, F.; Almairac, R.; Noe, L.; Serafinz, A. Evidence for Electro-chemical Interactions Between Multi-walled Carbon Nanotubes and Human Macrophages. *Carbon* **2009,** *47* (12), 2789–2804.
98. Dolatabadi, E. N.; Omidi, Y.; Losic, D. Carbon Nanotubes as an Advanced Drug and Gene Delivery Nanosystem. *Curr. Nanosci.* **2011,** *7* (3), 297–314.
99. Heister, E.; Neves, V. Drug Loading, Dispersion Stability, and Therapeutic Efficacy in Targeted Drug Delivery with Carbon Nanotubes. *Carbon* **2006,** *128* (2), 10568–10571.
100. Liu, Z.; Chen, K.; Davis, C.; Sherlock, S.; Cao, Q.; Chen, X.; Dai, H. Drug Delivery with Carbon Nanotubes for In Vivo Cancer Treatment. *Cancer Res.* **2008,** *16*, 6652–6660.
101. Modi, C. D.; Patel, S. J.; Desai, A. B.; Murthy, R. S. R. Functionalization and Evaluation of PEGylated Carbon Nanotubes as Novel Drug Delivery for Methotrexate. *J. Appl. Pharma. Sci.* **2011,** *01* (05), 103–108.
102. Worle-Knirsch, J. M.; Pulskamp, K.; Krug, H. F. Oops They did it Again! Carbon Nanotubes Hoax Scientists in Viability Assays. *Nano Lett.* **2006,** *6* (6), 1261–1268.
103. Wu, W.; Li, R.; Bian, X.; Zhu, Z.; Ding, D.; Li, X.; Jia, Z.; Jiang, X.; Hu, Y. Covalently Combining Carbon Nanotubes with Anticancer Agent: Preparation and Antitumor Activity. *ACS Nano.* **2009,** *3* (9), 2740–2750.
104. Wu, L.; Man, C.; Wang, H.; Lu, X.; Ma, Q.; Cai, Y.; Ma, W. PEGylated Multi-walled Carbon Nanotubes for Encapsulation and Sustained Release of Oxaliplatin. *Pharm. Res.* **2013,** *30* (2), 412–423.
105. Wang, H. F.; Wang, J.; Deng, X. Y.; Sun, H. F.; Shi, Z. J.; Gu, Z. N. Biodistribution of Carbon Single-wall Carbon Nanotubes in Mice. *J. Nanosci. Nanotechnol.* **2004,** *4* (8), 1019–1024.
106. Singh, R.; Pantarotto, D.; Lacerda, L.; Pastori, G.; Klumpp, C.; Prato, M. Tissue Biodistribution and Blood Clearance Rates of Intravenously Administered Carbon Nanotube Radiotracers. *Proc. Natl. Acad. Sci.* **2006,** *103* (9), 3357–3362.

CHAPTER 2

A REVIEW OF A COMPUTATIONAL STUDY OF CARBON NANOTUBES

HITESHI TANDON[1,2], SHALINI[1,2], PRABHAT RANJAN[3], VANDANA SUHAG[1], and TANMOY CHAKRABORTY[1*]

[1]*Department of Chemistry, Manipal University Jaipur, Jaipur 303007, India*

[2]*Department of Chemistry, Alankar P.G. Girls College, Jaipur 302012, India*

[3]*Department of Mechatronics Engineering, Manipal University Jaipur, Jaipur 303007, India*

[*]*Corresponding author. E-mail: tanmoychem@gmail.com; tanmoy.chakraborty@jaipur.manipal.edu*

ABSTRACT

Nanomaterials are extremely significant products of nanotechnologies. A material which has at least one dimension in the range of 1–100 nm is known as nanomaterial. The study of nanomaterials is of high importance due to their wide range of applications in industries such as manufacturing, semiconductor, healthcare, cosmetics, sports wearing, etc. Carbon nanotube (CNT) is the most vital amongst all such materials and it gains immense importance for nanotechnology, optics, nanoelectronics, and other fields of science and technology. The bonding pattern, durability, and high aspect ratios make it useful to prepare emitters, composites, sensors, wind blades, hydrogels, artificial implants, and so on. Though there exists a number of different computational approaches to study CNT, density functional theory (DFT) is very much relevant as it provides result closer to the experimental data. In this chapter, we have presented the study of carbon nanotubes and its applications based on DFT approach.

2.1 INTRODUCTION

Nanomaterial is defined as one which comprises at least one dimension of particles in the range of 1–100 nm. Nanomaterials offer exciting challenges and prospects in the field of chemical sciences, physical sciences, biological sciences, and material sciences. They act as building blocks in production of devices and help the downscale conventional technologies to a larger extent. They also offer an economical and environment-friendly manufacturing route because of utilization of a much less amount of raw material.[1] Besides this, when chemical and physical properties of materials are present on a nanoscopic level, they can be significantly transformed and adjusted, introducing the particle size as an effective novel parameter.[2] When the particle size is decreased to nanometer dimensions, specific surface area of the material is increased and as a result, reactions taking place at the liquid–solid or gas–solid interface get an advantage. These materials have received a lot of attention due to their exclusive properties and practical applicability. Characteristic examples are zero-dimensional nanoparticles,[3] one-dimensional nanowires,[4–9] and two-dimensional (2D) graphenes.[10–20] Since in these structures, electrons remain confined in one or more dimensions, different physicochemical properties, such as structural, optical, magnetic and electronic, can be attained in nanostructures. Exploitation of nanomaterials in environmental chemistry, electrochemistry, catalysis, and energy conversion enhances response time, sensitivity, and efficiency.[21] At this time, the most promising amongst nanoscale materials are carbon nanotubes (CNTs).

In 1991, multiwalled carbon nanotubes (MWNT) were discovered by Iijima[13] succeeding which other researches were carried out that resulted in the synthesis of single-walled carbon nanotubes (SWNTs).[14,19] Since then several theoretical and experimental studies have been performed to understand its electronic, chemical, and other properties. Nanotubes are now being considered as the fundamental building blocks in the field of nanotechnology due to their extreme strength, chemical stability, and diverse electronic properties. Various hypotheses have been put forward for the development of nanoscale devices such as chemical sensors, nanotube composite materials, gas sensors, electron emitters, and so on. Further, nanotubes have a wide variety of applications in genetic engineering, biomedical fields, drugs, and artificial implants. Thus, it is vital to meticulously understand the basic limits of various characteristics of nanotubes to understand the proposed devices and other concepts.

Since last few years, computational approaches have received a lot of attention to explain various physicochemical properties, as these results have been found to be consistent and close to experimental values.[22] Density functional theory (DFT) is a very promising approach to accomplish this purpose. In some instances, DFT has provided much efficient and better results as compared to the experimental studies. In-depth understanding and information of nanotubes provide ways by which novel devices utilizing nanotubes can be materialized. For this purpose, it is very important to understand the CNTs and its applications through DFT approach.[23]

In this chapter, a detailed analysis of CNTs based on DFT techniques has been presented. The chapter has been divided into four sections, in the first section, emergence of CNTs and in the second section structure and its properties have been discussed. In Section 3, detailed investigation of CNTs by using DFT method have been elaborated and in Section 4, recent works based on CNTs and their applications in numerous fields have been reported.

2.2 DISCOVERY OF CNTs

Bacon have identified images of hollow and straight nanowhiskers with carbon layers having the two layers separated by the spacing as present in the planar layers of graphite.[24] In 1970, Endo reported the production of carbon fibers with a hollow core presence of catalytic particles at the end by the pyrolysis of ferrocene and benzene at 1000°C.[25] Bochvar and Galpern suggested that hollow molecules having a relatively large band gap can be formed by curling up sheets of graphene through their molecular orbital energy level calculations.[26] The study of carbon-based nanomaterials gained attention when needles of carbon made of hollow tubes were synthesized by the electric-arc technique by Iijima in 1991.[13] Based on the same concept, Iijima[14] and Bethune[19] have discovered the SWNTs. In 1996, Nobel Prize in Chemistry was awarded for the discovery of fullerenes (C_{60}) after which these materials gained further interest.[5]

2.3 STRUCTURE AND PROPERTIES OF CNTs

CNTs have remarkable structural characteristics with high structural precision and unique electronic, thermal, optical, mechanical, and transport properties. It is composed of only carbon atoms which form a hollow cylinder that can be visualized as a graphene sheet rolled into a seamless tube, uncapped or capped at the ends. When one s-orbital and two p-orbitals of carbon hybridize, three sp^2 orbitals are formed at 120° to each other within a plane. When many such carbon atoms combine with each other in a similar pattern, graphite is formed. The high stiffness and strength of the CNTs result from the strong covalent in-plane σ bonds which bind the atoms in the plane.[27] The other p-orbital that is present perpendicular to the plane of the σ bonds is responsible for the interlayer interaction and forms weak delocalized π bonds out of the plane.[28] CNTs have diameters in nanometer scale with very large aspect ratios (length/diameter). Having a large aspect ratio makes CNTs characteristics comparable to single molecules or quasi-one-dimensional crystals which show translational periodicity along the tube axis. CNTs can be classified into two categories: the SWNTs and the MWNTs. Lucid schematic representation is reported in Ref. [29]. SWNT comprises a single rolled up tubular shell of graphene sheet[30,31] which has diameters between 0.4 and 4 nm.[32] SWNTs have a tendency to self-assemble in the form of bundles or ropes due to existence of van der Waals forces between them. MWNT has a number of concentric cylinders of graphene sheet placed around a hollow core. Its diameter ranges from 5 to 100 nm and a spacing of 0.34 nm is present between the layers.[29] There are various methods by which a graphene sheet can be rolled to form open-ended or closed types of CNTs. When a single graphitic layer is folded in the form of a cylinder and its open edges match each other completely to form a seamless structure, an open-ended CNT is formed. If pentagons are inserted on both the ends of the cylinder, a closed CNT is produced. Its tubular shell constitutes hexagonal rings of carbon in the form of a sheet while its ends contain dome-shaped half fullerene molecules which help in capping. The curve in the walls is because of the rolling of the sheet in a cylindrical pattern while the curve in the capped ends is because of the occurrence of pentagonal ring defects in the hexagonal arrangement of the lattice. The pentagonal ring defect helps in giving a convex (positive) curvature to the surface that supports closing of both the ends of the tube. The resulting apex angle is dependent on the number of inserted pentagons and chemical reactivity is increased due to enhanced strain at these positions. Symmetries are responsible for the properties of CNTs and they can be determined through their helicity and diameter. The helicity or chirality of a CNT is given by a chiral vector and angle.

In the literature, single-wall nanotubes are represented as (n, m) which corresponds to integer indices of two graphene unit lattice vectors related to the chiral vector of a nanotube.[33] The (n, n) type nanotubes are known as armchair nanotubes as they have _/‾_/ shape. They are perpendicular to the tube axis, and are symmetrical along the axis with a short unit cell of about 0.25 nm that can be repeated to construct the entire segment of a long nanotube. Other nanotube is of the type (n, 0) and is called as zigzag nanotube because of the /\/\/ shape. It is perpendicular to the axis, and has a short unit cell of approximately 0.43 nm along the axis. Apart from these, all other nanotubes are referred as chiral or helical nanotubes, and they have longer unit cell sizes along the tube axis. Detailed descriptions of their symmetrical properties with different chiralities have been also investigated.[33,34]

Both SWNTs and MWNTs are attractive candidates in nanoscale material domains because they possess excellent elastic and mechanical properties which arise due to their 2D arrangement of carbon atoms in a graphene sheet. This arrangement allows large out-of-plane distortions and at the same time, keeps the graphene sheet remarkably strong against any in-plane rupture or distortion because of the presence of strong carbon–carbon in-plane bonds. These structural and material characteristics would perhaps allow nanotubes to be used as composite materials in near future which would be exceptionally lightweight, highly elastic, and incredibly strong. A single-wall nanotube depending on its chiral vector (n, m), where n and m are two integers, can be either conducting or semiconducting. When the difference (n − m) is a multiple of three, we obtain a conducting nanotube while if the difference (n − m) is other than multiple of three, we get a semiconducting nanotube. Besides, it is also possible to connect nanotubes with different chiralities creating nanotube hetero-junctions, which can form a variety of nanoscale molecular electronic device components. They also have a high aspect ratio and hence own fine electronic and mechanical properties. As a result of these properties, nanotubes are being used in field-emission displays or scanning probe microscopic tips for metrological purposes. In addition, their materialization has started in commercial sector. It can be used in the casting structures for making nanowires and nanocapsulates, nanoscale containers for molecular drug delivery, fuel storage and gas-storage devices for hydrocarbons, and also as gas or liquid filtration devices due to their hollow, tubular, and caged structure.[23]

2.4 COMPUTATIONAL APPROACHES TO STUDY CNTs

In view of the fact that the majority of physical properties of nanomaterials can be well explained by quantum mechanics, and a few of them manifest relatively different properties from classical ones,[33] it has been believed that the computational approach has contributed significantly to this field of study. High accuracy exists between the theoretical and experimental results which have encouraged a further scientific research in the CNTs.[34,35] A simple application of the ab initio or first-principles method, though, often faces a serious constraint due to its high computational costs. Generally, much large-sized calculations are needed for the ab initio method based on the DFT[36,37] that has been effectively established aiming at the solid states, that is, nanoscale materials, whose structures do not show periodicity and can be better regarded as large molecules instead of a solid. Many researchers have developed large-sized computation methods for deep understanding of these nanomaterials.[38] The ab initio or first principle method is a technique carried out using numerical algorithms for directly solving the complex quantum many-body Schrödinger equation.[39] A precise description of quantum mechanical behavior of materials properties can be derived through ab initio method despite the fact that the system size is at present restricted to merely about a few hundred atoms.

The rigorous mathematical foundation of the DFT is the basis of current ab initio simulation methods.[36,37] DFT states that a functional of system density is its ground state total electronic energy.[36]

$$E_T[\rho] = T[\rho] + \int v\rho d^3r \tag{2.1}$$

Here, $T[\rho]$ is kinetic energy of electrons and v is the external potential. The expression for interaction energies between the electrons is given by

$$U[\rho] = \frac{1}{2}\iint \frac{e^2\rho(\vec{R}_1)\rho(\vec{R}_2)}{\left|\vec{R}_1 - \vec{R}_2\right|} d^3\vec{R}_1\, d^3\vec{R}_2 + E_{xc}[\rho] \tag{2.2}$$

Supposing that the ground state charge density can be expanded by N orbitals,

$$\rho = \sum_{i=1}^{N} |\phi_i|^2 \tag{2.3}$$

The construction of Eigenvalue equation for ψ_i is possible and thus the calculation for the total energy can be made; N single particle equations can be produced using the Euler–Lagrange equation which, with respect to the variation of $\{\psi_1.\psi_2.\cdots\cdots\cdots\psi_N\}$, minimizes the total energy[38] as

$$(-\vec{\nabla}^2 + u(\vec{r}) + v_h(\vec{r}) + \mu_{xc}[\rho]\psi_i = \varepsilon_i\psi_i \tag{2.4}$$

where $v_h(r)$ and $\mu_{xc}[\rho]$ are potentials that describe the electron–electron interaction energy $U[\rho]$. This equation is normally attacked by the most common method, that is, by expanding the Kohn–Sham orbital ψ_i with a set of plane waves (PWs).[40] PWs can efficiently explain the electronic states in solids, whose wave numbers are integer multiples of the reciprocal lattice vectors of the crystal. According to Kohn and Sham,[36,37] together with all the exchange-correlation effects of electronic interactions, the DFT can be reformulated as a single-electron problem with self-consistent effective potential:

$$H_1 = \frac{\rho^2}{2m_e} + v_h(r) + X_{xc}[\rho(r)] + v_{ion-el}(r) \tag{2.5}$$

$$H_1\psi(r) = \varepsilon\psi(r), \text{ for all atoms} \tag{2.6}$$

This single-electron Schrödinger equation is the well-known Kohn–Sham equation and to approximate the unknown effective exchange-correlation potential, the local density approximation (LDA) has been introduced. The DFT–LDA method requires only the identity of the constituent atoms for successfully predicting material properties without use of any experimental inputs. The DFT–LDA method is applied for practical purposes invoking a pseudopotential approximation and a PW basis expansion of single-electron wave functions.[39] Using these approximations, the electronic structure problem is reduced to a self-consistent matrix diagonalization problem.

DFT can provide highly accurate self-consistent structures. A vigilant selection of an appropriate sized system is recommended to get benefitted by the complete capacity of DFT methods.

2.5 RESEARCHES BASED ON THE COMPUTATIONAL STUDY OF CNTs

To examine the effect of CNT size on the properties of the electronic structure of different junction models constructed through covalent linking between (6, 0) CNT and graphene nanoribbon (GNR) units, Omidvar et al.[41] performed DFT calculations. They calculated the HOMO–LUMO energy gap and chemical shielding tensors for various models of the investigated hybrids of GNR and CNT. Their results exhibit that the tube length and number of atoms have controlling effect on the HOMO–LUMO gap. Decrease in the HOMO–LUMO gap was observed with an increase in tube length. Dai et al.[42] investigated the potential of secondary CNT reinforcement with advanced carbon or glass hybrid-reinforced composite for wind energy applications. They used multiscale 3D unit cells for the simulation of fatigue behavior of carbon and glass fiber reinforced composites as well as hybrid with and without secondary CNT reinforcement. They established that better fatigue performances resulted in the composites having secondary CNT reinforcements as compared to those without reinforcements. Sanchez-Portal et al.,[43] through ab initio DFT computations, observed that the single-wall nanotubes stiffness is very much comparable to the in-plane stiffness of graphite. They also found that carbon-made SWNTs are the strongest amongst all other SWNTs such as boron nitride (BN) or $B_xC_yN_z$. The advancements in ab initio method utilizing the localized basis set were described by Park.[44] He reported that if bond length change is assumed to be effectively limited to five bonds that form the pentagon, the pentagonal strain energy can be modeled. DFT is also applied in the study of adsorption of a variety of gas molecules on SWNTs through first-principles calculations. The adsorption of NO_2 on to SWNTs was studied by Peng and Cho[45] utilizing this method. Zhao et al.[46] also used first principles method for studying the adsorption of various gas molecules, such as NO_2, O_2, NH_3, N_2, CO_2, CH_4, H_2O, H_2, and Ar, on single SWNT as well as SWNT bundles. The self-consistent field electronic structure calculations are based on DFT and can be performed with localized basis (DMol) or else with PW basis Cambridge Serial Total Energy Package (CASTEP). The properties of nanotubes may be influenced by a number of local defects, such as Stone Waals defect or dislocation of carbon atoms, and these effects have been also observed.[47,48] Rols et al.[49] and Kuzmanyet et al.[50] carried out computational analysis of powder diffraction spectra of SWNTs bundles using the general formulas for X-ray diffraction to study the effect of a variety of parameters such as the finite size of the bundles, the mean tube diameter, and the diameter dispersivity of the tubes. They concluded that these parameters influence the thickness and position of the peak to a considerable level.

2.6 APPLICATIONS

CNTs have fascinated scientists and engineers ever since their discovery in 1991[13] and gained immense importance due to their exceptional structural, electronic, and mechanical properties, in addition to their large application potential.[51–54] These materials have a broad spectrum of potential applications ranging from the field of nanoelectronics to nanoscale biotechnology. For instance, they may be used as electron field emitters,[51,53] molecular field effect transistors, artificial muscles,[51,54] or even in DNA sequencing.[55] Due to the presence of strong in-plane graphitic C–C bonds, CNTs become free from defects and turn out to be extremely strong and rigid against axial strains and are extraordinarily flexible against nonaxial strains. Nanotubes, in addition, also have excellent thermal and electrical conduction capabilities. Consequently, nanotubes are being proposed as additive fibers in lightweight multifunctional composite materials in various applications. Numerous experiments have come into light on the preparation and characterization of nanotube–polymer composite materials.[56–58] An augment in the strength of CNT-embedded polymer matrices was observed as compared to the bare polymer matrices according to the measurements. Nanotube ribbons were also produced by Vigolo et al.[56] through condensing nanotubes in the flow of a polymer solution. Preliminary results of the structural, thermal, and mechanical characterization of nanotube polymer composites were obtained which revealed that important characteristics such as thermal expansion and diffusion coefficients from the processing and applications perspective can be simulated for computational design of nanotube composite materials. These simulations demonstrate the large potential of computational nanotechnology-based investigations. For large system sizes and realistic interface between nanotubes and polymer, the simulation techniques and underlying multiscale simulations and modeling algorithms need to be developed and improved significantly before high reliability simulations can be attempted in the near future.[23] The potential of using hybrid and nanoreinforced composites for wind turbine and other long-term cyclic, high-humidity service conditions can be analyzed using computational methods. Since superior fatigue lifetime exists in carbon-reinforced composites than glass-reinforced composites, those with CNT reinforcement in the fiber sizing are an excellent prospective for exploiting in the large wind turbine blades, particularly, off-shore turbines. A high potential for hybrid carbon/glass composites having secondary CNT reinforcement exists over common composites with vast improvement of the materials performances.[42] Since its discovery, CNT is considered to be an appropriate material which can be useful in electron field emitter. CNT, being a very good conductor having a high aspect ratio, consumes very less power in the process of electron field emission and thereby provides a big advantage to be used as an electron field emitter. This advantage has been demonstrated in various experiments with SWNTs[59,60] and MWCTs.[61–66] A 5 in. nanoemissive display has also been developed by Motorola.[67] At present, 20–30% power savings have been noticed with the use of these CNT-based displays over existing flat panel displays. The measurements of electron transfer in single SWNT were initiated by the Delft group. This group constructed the first single molecule field-effect transistor using only one semiconducting SWNT.[68] This device consists of two metal electrodes bridged through a nanotube and it functions at room temperature. Nanotube–PPV composite, SWNT- and MWNT-filled poly(*m*-phenylenevinylene-*co*-2,5-dioctoxy-*p*-phenylenevinylene) (a conjugated luminescent polymer),[69] allows an enhancement in electrical conductivity in comparison to the pure polymer bearing a small loss in photoluminescence or electroluminescence yield. Also, the strength of the composite is much better than the polymer in its pure state as regards photobleaching and its mechanical strength. Nanoprobes are another decisive application of nanotubes which can be materialized due to its very small size, high mechanical strength, stretchability, and superior conductivity. The exploitation of such probes in various areas such as nanoelectrodes, high-resolution imaging, sensors, nanolithography, field emitters, and drug delivery can be imagined without any doubt. It has been reported that a single SWNT or MWNT is used for lithography[70] as well imaging[71] by joining it to the end of a scanning tunneling microscope (STM) or atomic force microscope (AFM) tip. Ultimately, for the production and storage of energy, CNTs are being considered of as potential materials. Batteries, fuel cells and numerous other electrochemical applications are till now utilizing carbon fiber electrodes.[72,73] Since nanotubes possess high surface specificity and are greatly efficient in carrying out electron-transfer reactions, they are unique from all other available alternatives and consequently, can be applied in the energy production and storage.

Gas sensors are one of the rigorous research interest areas because of the growing demand of rapid response, sensitive, and stable sensors for environmental monitoring, industries, biomedicine, and so on. Nanotechnology has brought enormous possibility for constructing highly sensitive, low-cost, low power-consuming portable sensors.[74] Meyyappan[75] conducted theoretical and simulation works to comprehend this nanoscaled material and associated phenomenon.

Functionalized CNTs are exploited for targeting of amphotericin B to cells.[76] Nanotubes, when given with antibiotic doxorubicin, are reported to improve intracellular penetration of doxorubicin. The gelatin CNT mixture, a hydrogel, has been used in biomedicine as a potential carrier system.[77] Due to gastric environmental conditions and presence of enzymes, denaturation of erythropoietin (EPO) takes place and hence it has not been possible so far to administer EPO orally. Use of CNT-based carrier system can facilitate oral administration of EPO effectively.[77] Moreover, CNTs, because of their nanosize and sliding nature of graphite layers bound with van der Waals forces, can be used as lubricants in tablet manufacturing.[77] CNTs are being used in genetic engineering to control genes and atoms in the development of bioimaging genomes, proteomics, and tissue engineering, and also in artificial implants.[78]

2.7 CONCLUSION

Ever since the unearthing of CNTs in 1991, they have been rising immensely and progressively and have found applications in diverse fields. The structure–topology–property relationship studies in nanotubes have been highly complemented by theoretical and computational modeling. Among all the computational approaches, the analysis invoking DFT is very popular. Theoretical and simulation research, involving DFT, not only elucidate the observed experimental results with in-depth mechanism but can also simulate the CNTs behaviors under certain assumptions of interest before carrying out the actual experiment. DFT methods are extremely competent and afford highly precise self-consistent structures. The best structure that could be constructed from the graphitic arrangement is a nanotube and it reproduces the physical properties of in-plane graphite. Due to the presence of superior in-plane graphite properties, some exclusive properties owing to the small dimensions, lattice helicity, and closed topology, nanotubes serve as a well of properties making it extraordinary. It is very difficult to trace any other material that encompasses such a wide range of potential properties. In the future, many of these remarkable properties can be utilized for realistic applications. These minute structures that self-assemble with flawlessness could one day become an imperative material in our daily lives.

KEYWORDS

- **nanomaterials**
- **carbon nanotubes**
- **density functional theory**
- **composite materials**
- **computation**

REFERENCES

1. Roco, M. C.; Williams, R. S.; Alivisatos, A. P. *Nanotechnology Research Directions*; Kluwer Academic Publishers: Boston, 2000.
2. Rao, C. N. R.; Cheetham, A. K. *J. Mater. Chem.* **2001,** *11*, 2887.
3. Dresselhaus, M. S.; Dresselhaus, G.; Eklund, P. C. *Science of Fullerenes and Carbon Nanotubes*; Academic Press: New York, 1996.
4. Rohlfing, E. A.; Cox, D. M.; Kaldor, A. *J. Chem. Phys.* **1984,** *81*, 3322.
5. Kroto, H. W.; Heath, J. R.; O'Brien, S. C.; Curl, S. C.; Smalley, R. E. *Nature* **1985,** *318*, 162.
6. Ebbesen, T. W. *Carbon Nanotubes: Preparation and Properties*; CRC Press: Boca Raton, FL, 1997.
7. Kiang, C. H.; Goddard, W. A.; Beyers, R.; Bethune, D. S. *Carbon* **1995,** *33*, 903.
8. Ajayan, P. M.; Ebbesen, T. W. *Rep. Prog. Phys.* **1997,** *60*, 1025.
9. Yakabson, B. I.; Smalley, R. E. *Am. Sci.* **1997,** *85*, 324.
10. Saito, R.; Dresselhaus, M. S.; Dresselhaus, G. *Physical Properties of Carbon Nanotubes*; World Scientific: New York, 1998.
11. Mintmire, J. W.; Dunlap, B. I.; Carter, C. T. *Phys. Rev. Lett.* **1992,** *68*, 631.
12. Hamada, N.; Sawada, S.; Oshiyama, A. *Phys. Rev. Lett.* **1992,** *68*, 1579.
13. Iijima, S. *Nature* **1991,** *354*, 56.
14. Iijima, S.; Ichihashi, T. *Nature* **1993,** *363*, 603.
15. Ge, M.; Sattler, K. *Science* **1993,** *260*, 515.
16. Wildoer, J. W. G.; Venema, L. C.; Rinzler, A. G.; Smalley, R. E.; Dekker, C. *Nature* **1998,** *391*, 59.
17. Odom, T.; Huang, J.; Kim, P.; Lieber, C. *Nature* **1998,** *391*, 62.
18. Dresselhaus, M. S.; Dresselhaus, G.; Sugihara, K.; Spain, I. L.; Goldberg, H. A. *Graphite Fibers and Filaments*; Springer-Verlag: New York, 1988.
19. Bethune, D. S.; Kiang, C. H.; de Vries, M. S.; Gorman, G.; Savoy, R.; Vazquez, J.; Beyers, R. *Nature* **1993,** *363*, 605.
20. Thess, A.; Lee, R.; Nikolaev, P.; Dai, H.; Petit, P.; Robert, J.; Xu, C.; Lee, Y. H.; Kim, S. G.; Rinzler, A. G.; Colbert, D. T.; Scuseria, G. E.; Tomanek, D.; Fischer, J. E.; Smalley, R. E. *Science* **1996,** *273*, 483.

21. On, D. T.; Desplantier-Giscard, D.; Danumah, C.; Kaliaguine, S. *Appl. Catal. A Gen.* **2001,** *222*, 299.
22. Ribeiro, M. S.; Pascoini, A. L.; Knupp, W. G.; Camps, I. *Appl. Surf. Sci.* **2017,** *426*, 781.
23. Srivastava, D.; Wei, C.; Cho, K. *Appl. Mech. Rev.* **2003,** *56* (2), 215.
24. Bacon, R. J. *Appl. Phys.* **1960,** *31*, 283.
25. Oberlin, A.; Endo, M.; Koyoma, T. *J. Cryst. Growth* **1976,** *32*, 335.
26. Bochvar, D. A.; Gal'pern, E. G. *Proc. Acad. Sci. USSR* **1973,** *209*, 610.
27. Dumitrica, T.; Hua, M.; Yakobson, B. I. *Proc. Natl. Acad. Sci. USA* **2006,** *103* (16), 6105.
28. Qian, D.; Wagner, G. W.; Liu, W. K. *Appl. Mech. Rev.* **2002,** *55* (6), 495.
29. Basu-Dutt, S.; Minus, M. L.; Jain, R.; Nepal, D.; Kumar, S. *J. Chem. Educ.* **2012,** *89*, 221.
30. Saito, R.; Dresselhaus, G.; Dresselhaus, M. S. *Physical Properties of Carbon Nanotubes*; Imperial College Press: London, 1998.
31. Dresselhaus, M. S.; Dresselhaus, G.; Avouris, Ph. *Carbon Nanotubes: Synthesis, Structure, Properties, and Applications*; Springer-Verlag: Berlin, Heidelberg, 2001.
32. Rao, C. N. R.; Satishkumar, B. C.; Govindaraj, A.; Nath, M. *Chem. Phys. Chem.* **2001,** *2*, 78.
33. Frank, S.; Poncharal, P.; Wang, Z. L.; de Heer, W. A. *Science* **1998,** *280*, 1744.
34. Wildöer, J. W. G.; Venema, L. C.; Rinzler, A. G.; Smalley, R. E.; Dekker, C. *Nature* **1998,** *391*, 59.
35. Odom, T. W.; Huang, J.; Kim, P.; Lieber, C. M. *Nature* **1998,** *391*, 62.
36. Hohenberg, P.; Kohn, W. *Phys. Rev.* **1964,** *136*, B864.
37. Kohn, W.; Sham, L. J. *Phys. Rev.* **1965,** *140*, A1133.
38. Park, N.; Lee, K.; Han, S.; Yu, J.; Ihm, J. *Phys. Rev. B* **2002,** *65*, 121405.
39. Payne, M. C.; Teter, M. P.; Allan, D. C.; Arias, T. A.; Joannopoulos, J. D. *Rev. Mod. Phys.* **1992,** *68*, 1045.
40. Ihm, J.; Zunger, A.; Cohen, M. L. *J. Phys. C* **1979,** *12*, 4409.
41. Omidvar, A; Anafcheh, M.; Hadipour, N. L. *Sci. Iran. Trans. F Nanotechnol.* **2013,** *20*, 1014.
42. Dai, G.; Mishnaevsky, L. Jr. *Composites Part B* **2015,** *78*, 349.
43. Sanchez-Portal, D.; Artacho, E.; Solar, J. M.; Rubio, A.; Ordejon, P. *Phys. Rev. B* **1999,** *59*, 12678.
44. Park, N. *J Nano Bio Tech* **2005,** *2* (2), 59.
45. Peng, S.; Cho, K. *Nanotechnology* **2000,** *11* (2), 57.
46. Zhao, J.; Buldum, A.; Han, J.; Lu, J. P. *Nanotechnology* **2002,** *13* (2), 195.
47. Belytschko, T.; Xiao, S. P.; Schatz, G. C.; Ruoff, R. S. *Phys. Rev. B* **2002,** *65*, 235430.
48. Maiti, A.; Svizhenko, A.; Anantram, M. P. *Phys. Rev. Lett.* **2002,** *88* (12), 126905.
49. Rols, S.; Almairac, R.; Henrard, L.; Anglaret, E.; Sauvajol, J. *Eur. Phys. J. B* **1999,** *10*, 263.
50. Kuzmany, H.; Plank, W.; Hulman, M.; Kramberger, Ch.; Gruneis, A.; Pichler, Th.; Peterlik, H.; Kataura, H.; Achiba, Y. *Eur. Phys. J. B* **2001,** *22*, 307.
51. Minett, A.; Schüth, F.; Sing, S. W.; Weitkapmp, J.; Atkinson, K.; Roth, S. Carbon Nanotubes. In *Handbook of Porous Solids*; Wiley-VCH: Weinheim, Germany, 2002.
52. Yao, Z.; Postma, H. W. C.; Balents, L.; Dekker, C. *Nature* **1999,** *402*, 273.
53. Zhou, O.; Shimoda, H.; Gao, B.; Oh, S.; Fleming, L.; Yue, G. *Acc. Chem. Res.* **2002,** *35*, 1045.
54. Baughman, R. H.; Cui, C.; Zakhidov, A. A.; Iqbal, Z.; Barisci, J. N.; Spinks, G. M.; Wallace, G. G.; Mazzoldi, A.; De Rossi, D.; Rinzler, A. G.; Jaschinski, O.; Roth, S.; Kertesz, M. *Science* **1999,** *284*, 1340.

55. Gao, H.; Kong, Y.; Cui, D.; Ozkan, C. S. *Nano Lett.* **2003,** *3*, 471.
56. Vigolo, B.; Penicaud, A.; Coulon, C.; Sauder, C.; Pailler, R.; Journet, C.; Bernier, P.; Poulin, P. *Science* **2000,** *290*, 1331.
57. Schadler, L. S.; Giannaris, S. C.; Ajayan, P. M. *Appl. Phys. Lett.* **1998,** *73*, 3842.
58. Andrews, R.; Jacques, D.; Rao, A. M.; Rantell, T.; Derbyshire, F.; Chen, Y.; Chen, J.; Haddon, R. C. *Appl. Phys. Lett.* **1999,** *75*, 1329.
59. Saito, Y.; Hamaguchi, K.; Nishino, T.; Hata, K.; Tohji, K.; Kasuya, A.; Nishina, Y. *Jpn. J. Appl. Phys.* **1997,** *36*, L1340.
60. Bonard, J.; Salvetat, J.; Stockli, T.; de Heer, W. A. *Appl. Phys. Lett.* **1998,** *73*, 918.
61. Rinzler, A. G.; Hafner, J. H.; Nikolaev, P.; Lou, L.; Kim, S. G.; Tománek, D.; Nordlander, P.; Colbert, D. T.; Smalley, R. E. *Science* **1995,** *269*, 1550.
62. De Heer, W. A.; Chatelain, A.; Ugarte, D. *Science* **1995,** *270*, 1179.
63. Saito, Y.; Hamaguchi, K.; Uemura, S.; Uchida, K.; Tasaka, Y.; Ikazaki, F.; Yumura, M.; Kasuya, A.; Nishina, Y. *Appl. Phys. A* **1998,** *67*, 95.
64. Wang, Q. H.; Corrigan, T. D.; Dai, J. Y.; Chang, R. P. H. *Appl. Phys. Lett.* **1997,** *70*, 3308.
65. Collins, P. G.; Zettl, A. *Appl. Phys. Lett.* **1996,** *69* (13), 1969.
66. Collins, P. G.; Zettl, A. *Phys. Rev. B* **1997,** *55*, 9391.
67. Motorola. http://www.motorola.com/content.jsp?globalObjectId=8206 (accessed 14 May, 2018).
68. Tans, S. J.; Verschueren, A. R. M.; Dekker, C. *Nature* **1998,** *393*, 49.
69. Curran, S.; Ajayan, P. M.; Blau, W.; Carroll, D.; Coleman, J.; Dalton, A.; Davey, A. P.; McCarthy, B.; Strevens, A. *Adv. Mater.* **1998,** *10*, 1091.
70. Dai, H.; Franklin, N.; Han, J. *Appl. Phys. Lett.* **1998,** *73*, 1508.
71. Dai, H. J.; Hafner, J. H.; Rinzler, A. G.; Colbert, D. T.; Smalley, R. E. *Nature* **1996,** *384*, 147.
72. Dresselhaus, M. S.; Dresselhaus, G.; Sugihara, K.; Spain, I. L.; Goldberg, H. A. *Graphite Fibers and Filaments*; Springer-Verlag: New York, 1988.
73. Che, G.; Lakshmi, B. B.; Fisher, E. R.; Martin, C. R. *Nature* **1998,** *393*, 346.
74. Wang, Y.; Yeow, J. T. W. *J. Sens.* **2009,** *2009*, 1. DOI:10.1155/2009/493904.
75. Meyyappan, M. *Carbon Nanotubes: Science and Applications*; CRC Press: Boca Raton, FL, USA.
76. Barroug, A.; Glimcher, M. *J. Orthop. Res.* **2002,** *20*, 274.
77. Pai, P.; Nair, K.; Jamade, S.; Shah, R.; Ekshinge, V.; Jadhav, N. J. *Curr. Pharm. Res.* **2006,** *1*, 11.
78. Deng, P.; Xu, Z.; Li, J. *J. Pharm. Biomed. Anal.* **2013,** *76*, 234.

CHAPTER 3

CARBON NANOTUBE COMPOSITES AS PHOTOCATALYTIC MATERIALS

RAKSHIT AMETA[1*], NEHA KAPOOR[2], BHAVYA PATHAK[2], JAYESH BHATT[2], and SURESH C. AMETA[2]

[1]*Department of Chemistry, J.R.N. Rajasthan Vidyapeeth, Udaipur 313003, Rajasthan, India*

[2]*Department of Chemistry, PAHER University, Udaipur 313003, Rajasthan, India*

Corresponding author. E-mail: rakshit_ameta@yahoo.in

ABSTRACT

In last few decades, the growing industrial activities have resulted in an increasing amount of pollutants in the environment, which has become a burning problem all over the globe and it still is a challenge for many countries. Carbon nanotubes (CNTs) have attracted considerable attention because of their special structure and high mechanical strength, which makes them potential candidates for advanced composites. They can be semiconducting, semimetallic, or metallic in nature. CNTs are also efficient adsorbents due to their large specific surface area and hollow and layered structures. The chemical, physical, and mechanical properties of CNTs find applications in diverse fields. There are different types of nanotubes including single-walled nanotubes, double-walled nanotubes, and the multi-walled nanotubes. Apart from their different applications, these can be used as a photocatalytic material in pure form or in the form of a composite with some photocatalyst. Thus, CNTs can be used as a promising material in environmental cleaning. Here, use of CNTs as a photocatalytic material has been discussed.

3.1 CARBON NANOTUBES

Naming a discoverer of carbon nanotube (CNT) had been a controversial subject. Monthioux and Kuznetsov[36] in their editorial have given an interesting origin of the CNT, which is often misstated. Major literature attributes the discovery of CNT to Iijima,[20] but it seems that initial discovery of CNT goes back much further than that. Radushkevich and Lukyanovich[46] published images of 50 nm diameter tubes made of carbon, but it remained largely unnoticed. They should be given the credit for the CNTs. Later, Oberlin et al.[40] observed hollow tubes of rolled up graphite sheets synthesized by a chemical vapor-growth technique. Endo[17] reported in a review that he has observed a hollow tube, which is linearly extended with parallel carbon layer faces near the fiber core and it is multiwalled carbon nanotube (MWCNT) at the center of the fiber.

Abrahamson et al.[2] presented an evidence of CNTs at the 14th Biennial Conference of Carbon in 1979 and described CNTs as carbon fibers, which were produced on carbon anodes during arc discharge. Characterization of these fibers and hypotheses for their growth in a nitrogen atmosphere at low pressures were given. A group of Soviet scientists (1981) observed that carbon multilayer tubular crystals were formed by rolling graphene layers into cylinders, many different arrangements of graphene hexagonal nets are possible, but two possibilities are common. These are circular arrangement (armchair nanotube) and a spiral, helical arrangement (chiral tube). Tennent[53] got a patent for the production of cylindrical discrete carbon fibrils with a constant diameter between about 3.5 and 70 nm with lengths about 100 times its diameter. Mintmire et al.[34] predicted that if single-walled carbon nanotubes (SWCNTs) could be made, then they would show remarkable conducting properties. Bethune et al.[6] and Iijima and Ichlhashi[21] developed methods to produce SWCNTs.

Depending on the diameter of nanoparticle, an individual SWCNT may be there or in the form of a bundle of tightly packed SWCNTs, where a different structure is obtained as MWCNTs. MWCNTs consist of a concentric set of SWCNTs. The spacing between the walls in a MWCNT is almost 0.34 nm, which is the basal plane separation in graphite. These concentric tube walls are not chemically bonded to each other, but are weakly held by the van der Waals force and one tube may slip within another without any significant damage to its structure. MWCNTs can have many cylindrical shells (as many as 50) with an inner diameter (~1–2 nm) and lengths in the tens of millimeters. These are shown in Figures 3.1 and 3.2.

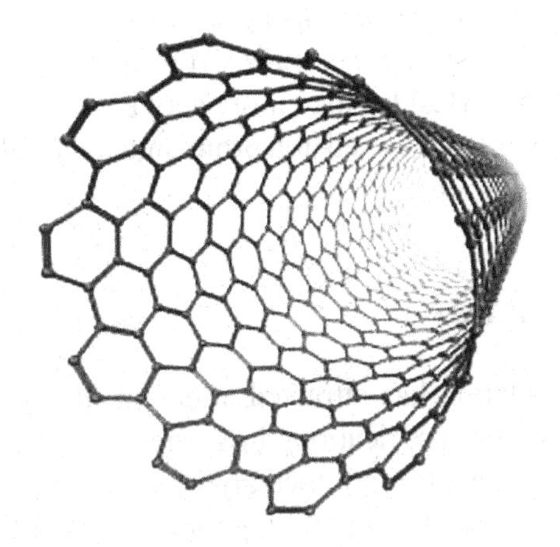

FIGURE 3.1 Single-walled carbon nanotube (SWCNT).

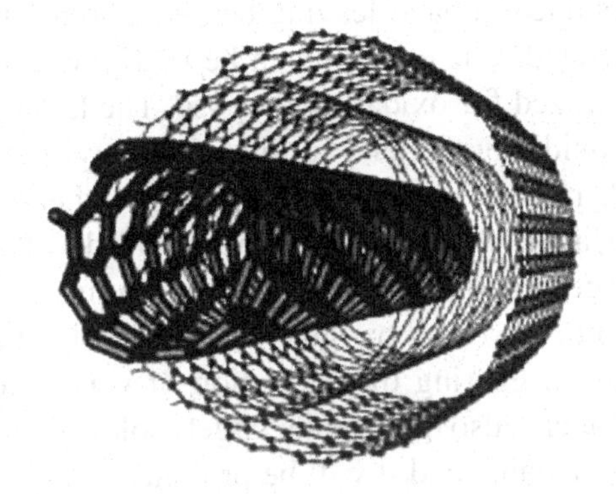

FIGURE 3.2 Multiwalled carbon nanotube (MWCNT).

CNTs are the strongest (very high tensile strength) and stiffest materials (elastic modulus). These are either metallic or semiconducting along the tubular axis.

Several methods have been used to prepare high-purity CNTs so as to have a control on wall thickness, length, and acceptable price. Some of the important ones are[33]:

- arc discharge,
- chemical vapor deposition (CVD),
- plasma-enhanced CVD method, and
- laser ablation of graphite.

CNTs are quite efficient adsorbents because of their large specific surface area, hollow and layered structures, and apart from it, the presence of π-bond electrons on the surface. Besides that, more active sites can be created on the nanotubes. Thus, CNTs can be used as a promising material in environmental cleaning also.

3.2 PHOTOCATALYSIS

Photocatalytic degradation has attracted attention of scientists all over the globe due to its potential applications in wastewater treatment assisting in conserving the environment. A photocatalytic reaction is the process, which is carried out in presence of light and a photocatalyst (PC) (semiconductor material). When a photocatalytic material is exposed to the light of a suitable wavelength (depending on its band gap) it will excite an electron from its valence band to conduction band leaving behind a hole. Thus, electron and hole pairs are generated. The electron can be used for reducing a substrate while the hole is utilized for oxidation purpose. The holes react with water molecules or hydroxide ions (OH^-) forming hydroxyl radicals ($^{\bullet}OH$). The generation of such radicals depends on the pH of the system. Pollutants adsorbed on the surface of the catalyst will be oxidized by these $^{\bullet}OH$ radicals, which are the strongest oxidizing agent next to fluorine.

On the other hand, dissolved oxygen may also abstract an electron form the conduction band triggering the formation of very reactive superoxide radical ion ($O_2^{\bullet-}$) that can also oxidize the target molecule. In acidic medium, this radical ion is not stable and it will be protonated to form hydroperoxyl radical ($HO^{\bullet}_2$), of course it is relatively a mild oxidant. This oxidation/reduction is responsible for degrading organic pollutants, sometimes even recalcitrant molecules. Organic pollutants are efficiently degraded producing harmless or less harmful compounds. Main products of this type of degradation are water, carbon dioxide, some inorganic anions, etc.

3.3 CNT-PC COMPOSITES

3.3.1 TITANIUM DIOXIDE

Titanium dioxide has been commonly used as a photocatalytic material but its photocatalytic activity is suppressed by its large band gap and the relatively higher recombination rate of the electron–hole (e^- and h^+) pairs.

Kasuga et al.[22] have successfully modified titania nanoparticles (n-TiO_2) into nanotubes, which have good electronic and mechanical characteristics, larger surface area and pore volume apart from photocatalytic activity. Titania may be able to interact relatively more conveniently with other materials in the form of nanotubes than nanoparticles. Use of PC has its own limitations decreasing its efficiency in wastewater treatment and these are:

- low adsorption in visible range and
- high rate of electron–hole recombination.

Different efforts were made to resolve these problems using doping, metallization, sensitization, composite formation, etc. In this context, PCs were also combined with some other materials so as to enhance their photocatalytic performance. One attempt in this direction has been made by adding a coadsorbent such as CNTs, because it has been reported that CNT has many special characteristics, such as unique electrical properties, high chemical stability, and high adsorption ability. Yu et al.[65] observed that the adsorption ability of CNT may increase the photocatalytic activity of TiO_2 (P25). In addition to this, the presence of CNT on the surface of the PC may trap electron released by titania (available in its conduction band). In other words, CNTs can act as electron sink. This process will prevent the process of electron–hole recombination process. Eder and Windle[15] reported that the interaction between n-TiO_2 and CNT in CNT/TiO_2 composite may enhance the photocatalytic activity in three ways, which are:

- storing or trapping electrons,
- creating desired band gap energy, and
- allowing visible light absorption.

The degradation rate of pollutant can be enhanced by reducing the electron–hole recombination rate; preventing the particles agglomeration; and increasing the adsorption capacity. Several methods have been tried to improve the photocatalytic efficiency of a PC. These include:

- increasing the surface area of the metal oxide by synthesizing nano-sized materials,
- generation of defect structures to induce space-charge separation and thus, reducing the recombination of electron–hole,
- modification of the semiconductors with metal or other semiconductor, and
- adding a cosorbent such as silica, alumina, zeolite, or clay.

TiO_2-developed charge separation on excitation by light resulted in generation of electron–hole pairs. Normally, these electron–hole pairs recombine quickly and only a fraction of the electrons and pairs are consumed in the photocatalytic reaction. Such rapid recombination rate of electron–hole pairs leads to low photocatalytic activity.[62] With the presence of CNT in the composites, the photoexcited electrons are transported to the neighboring CNT, prolonging the lifetime of electron–hole pairs, and thus, enhancing the photocatalytic activity for CO_2 reduction. Furthermore, CNT plays the role of an electron sink, which enhances the density of photoexcited electrons and encourages the formation of CH_4.

CNT/TiO_2 nanocomposite PC has been prepared by a simple impregnation method by Xu et al.[60] and used for gas-phase degradation of benzene. It was found that the as-prepared CNT/TiO_2 nanocomposite exhibits higher photocatalytic activity for benzene degradation, as compared with commercial titania (Degussa P25). It was also found true for liquid-phase degradation of methyl orange. Here, CNTs play two kinds of crucial roles in enhancement of photocatalytic activity of TiO_2 and these are:

- To act as electron reservoir and assist in trapping electrons emitted in conduction band of TiO_2 particles generated irradiation with UV light; consequently, electron–hole pair's recombination is hindered.
- To act as a dispersing template or support to control the morphology of TiO_2 particles in the CNT/TiO_2 nanocomposite.

Yu et al.[63] synthesized titanium dioxide/CNTs composite (TiO_2/CNTs) in presence of ultrasonic irradiation. The photocatalytic activity was evaluated for the degradation of acetone. It was found that the crystalline TiO_2 was composed of both anatase and brookite phases. The agglomerated morphology and the particle size of TiO_2 in the composites change in the presence of CNTs. The addition of CNTs was not found to affect the mesoporous nature of the TiO_2, but it increases the surface area. It was observed that higher is the content of CNTs, the suppression of the recombination of photogenerated e^-/h^+ pairs was more, but excessive CNTs may also shield TiO_2 from absorbing UV light. The optimal amount of TiO_2 and CNTs was observed in the range of 1:0.1 and 1:0.2 (feedstock molar ratio).

Djokic et al.[13] reported the preparation of TiO_2/MWCNTs nanocomposite by precipitation of anatase TiO_2 nanoparticles onto differently oxidized CNTs. Titanium (IV) bromide was hydrolyzed producing pure anatase phase TiO_2 nanoparticles decorated on the surface of the oxidized

CNTs. The oxidative treatment of these CNTs influenced the type, quantity, and distribution of oxygen-containing functional groups, which had a significant influence on the electron transfer properties. Photocatalytic activity of the synthesized nanocomposites was evaluated in photodegradation of reactive orange 16, which showed that it has better photocatalytic activity as compared to the commercial PC Degussa P25. Current process of cyclohexane oxidation suffers from low conversion, poor selectivity, and excessive production of waste.

Mahmoodi[30] used CNT and n-TiO_2 to degrade dyes using single (UV/CNT/H_2O_2 and UV/n-TiO_2/H_2O_2) and binary catalyst (UV/CNT/n-TiO_2 and UV/CNT/n-TiO_2/H_2O_2) systems. They used direct red 23 and direct red 31 as model systems. Photocatalytic dye degradation was monitored using UV–Vis spectrophotometry and ion chromatography. The effects of initial dye concentration and salt on dye degradation were investigated. Formate, acetate, oxalate, nitrate, and sulfate were detected as dye mineralization products.

Dai et al.[10] prepared nanocomposite catalysts (MWCNT–TiO_2) hydrothermally from MWCNTs and titanium sulfate as the titanium source. Photocatalytic activity of these composites on splitting water was observed using triethanolamine as an electron donor. Hydrogen was successfully produced over the Pt/MWCNT–TiO_2 under visible light irradiation ($\lambda > 420$ nm) while no water splitting was observed on the Pt-loaded pristine TiO_2 and MWCNTs. Hydrogen generation rate using full spectral irradiation of a Xe lamp of up to 8 mmol g^{-1} h^{-1} or more was achieved. This significant photocatalytic activity of the nanocomposites was attributed to the synergetic effect of the intrinsic properties of its components such as excellent light absorption and charge separation on the interfaces between the modified MWCNTs and TiO_2.

Composites of CNTs–titanate nanotubes (MWCNTs/TNTs) were successfully synthesized by Vu et al.[55] the hydrothermal reaction process. The charge recombination of MWCNTs/TNTs composites was examined and the photocatalytic activity of TNTs and MWCNTs/TNTs composites was evaluated on the basis of the model oxidation of H_2S.

A novel approach has been presented by Pyrgiotakis et al.[45] for enhancing photocatalytic activity in a nanocomposite with high aspect ratio of CNT. Composite nanoparticles were synthesized with sol–gel nanocoating on MWCNTs and their photocatalytic efficiencies (Degussa P25 and TiO_2 nanocoated MWCNTs) were evaluated by azo dye degradation. It was observed that the nanocomposite was having superior photocatalytic activity with UVA and visible light.

Yu et al.[64] observed the effect of CNTs on the adsorption and the photocatalytic properties of TiO_2 (P25) for the treatment of monoazo dye, procion red MX-5B, diazo dyes, procion yellow HE4R, and procion red HE3B. The results showed that CNTs can better improve the adsorption of the dyes onto P25 compared with activated carbon (AC). It may be due to the strong interaction between P25 and CNTs. Residual total organic carbon (TOC) in the solutions and the amount of cyanuric acid evolved after degradation indicated that the adsorption ability of P25 was enhanced by CNTs. CNTs also facilitate the photocatalytic activity of P25 in the degradation of these azo dyes more effectively than AC. The excited electron in conduction band of TiO_2 may migrate into CNTs because of special structure and the ability for electron transport. It resulted in the possibility of a decrease in the recombination of e^-/h^+ pairs. O_2 adsorbed on the surface of CNTs may also accept electron to form oxygen anion radical, $O_2\bullet^-$, which also leads to the formation of •OH in the system.

Kuo[25] also presented an application of CNTs to increase the photocatalytic activity of TiO_2. The effect of CNTs dose on the decolorization efficiency of aqueous reactive red 2 (RR2) was observed. The effects of SO_4^{2-} formation and removal of TOC were also determined. Decolorization of RR2 was monitored spectrophotometrically. It was indicated that photodegradation in the presence of combination of TiO_2 with CNTs as well as electron transfer was higher for 410 nm irradiation than for 365 nm. The electron transfer in the TiO_2/CNTs composites reduced the electron/hole recombination and increased the photodegradation efficiency.

MWCNT/TiO_2 nanocomposites with different crystallographic forms of TiO_2 (anatase, A-TiO_2; and rutile, R-TiO_2) were synthesized by Chaudhary et al.[7] via hydrothermal method. Photocatalytic activity of as-prepared TiO_2 nanocomposites was investigated for the degradation of methylene blue under UV irradiation. Enhanced photocatalytic activity of MWCNT/TiO_2 nanocomposites was due to efficient transfer of photogenerated electron from TiO_2 to MWCNT. The MWCNT/A-TiO_2 nanocomposite also showed three times higher photocatalytic activity and faster degradation rate than MWCNT/R-TiO_2. The difference has been attributed to efficient generation of •OH radicals in A-TiO_2 as compared to R-TiO_2.

Photocatalytic degradation of tetracycline using MWCNT/TiO_2 nanocomposite was investigated by Ahmadi et al.[3] under UVC irradiation. The effect of different operational parameters was studied such as pH, irradiation time, PC dosage, weight ratio of MWCNT to TiO_2, and concentration of tetracycline. Complete removal of tetracycline concentrations up to 10 mg L^{-1} was obtained at MWCNT:TiO_2 ratio 1.5 (w/w%); pH = 5; and PC dosage 0.2 g L^{-1}. Mineralization was about 37% in these reaction conditions for initial tetracycline concentration of 10 mg L^{-1}, which reached to 83% after 300 min based on TOC analysis. The chemical oxygen demand (COD) concentration of 2267 mg L^{-1} decreased to 342 mg L^{-1} after 240 min in the same operational conditions for real pharmaceutical wastewater samples.

Highly active MWCNTs–TiO_2 composite PCs were prepared by sol–gel method by Wang et al.[57] Its photocatalytic activity for the degradation of 2,6-dinitro-p-cresol (DNPC) was obtained in aqueous solution under solar irradiation. They showed that there was a red shift in UV–Vis spectrum compared with pure TiO_2. The degradation of DNPC by this composite under solar irradiation was studied by varying the parameters such as pH, irradiation time, the initial substrate concentration, reaction temperature, catalyst concentration, etc. The optimal conditions were obtained as DNPC concentration = 33.4 mg L^{-1}; pH = 6.0; and MWCNTs–TiO_2 concentration = 6.0 g L^{-1}; under solar irradiation for 150 min. The presence of MWCNTs was found to enhance the photoefficiency of TiO_2. The highest efficiency on photodegradation of DNPC was achieved with an optimal MWCNTs/TiO_2 mass ratio as 0.05%. The PC can be used for five cycles with photocatalytic degradation efficiency maintained higher than 96%.

Wang et al.[56] also synthesized nanosized MWCNTs/TiO_2 composite and neat TiO_2 PCs via sol–gel technique using tetrabutyl titanate as a precursor. These samples were evaluated for their photocatalytic activity toward the degradation of2,4-dinitrophenol (DNP) under solar irradiation. It was observed that the addition of MWCNTs could remarkably improve the photocatalytic activity of TiO_2. The maximum rate of DNP degradation was found with MWCNTs:TiO_2 ratio of 0.05% (w/w). The effects of different parameters such as pH, irradiation time, catalyst concentration, DNP concentration, etc., were studied on the photocatalytic activity of the composite. The optimal conditions obtained were: initial DNP concentration = 38.8 mg L^{-1}; pH = 6.0; and catalyst concentration = 8 g L^{-1}; under solar irradiation for 150 min. The degree of photocatalytic degradation of DNP was found to increase with an increase in temperature. The MWCNTs/TiO_2 composite was also found to be quite effective in the decolorization and COD reduction of real wastewater samples from DNP manufacturing units. The composite has good recyclization ability.

A visible light PC of MWCNTs decorated with TiO_2 nanoparticles (MWCNT/TiO_2) was also synthesized by a two-step method,[68] where TiO_2 was first mounted on MWCNT surfaces by hydrolysis of tetrabutyl titanate and then was crystallized into anatase nanocrystal in a vacuum furnace at

500°C. The MWCNT/TiO_2 composite was able to absorb large amount of photoenergy in the visible light region, driving photochemical degradation of methylene blue effectively. More •OH radicals were produced by the MWCNT/TiO_2 (1:3) than pure TiO_2 under UV and visible light exposure. A signification enhancement in the reaction rate was observed with the MWCNT/TiO_2 (1:3) composite as compared to bare TiO_2 and the physical mixture of MWCNTs and TiO_2. MWCNTs can improve the photocatalytic activity of TiO_2 in two aspects, namely, e− transportation and adsorption.

Dalt et al.[11] prepared CNTs/TiO_2 composites using MWCNTs, titanium (IV) propoxide, and commercial TiO_2 as titanium oxide sources. These were used to degrade methyl orange solution through photocatalytic degradation using UV irradiation. These composites were prepared by solution processing method followed by a thermal treatment at 400°C, 500°C, and 600°C. The heterojunction between nanotubes and TiO_2 was also confirmed by XRD and specific surface area.

Titanium dioxide (anatase) nanoparticles have also been successfully deposited onto MWCNTs by An et al.[4] by hydrolysis of titanium isopropoxide in supercritical ethanol. It was found that MWCNTs were decorated with well-dispersed anatase nanoparticles (>7 nm diameter). The size as well as loading content of the nanoparticles on MWCNTs could be tuned by changing the ratio of precursor to MWCNTs. The absorbance spectrum of the as-prepared TiO_2/MWCNT composites was extended to the whole UV–visible region due to this decoration of TiO_2. These composites were used as a PC for degradation of phenol under visible light with higher efficiency as compared to a mixture of TiO_2 and MWCNTs.

CNT/TiO_2 composites were prepared using surface-modified MWCNT and titanium n-butoxide (TNB) with benzene.[9] The UV-induced photoactivity of the CNT/TiO_2 composites was tested using a fixed concentration of methylene blue in an aqueous solution. It was concluded that the removal of methylene blue by CNT/TiO_2 composites was not only due to the adsorption effect of MWCNT and photocatalytic nature of TiO_2 but also due to electron transfer between MWCNT and TiO_2.

The high rate of electron/hole pair recombination reduces the quantum yield of the photocatalytic processes and is one of its major drawbacks. Addition of a coadsorbent is likely to increase its photocatalytic efficiency. Efforts have been made to hybridize the photocatalytic activity of TiO_2 with the adsorptivity of CNT. In this context, composite of MWCNTs and different PC (MWCNT/PC) has been synthesized by Saleh and Gupta.[49] The catalytic activity of this MWCNT/TiO_2 was investigated for the degradation of methyl orange. It was observed that the composite material exhibits an enhanced photocatalytic activity as compared to pure TiO_2. This enhancement in performance of the MWCNT/TiO_2 composite may be explained in terms of putting a check on recombination of photogenerated electron–hole pairs. MWCNT acts as a dispersing agent, which prevents TiO_2 from agglomerating activity during the catalytic process; thus, providing a high catalytically active surface area.

CNT/TiO_2 composites were prepared by Chen et al.[8] using MWCNTs as a starting material. Titanium (IV) isopropoxide, titanium (IV) propoxide, and titanium (IV) n-butoxide were used as titanium sources and benzene as a solvent. The photoactivity of as-prepared materials was evaluated by the degradation of methylene blue in aqueous solution under UV irradiation. It was reported that the removal of methylene blue in presence of CNT/TiO_2 composites is not only due to the adsorption of MWCNTs and the photocatalytic degradation of TiO_2 but it is also due to electron transfer between MWCNTs and TiO_2.

MWCNTs are considered a good catalyst support and they can effectively improve the performance of the TiO_2 PCs. Unique one-dimensional TiO_2@MWCNTs nanocomposites have been prepared by Zhang et al.[66] using a facile hydrothermal method. The TiO_2 coating layers are extremely uniform and their thickness can be adjusted. An intimate electronic interaction existed between MWCNTs and TiO_2 via interfacial Ti–O–C bond and the fast electron transfer rates. It was observed that the thickness of TiO_2 coating layers in nanocomposites plays a significant role in the photocatalytic degradation of methylene blue and Rhodamine B (RhB) and also in photocatalytic hydrogen evolution from water. It was found that the photocatalytic activity of the system is significantly improved due to the formation of one-dimensional heterojunction of TiO_2@MWCNTs nanocomposites and the positive synergistic effect between TiO_2 and CNTs.

Pajootan et al.[43] utilized the substrate to immobilize TiO_2 nanoparticles using solvent evaporation method and CNTs-coated carbon plates were fabricated with large surface area and high adsorption capacity without any agglomeration. The photocatalytic activity has been observed in a continuous photocatalytic reactor to degrade three acid dyes and it was found that PC was stable for 16 successive cycles without any reasonable loss in activity. A fast and complete decolorization of real industrial wastewater was achieved in 100 min by adding H_2O_2 to the system.

Orge et al.[42] carried out photocatalytic ozonation of aniline in the presence of neat titanium dioxide, MWCNTs, and their composites. Independent tests for catalytic ozonation and photocatalysis were also carried out in order to explore the potential occurrence of a synergetic effect. Catalytic ozonation was carried out with ozone dose of 50 g m^{-3} converting aniline in 15 min. Photocatalysis using P25, commercial TiO_2, and an 80:20 (w/w) composite of P25 and MWCNT also led to total aniline conversion, but these took relatively longer reaction times. Removal of TOC was higher than 70% for all photocatalytic ozonation systems in 1 h with the exception of neat MWCNT, where photocatalytic ozonation in the presence of the selected samples led to nearly complete mineralization after 3 h. Photocatalytic ozonation completely removed oxalic acid (by-product) formed during degradation of aniline. However, the concentration of oxamic acid, other by-product is more refractory to increase with time.

A conductive and photocatalytic nanocomposite thin film comprising MWCNTs and TiO_2 nanoparticles was prepared based on layer-by-layer assembly in toluene.[54] They used an amphiphilic surfactant, aerosol OT (AOT), to impart opposite surface charge onto MWCNTs and TiO_2. This technique enables the incorporation of unoxidized MWCNTs into the nanocomposite thin films and also provides method of forming conformal thin films over a large area. The physicochemical properties of MWCNT/TiO_2 nanocomposite thin films can be varied by changing concentration of AOT during assembly. The electrical properties of the nanocomposite film, sheet resistance and conductivity, can also be tuned through change in assembly conditions. The incorporation of MWCNTs within films will lead to a significant enhancement of the photocatalytic activity of TiO_2.

Natarajan et al.[38] synthesized MWCNT-loaded TNT composites via hydrothermal method. The tubular morphology of as-prepared TNT and MWCNT/TNT composites was supported by TEM and SEM, which was confirmed by the increase in surface area. The MWCNT/TNT PCs show high photocatalytic decomposition efficiency (PDE) for the degradation of rhodamine 6G (RhB-6G), with excellent stability and reusability. Loading of MWCNT (10%) results in a significantly higher PDE (89%) as compared with that of bare TNT (78%), Degussa P25 TiO_2 (P25, 60%), and TiO_2 nanoparticles (56%). This enhancement was due to the ability of the MWCNTs to promote the electron transfer process and reduce the electron–hole pair recombination rate. Moreover, COD and TOC analyses were performed to verify RhB-6G degradation. Tubular morphology, enriched adsorption, synergic effect, and efficient separation of photogenerated electron–hole pairs were considered responsible for increased PDE.

MWCNTs/Ag/TNTs plates were synthesized by Mohaghegh *et al.*[35] *via* electrochemical reduction of functionalized MWCNTs on Ag/TiO_2NTs. They carried out loading of silver nanoparticles by electroless reduction of Ag^+ onto TiO_2 nanotubes. SEM analysis revealed that nanoparticles of Ag had grown on the walls of TiO_2NTs and MWCNTs were loaded significantly on the as-prepared Ag/TiO_2NTs. The photocatalytic activity as synthesized plates was evaluated by the photodegradation of methyl orange under UV light irradiation. It was observed that the MWCNTs/Ag/TiO_2NTs showed improved efficiency for photodegradation of methyl orange as compared to Ag/TiO_2NTs and TiO_2NTs. This is attributed to the efficient separation and transfer of photogenerated electron–hole pairs. The COD of the dye solution was measured at specified time intervals, which indicated the degradation of methyl orange.

An enhancement in photocatalytic activity of mesoporous TiO_2 modified by the addition of CNTs and Cu has been reported.[39] Nanocomposites of CNTs containing varying amounts of Cu were prepared by treatment with Cu^{2+}, which was then reduced to Cu^0 by $NaBH_4$ as the reducing agent. The mesoporous TiO_2 was synthesized by a sol–gel method using titanium isopropoxide as a precursor. It was then combined with the CNT/Cu nanocomposites to form the desired PCs. The photocatalytic properties of as-prepared mesoporous TiO_2 composites were studied by observing the degradation of methyl orange. It was found to be maximum with the sample containing 20 wt% of the Cu–CNT nanocomposite. The higher degradation efficiency was a synergistic effect may be due to improved electrical conductivity of the system by the presence of the CNT/Cu networks. It was observed that the photodegradation of methyl orange and photocatalytic activity of the photoactive systems increases as the copper content was increased.

MWCNT/N, Pd codoped TiO_2 nanocomposites were prepared by calcining the hydrolysis products of the reaction of titanium isopropoxide, containing MWCNTs with aqueous ammonia.[26] The red shift was observed in the absorption edge at lower MWCNT percentages. Complete coverage of the MWCNTs with clusters of anatase TiO_2 was observed at low MWCNT percentages while higher MWCNT levels led to their aggregation and as a result, poor coverage by N, Pd codoped TiO_2. The photocatalytic activities of the nanocomposites were observed by photodegradation of eosin yellow under simulated solar and visible light irradiation ($\lambda > 450$ nm). The optimum MWCNT weight percentage in the composites was found to be 0.5.

Masoud et al.[32] synthesized mesoporous TiO_2 by sol–gel method and mixed it with functionalized MWCNTs and decorated it by different amounts of Ag. The photoactivity of this nanocomposite containing mesoporous TiO_2, MWCNT, and Ag was evaluated for the degradation of methyl orange in aqueous solution under UV irradiation. They showed that the MWCNT decorated with the 15.5% of Ag shows enhanced photocatalytic activity of mesoporous TiO_2.

Synthesis of gold nanoparticle/CNT composites has been reported by Liu et al.[28] for oxidation of cyclohexane with high efficiency and high selectivity. Au nanoparticles confined in CNTs (Au-in-CNTs) were found photocatalytically active for the oxidation of cyclohexane. Using these, 14.64% conversion of cyclohexane and a high selectivity of 86.88% of cyclohexanol was achieved using air and visible light at room temperature. This can be considered a green chemical pathway.

Highly photoactive metal oxide catalyst can be prepared with efficient charge carrier separation using low-cost electron scavengers as an electron sink. Nair et al.[37] developed MWCNT decorated V-doped titania (TiV) via sol–gel technique. MWCNT/TiV sample showed 18 times enhancement in the photocatalytic performance compared to Degussa P25. It was revealed that there is a close bonding of TiV nanoparticles on the surface of MWCNT. The optical analyses confirm the extended and elevated absorption of the sample in visible range compared to the reference catalyst Degussa P25. Methylene blue was used as probe pollutant. They showed that TiV (mixed) was highly active as compared to MWCNT/TiV, TiV (anatase), and Degussa P25.

The preparation of hedgehog such as F-doped titanium dioxide bronze F-TiO_2 (B) and its nanocomposites containing SWCNTs and MWCNTs was reported out by Panahian and Arsalani[44] via combined ball milling–hydrothermal processes. Photocatalytic and sonophotocatalytic degradation of malachite green from aqueous solution was performed in presence of as-prepared materials. They showed that F-TiO_2 (B)/SWCNT displayed a good photocatalytic as well as sonophotocatalytic performance under visible light. The efficiency of malachite green degradation was found to be more than 91% and 95% for photocatalytic and sonophotocatalytic methods, respectively. The photocatalytic efficiency of catalysts increases in the presence of ultrasound. The presence of CNTs and fluorine as dopant was confirmed in all nanocomposites. This hybrid method reduced the band gap of F-TiO_2 (B) from 3.02 to 2.7 eV for F-TiO_2 (B)/SWCNT nanocomposite.

Gadolinium oxide nanoparticles (diameters < 5 nm) were uniformly decorated on the surfaces of MWCNTs by Mamba et al.[31] Later on, these were used as templates to fabricate gadolinium oxide nanoparticle-decorated

MWCNT/titania nanocomposites. As-prepared nanocomposites were evaluated for the photocatalytic degradation of methylene blue under simulated solar light irradiation. It was observed that photocatalytic activity for the gadolinium oxide-decorated MWCNT-based nanocomposites was higher as compared to the neat MWCNT/titania nanocomposite as well as commercial titania. Enhancement in photocatalytic activity of gadolinium oxide nanoparticles supported at the interface of the CNTs and titania was attributed to efficient electron transfer between the two components of the composite. A higher degree of mineralization of methylene blue (80.0% TOC removal) was achieved as evident from TOC analysis. This PC can be recycled for five times and a degradation efficiency of 85.9% was observed after the five cycles.

The importance of the order of the semiconductor layers in WO_3-TiO_2/MWCNT composite materials was studied by Bárdos et al.[5] The morphological parameters such as mean crystallite size, crystal phase composition, as well as photocatalytic efficiency were also observed. Oxalic acid was used as a model pollutant. The appearance of TiWOx phase in the composites contributed to the higher photocatalytic efficiencies. On using excess of MWCNT or WO_3, active composites were obtained, but TiWOx disappeared.

A new embedded heterogeneous PC on polyether block amides/polyvinyl alcohol/titanium dioxide/MWCNT (PEBAX/PVA/TiO_2/MWCNT) composite was used in advanced oxidation process by Shakouri et al.[50] The effect of functionalization, structure, surface morphology, optical properties of composites, and mineralization of pollutants were observed. MWCNT not only activated the catalyst with visible light but it also increases the active surface area and consequently, adsorption of species. Oxalic acid-treated composite showed high potential for the mineralization. A complete loss of remazol black brilliant under UV irradiation was observed by utilizing composite films in 240 min.

Diethyl phthalate (DEP) is quite commonly used phthalate in industries and has a detrimental effect in environment. A series of polyaniline (PANI/CNT/TiO_2) PCs were immobilized on glass plate and used for degrading DEP with visible light.[19] The PANI/CNT/TiO_2 PCs were fabricated by codoping with PANI and two functionalized CNT (CNT–COCl and CNT–COOH) onto TiO_2 followed by a hydrothermal synthesis and sol–gel hydrolysis. Doping of PANI shifted absorption edge of this composite to 421–437 nm. The highest DEP degradation of about 41.5–59.0% and 44.5–67.4% was found on exposure to simulated sunlight for 120 min with composite obtained from sol–gel and hydrothermal synthesis, respectively.

The optimum pH was determined as 5.0 and 7.0 for these two PANI/CNT/TiO_2 PCs. The reusability of the sol–gel hydrolyzed PCs can be reused up to five times without any significant decline in the photodegradation efficiency but the photocatalytic stability of the hydrothermal synthesized composite was relatively less.

Photocatalytic activity of TiO_2 has emerged as a promising technology to combat biological warfare agents even anthrax spores, but the present challenge in using photocatalysis for degradation of bacterial endospores is the longer time required for disinfection (in hours) as compared to bacteria (in minutes). This rate of disinfection can be enhanced by controlling the rate of recombination of electron–hole pairs and also the rate of generation of reactive species. PCs have been synthesized by coating titanium dioxide (in anatase polymorph) on MWCNTs for increasing the disinfection rate of bacterial endospores. Anatase-coated MWCNTs were tested by Krishna et al.[24] for disinfection of *Bacillus cereus* spores, which were used as surrogate for anthrax spores. It was shown that disinfection rate was almost twice as compared to the commercial titanium dioxide.

TiO_2 nanoparticles were also decorated on the surface of MWCNTs by Koli et al.[23] A red shift of nanocomposites was observed with respect to the increasing content of MWCNTs. Its photocatalytic activity has been investigated for degradation of methyl orange under ultraviolet as well as sunlight irradiation. Photoinactivation of *Bacillus subtilis* under visible light irradiation was also observed. Significant enhancement in the degradation and inactivation reaction rate was observed with the TiO_2–MWCNTs (0.5 wt%) in nanocomposite as compared to TiO_2NPs. The COD observations confirmed environmentally benign nature of photodegraded dye solution.

The effect of variation of pH on photodegradation of methyl orange was studied using three different kinds of MWCNTs–ZnO hybrids.[1] They showed that the augmentation of the MWCNTs content in the synthesized CPs from 0.02 to 0.06 g leads to increased photodecomposition of methyl orange. It was also observed that photocatalytic activity of decorated MWCNTs was higher for the removal efficiency of methyl orange at acidic condition (pH = 4) than under neutral (pH = 7) and basic (pH = 10) conditions.

3.3.2 ZINC OXIDE

The photodecomposition of methyl orange was also studied by Roozban et al.[47] using decorated MWCNTs with different amounts of ZnO nanoparticles

(MWCNT–ZnO). The outer surface of MWCNTs has been successfully decorated with ZnO nanoparticles. It was reported that the photodegradation of methyl orange increases with increase in the UV irradiation time and weight fraction in different MWCNT–ZnO samples. The influence of weight fraction (0.1, 0.2, and 0.5 wt%) on the decomposition of methyl orange was found better than the UV irradiation time (5–30 min).

MWCNTs coated by zinc oxide nanoparticles were synthesized and then blended in polyethersulfone (PES) membrane casting solution.[70] The influence of as-prepared nanocomposite membrane on the performance and antibiofouling properties was evaluated. The water flux of the nanocomposite membranes was improved on addition of ZnO-coated MWCNTs because of their higher hydrophilicity. The 0.5 wt% ZnO/MWCNTs membrane was found to have the best antifouling capacity. Nanofiltration performance was also carried out, which revealed that the ZnO/MWCNTs membranes have greater direct red 16 removal ability rather than the PES alone. These ZnO/MWCNTs membranes presented the best antibiofouling characteristics, which may be due to high hydrophilicity and low surface roughness induced by the embedded nanoparticle.

ZnO/CNT nanocomposite was converted to ZnS/CNT by treating it with hydrogen sulfide using thioacetamide as a precursor.[41] The composition of this resulting nanocomposite can be tuned from a mixed ternary ZnS/ZnO/CNT nanocomposite to a pure ZnS/CNT nanocomposite. It was observed that the amount of wurtzite and sphalerite phases varies in the ZnS/CNT nanocomposite. Different nanocomposites were evaluated for their photocatalytic activity by the decomposition of methyl orange under visible light.

An additive free and fast energy-efficient microwave-assisted synthesis of nanocrystalline Ce-doped ZnO/CNT composite without calcinations was reported by Elias et al.[16] It was reported that all the samples were free of any impurity phase and crystallized in the hexagonal structure only. The 96.4% photocatalytic efficiency for methylene blue degradation was achieved under UV irradiation, which was only 34.6% with pure ZnO nanoparticles. The enhanced photocatalytic activity of the composite was due to Ce doping and insertion of CNTs into ZnO matrix resulting in reduction of the charge recombination.

Sobahi et al.[51] used a stepwise method to prepare a ternary nanocomposite comprising Pt/ZnO–MWCNT. They used sol–gel method to prepare ZnO–MWCNT nanocomposite followed by photo-assisted deposition of platinum nanoparticle over ZnO–MWCNT nanocomposite. The photocatalytic activity of this nanocomposite was observed for degradation of methylene blue under visible light. The results revealed that the photocatalytic activity of Pt/ZnO–MWCNT nanocomposites was higher than ZnO–MWCNT and pure ZnO. This enhanced photocatalytic activity of ternary nanocomposites which was attributed to synergetic effect of Pt nanoparticles and MWCNT. It leads to high absorption ability as well as separation ability between electron and hole.

3.3.3 OTHER BINARY CHALCOGENIDES

Saleh and Gupta[48] successfully synthesized a composite of MWCNT/tungsten oxide (MWCNT/WO_3). The photocatalytic activity was investigated for RhB degradation under solar insulation. The influence of various parameters such as illumination time, initial dye concentration, dosage, and pH was also investigated. It was observed that the composite exhibited an enhanced photocatalytic activity as compared to WO_3 and a mechanical mixture of MWCNTs and WO_3. This enhancement in photocatalytic performance of the MWCNT/WO_3 composite may be due to adsorption ability and electron transportation as a result of a strong interaction between WO_3 and MWCNTs. MWCNTs also acts as a dispersing agent preventing WO_3 from agglomeration during the catalytic process; thus, providing a high active surface area of the catalyst.

Flowerlike WSe_2 and MWCNTs-modified WSe_2 (CNT/WSe_2) composites have been successfully synthesized by Wang et al.[58] via a facile one-pot solvothermal reaction. The CNT/WSe_2 composite exhibits enhanced photocatalytic activity in photocatalytic degradation of methyl orange under visible light irradiation as compared to bare WSe_2. This enhanced photocatalytic activity has been attributed to the fact that CNTs can reduce electron–hole pair recombination and efficient inhibition of the aggregation of WSe_2 for fully exposing the active edges.

MWCNT/CdS-diethylenetriamine (DETA) composite PC was synthesized by Lv et al.[29] via hydrothermal method. CdS-DETA was uniformly dispersed on the surface of MWCNT. The photocatalytic activity of as-prepared PCs was investigated under 410 nm (LED light irradiation) for photodegradation of methylene blue. The rate of degradation in presence of MWCNT/CdS-DETA was found to be about six times more than pure CdS-DETA. The MWCNT/CdS-DETA hybrid can be reused again for degradation of organic pollutant and it indicates its photostability.

Zhao et al.[67] prepared a series of MWCNTs-loaded Bi_2S_3 nanomaterials composites via hydrothermal method. The photocatalytic activity of as-prepared composite was evaluated for degradation of methylene blue under visible light irradiation and compared it with others. This composite showed excellent photocatalytic performance and also maintained a good stability during the recycling experiments.

3.3.4 TERNARY PCs

Wang et al.[59] synthesized a novel Ag_3PO_4/MWCNT composite by a simple precipitant approach. It was found that the MWCNTs were well deposited on Ag_3PO_4 nanoparticles. The composite with MWCNT content of 1.4 wt% exhibited maximum photocatalytic activity under visible light. The rate constant of methyl orange degradation over Ag_3PO_4/1.4 wt% MWCNT was 5.84 times than that with pure Ag_3PO_4. The incorporation of MWCNTs not only enhances the photocatalytic activity significantly but it also improves the structural stability of Ag_3PO_4, which suggests that there are synergistic effects taking place in the composite. The improvement in activity was attributed to the conductive structure supported by MWCNTs, which not only favors electron–hole separation but the removal of photogenerated electrons from the decorated Ag_3PO_4 as well.

A facile approach has been reported by Yin et al.[61] for the preparation of MWCNT/BiOCl microspheres via an ionic liquid-assisted solvothermal method. These heterostructures were fabricated by dispersing MWCNT onto the surface of BiOCl. As-synthesized sample microspheres showed higher photocatalytic activity for the degradation of RhB as compared to pure BiOCl under visible light irradiation. The photocatalytic activity of the MWCNT/BiOCl microspheres was found to be highest for MWCNT (0.1 wt%).

A search is still on for developing efficient, newer, and visible-light-driven (VLD) PCs for environmental decontamination. Li et al.[27] reported the preparation of novel heterostructure of MWCNTs coated with BiOI nanosheets as an efficient VLD PC via a simple solvothermal method. They showed that BiOI nanosheets were well deposited on MWCNTs. The MWCNTs/BiOI composites exhibited significant enhancement in photocatalytic activity for the degradation of RhB, methyl orange, and parachlorophenol under visible light, compared with pure BiOI. The MWCNTs/BiOI composite achieves the highest activity when the MWCNTs content is 3 wt%. It is higher than that of a mechanical mixture of 3 wt% MWCNTs + 97 wt% BiOI. This higher photocatalytic activity is mainly due to the strong coupling interface between MWCNTs and BiOI, which significantly promotes the efficient electron–hole separation.

MWCNTs/BiOCOOH composites were prepared by Hu et al.[18] via a facile solvothermal method. This composite composed of BiOCOOH nanosheets and MWCNTs. Photocatalytic properties of all catalysts were evaluated by degrading RhB and 4-chlorophenol under simulated sunlight irradiation. The MWCNTs/BiOCOOH composites were found to be much more active than BiOCOOH and a mechanical mixture of MWCNTs and BiOCOOH. The MWCNTs/BiOCOOH composite with 3 wt% MWCNTs (3% M-B) had the highest activity and it also displayed good stability. The higher photocatalytic activity was attributed to the effective electron–hole separation and improved visible light harvesting. They revealed that the holes and superoxide radicals were the primary active species for this degradation using trapping experiments.

A composite material (MWCNT@$BiVO_4$) was prepared by Zhao et al.[69] using a one-step hydrothermal method. They showed that MWCNTs were successfully embedded into $BiVO_4$ and composite showed a strong visible light absorption capacity, high efficiency for electron–hole separation, and excellent stability. The degradation of RhB was observed under visible light irradiation. It was almost 10 times to that with $BiVO_4$ and 3 times with P25. The stability of MWCNT@$BiVO_4$ was also confirmed via recycling experiments and it was noted that even after five cycles, MWCNT@$BiVO_4$ could still maintain high removal rate of RhB (95.96%).

Sun et al.[52] synthesized a series of novel and facile quaternary composite $BiVO_4$/MWCNT/Ag@AgCl via hydrothermal and oxidization method. The catalytic performance of the composite was investigated in degrading RhB (5 mg L^{-1}) under visible light irradiation ($\lambda \geq 420$ nm). It was observed that the weight ratio of Ag and the mole ratios of Ag to AgCl in the composite can influence the photocatalytic performance. The catalyst exhibited much superior visible light photocatalytic activity than pure $BiVO_4$. This composite can decompose 99% RhB in 100 min only.

Datta et al.[12] prepared pure and CNT-attached $BiFeO_3$ and $Bi_2Fe_4O_9$ via hydrothermal route. It was revealed that as-synthesized materials exhibited various morphologies, which depend on the mineralizer used in the synthesis. The attachment of CNT reduces the band gap, and as a result, photocatalytic activity was enhanced due to the electron transfer from $BiFeO_3$ to CNT.

The Zn_2SnO_4/MWCNT nanocomposite was synthesized by Dorraji et al.[14] by a facile two-step process. The as-prepared nanocomposite exhibited much higher photocatalytic activity (about 94% removal in 120 min) as compared to bare zinc stannate (17% removal in 120 min) under UVA irradiation in degrading basic red 46 (BR46). The effect of radical scavengers on degradation efficiency was observed. The addition of benzoquinone had no effect on photocatalytic efficiency while t-butanol decreased it to some extent. The addition of iodide ion, as a hole scavenger, inhibited the photocatalytic degradation of BR46 completely, indicating that the degradation occurred by direct oxidation using photogenerated holes.

Immobilized surfactant-modified PANI–CNTs/TiO_2 PCs were prepared by Yuan et al.[65] via hydrothermally and sol–gel method. These were used in the degradation of DEP under visible light at 410 nm. The TiO_2 surface was modified with both sodium dodecyl sulfate (SDS) and functionalized CNTs (CNT–COOH and CNT–COCl) just to improve the dispersion of nanoparticles and the transfer of electrons. The adsorption edge of the prepared PCs shifted to 442 nm on addition of PANI (1–5%). The SDS linked the PANI polymers so that a coating of the film of up to 314–400 nm and 1301–1600 nm were achieved in sol–gel hydrolysis and hydrothermally-synthesized PCs, respectively. An appropriate film thickness will extend the transfer path of the electrons inhibiting recombination of the electrons and holes. It was observed that photodegradation performance of DEP was better on using hydrothermally-synthesized PCs samples obtained by sol–gel hydrolysis.

Some PCs are associated with some disadvantages such as they do not absorb in visible range and undergo relatively fast electron–hole recombination. Some supports such as CNTs can be used to overcome these problems by providing more surface area and acting as an electron sink; thus, putting a check on electron–hole recombination. As a result, photocatalytic activity of that semiconductor is enhanced. CNTs–PC composites have opened new avenues for photocatalytic degradation of contaminants in wastewaters.

KEYWORDS

- **carbon nanotube**
- **photocatalytic activity**
- **nanoparticles**
- **adsorption ability**
- **nanocomposites**
- **wastewater treatment**
- **multiwalled carbon nanotube**

REFERENCES

1. Abbasi, S.; Hasanpour, M. Variation of the Photocatalytic Performance of Decorated MWCNTs (MWCNTs-ZnO) with pH for Photo Degradation of Methyl Orange. *J. Mater. Sci. Mater. Electron.* **2017,** *28* (16), 11846–11855.
2. *Abrahamson, J.; Wiles, P. G.; Rhoades, B. L. Structure of Carbon Fibers Found on Carbon Arc Anodes. Carbon* **1999,** *37 (11), 1873–1874.*
3. Ahmadi, M.; Jaafarzadeh, N.; Mostoufi, A.; Saeedi, R.; Barzegar, G.; Jorfi, S. Enhanced Photocatalytic Degradation of Tetracycline and Real Pharmaceutical Wastewater Using MWCNT/TiO_2 Nano-composite. *J. Environ. Manage.* **2017,** *186* (1), 55–63.
4. An, G.; Ma, W.; Sun, Z.; Liu, Z.; Han B,; Miao, S.; Miao, Z.; Ding, K. Preparation of Titania/Carbon Nanotube Composites Using Supercritical Ethanol and Their Photocatalytic Activity for phenol Degradation Under Visible Light Irradiation. *Carbon* **2007,** *45* (9), 1795–1801.
5. Bárdos, E.; Kovács, G.; Gyulavári, T.; Németh, K.; Kecsenovity, E.; BerkI, P.; Baia, L.; Pap, Z.; Hernádi, K. N. Highlights Novel Synthesis Approaches for WO_3-TiO_2/ MWCNT Composite Photocatalysts: Problematic Issues of Photoactivity Enhancement Factors. *Catal. Today* **2017**. doi.org/10.1016/j.cattod.2017.03.019.
6. Bethune, D. S.; Kiang, C. H.; Devries, M. S.; Gorman, G.; Savoy, R.; Vazquez, J.; Beyers, R. Cobalt-catalysed Growth of Carbon Nanotubes with Single-atomic-layer Walls. *Nature* **1993,** *363*, 605–607.
7. Chaudhary, D.; Zubair, A. M.; Neeraj, K.; Vankar, V. D. Preparation, Characterization and Photocatalytic Activity of Anatase, Rutile TiO_2/Multiwalled Carbon Nanotubes Nanocomposite for Organic Dye Degradation. *J. Nanosci. Nanotechnol.* **2017,** *17* (3), 1894–1900.
8. Chen, M. L.; Zhang, F. J.; Oh, W. C. Synthesis, Characterization, and Photocatalytic Analysis of CNT/TiO_2 Composites Derived from MWCNTs and Titanium Sources. *New Carbon Mater*. **2009,** *24* (2), 159–166.
9. Chen, M. L.; Zhang, F. J.; Oh, C. W. Photocatalytic Degradation of Methylene Blue by CNT/TiO_2 Composites Prepared from MWCNT and Titanium n-Butoxide with Benzene. *J. Korean Ceram. Soc.* **2008,** *45* (11) 651–657.
10. Dai, K.; Peng, T; Ke, D.; Wei, B. Photocatalytic Hydrogen Generation Using a Nanocomposite of Multi-walled Carbon Nanotubes and TiO_2 Nanoparticles Under Visible Light Irradiation. *Nanotechnology* **2009,** *20* (12). doi.org/10.1088/ 0957-4484/20/12/125603.
11. Dalt, S. D.; Alves, A. K.; Bergmann, C. P. Photocatalytic Degradation of Methyl Orange Dye in Water Solutions in the Presence of MWCNT/TiO_2 Composites. *Mater. Res. Bull.* **2013,** *48* (5), 1845–1850.
12. Datta, A.; Chakraborty, S.; Mukherjee, S.; Mukherjee, S. Synthesis of Carbon Nanotube (CNT)-$BiFeO_3$ and (CNT)-$Bi_2Fe_4O_9$ Nanocomposites and Its Enhanced Photocatalytic Properties. *Int. J. Appl. Ceram. Technol.* **2017,** *14* (4), 521–531.
13. Djokic, V. R.; Marinkovic, A. D.; Mitric, M.; Uskokovic, P. S.; Petrovic, R. D.; Radmilvic, V. R.; Janackovic, D. T. Preparation of TiO_2/Carbon Nanotubes Photocatalysts: The Influence of the Method of Oxidation of the Carbon Nanotubes on the Photocatalytic Activity of the Nanocomposites. *Ceram. Int.* **2012,** *38*, 6123–6129.
14. Dorraji, M. S. S.; Ghadim, A. R. A.; Rasoulifard, M. H.; Taherkhani, S.; Daneshvar, H. The Role of Carbon Nanotube in Zinc Stannate Photocatalytic Performance Improvement: Experimental and Kinetic Evidences. *Appl. Catal. B* **2017,** *205*, 559–568.

15. Eder, D.; Windle, A. H. Carbon-inorganic Hybrid Materials: The Carbon-nanotube/TiO_2 Interface. *Adv. Mater.* **2008,** *20* (9), 1787–1793.
16. Elias, M. Md.; Amin, M. K.; Firozd, S. H.; Hossain, M. A.; Akter, S.; Hossaina, M. A.; Uddin, M. N.; Siddiquey, I. A. Microwave-assisted Synthesis of Ce-doped ZnO/CNT Composite with Enhanced Photo-catalytic Activity. *Ceram. Int.* **2017,** *43* (1), 84–91.
17. Endo, M. Grow Carbon Fibers in the Vapor Phase. *ChemTech* **1988,** *18*, 568–576.
18. Hu, S.; ShijieLi, S.; Xu, K.; Jiang, W.; Zhang, J.; Liu, J. MWCNTs/BiOCOOH Composites with Improved Sunlight Photocatalytic Activity. *Mater. Lett.* **2017,** *191*, 157–160.
19. Hung, C. H.; Yuan, C.; Li, H. W. Photodegradation of Diethyl Phthalate with PANi/CNT/TiO_2 Immobilized on Glass Plate Irradiated with Visible Light and Simulated Sunlight—Effect of Synthesized Method and pH. *J. Hazard. Mater.* 2017, *322* (A), 243–253.
20. Iijima, S. Helical Microtubules of Graphitic Carbon. *Nature* **1991,** *354* (6348), 56–58.
21. Iijima, S.; Ichlhashi, T. Single-shell Carbon Nanotubes of 1-nm Diameter. *Nature* **1993,** *363*, 603–605.
22. Kasuga, T.; Hiramatsu, M.; Hoson, A.; Sekino, T.; Niihara, K. Titania Nanotubes Prepared by Chemical Processing. *Adv. Mater.* **1999,** *11*, 1307–1311.
23. Koli, V. B.; Dhodamani, A. G.; Delekar, S. D.; Pawar, S. H. *In Situ* Sol–Gel Synthesis of Anatase TiO_2-MWCNTs Nanocomposites and Their Photocatalytic Applications. *J. Photochem. Photobiol. A* **2017,** *333*, 40–48.
24. Krishna, V.; Pumprueg, S.; Lee, S. H.; Zhao, J.; Sigmund, W.; Koopman, B.; Moudgi, B. M. Photocatalytic Disinfection with Titanium Dioxide Coated Multi-wall Carbon Nanotubes. *Proc. Safety Environ. Prot.* **2005,** *83* (4), 393–397.
25. Kuo, C. Y. Prevenient Dye-degradation Mechanisms Using UV/TiO_2/Carbon Nanotubes Process. *J. Hazard. Mater.* **2009,** *163* (1), 239–244.
26. Kuvarega, A. T.; Krause, R. W. M.; Mamba, B. B. M. Multiwalled Carbon Nanotubes Decorated with Nitrogen, Palladium Co-doped TiO_2 (MWCNT/N, Pd Co-doped TiO_2) for Visible Light Photocatalytic Degradation of Eosin Yellow in Water. *J. Nanopart. Res.* **2012**. doi.org/10.1007/s11051-012-0776-x.
27. Li, S.; Hu, S.; Xu, K.; Jiang, W.; Jianshe, L. J.; Wang, Z. A Novel Heterostructure of BiOI Nanosheets Anchored onto MWCNTs with Excellent Visible-light Photocatalytic Activity. *Nanomaterials* **2017,** *7* (1). DOI: 10.3390/nano7010022.
28. Liu, J; Liu, R; Li, H; Kong, W; Huang, H; Liu, Y; Kang, Z. Au Nanoparticles in Carbon Nanotubes with High Photocatalytic Activity for Hydrocarbon Selective Oxidation. *Dalton Trans.* **2014,** *43*, 12982–12988.
29. Lv, J.; Li, D.; Dai, K.; Liang, C.; Jiang, D.; Lu, L.; Zhu, G. Multi-walled Carbon Nanotube Supported CdS-DETA Nanocomposite for Efficient Visible Light Photocatalysis. *Mater. Chem. Phys.* **2017,** *186*, 372–381.
30. Mahmoodi, N. M. Photocatalytic Degradation of Dyes Using Carbon Nanotube and Titania Nanoparticle. *Water Air Soil Pollut.* **2013,** *224*. doi.org/10.1007/s11270-013-1612-3.
31. Mamba, G.; Mbianda, X. Y.; Mishra, A. K. Gadolinium Nanoparticle-decorated Multiwalled Carbon Nanotube/Titania Nanocomposites for Degradation of Methylene Blue in Water Under Simulated Solar Light. *Environ. Sci. Poll. Res.* **2014,** *21* (8), 5597–5609.
32. Masoud, M.; Nourbakhsh, A.; Tabrizi, S. A. H. Influence of Modified CNT-Ag Nano-composite Addition on Photocatalytic Degradation of Methyl Orange by Mesoporous TiO_2. *Inorg. Nano-Metal Chem.* **2017,** *47* (8), 1168–1174.
33. Meyyappan, M. *Carbon Nanotubes: Science and Applications*; CRC Press: Boca Raton, 2004.

34. Mintmire, J. W.; Dunlap, B. I.; White, C. T. Are Fullerene Tubules Metallic? *Phys. Rev. Lett.* **1992,** *68*, 631–634.
35. Mohaghegh, N.; Faraji, M.; Gobal, F.; Gholami, M. R. Electrodeposited Multi-walled Carbon Nanotubes on Ag-loaded TiO_2 Nanotubes/Ti Plates as a New Photocatalyst for Dye Degradation. *RSC Adv.* **2015,** *5*, 44840–44846.
36. Monthioux, M.; Kuznetsov, V. Who Should be Given the Credit for the Discovery of Carbon Nanotubes? *Carbon* **2006,** *44* (9), 1621–1623.
37. Nair, R. G.; Das, A.; Paul, S.; Rajbongshi, B.; Samdarshi, S. K. MWCNT Decorated V-Doped Titania: An Efficient Visible Active Photocatalyst. *J. Alloys Compd.* **2017,** *695*, 3511–3516.
38. Natarajan, T. S.; Lee, J. Y.; Bajaj, H. C.; Jo, W. K.; Tayade, R. J. Synthesis of Multiwall Carbon Nanotubes/TiO_2 Nanotube Composites with Enhanced Photocatalytic Decomposition Efficiency. *Catal. Today* **2017,** *282* (1), 13–23.
39. Nourbakhsh, A.; Abbaspour, S.; Masood, M.; Mirsattari, S. N.; Vahedi, A.; K. Mackenzie, J. D. Photocatalytic Properties of Mesoporous TiO_2 Nanocomposites Modified with Carbon Nanotubes and Copper. *Ceram. Int.* **2016,** *42* (10), 11901–11906.
40. *Oberlin, A.; Endo, M.; Koyama, T. Filamentous Growth of Carbon Through Benzene Decomposition. J. Crystal Growth* **1976,** *32 (3), 335–349.*
41. Okeil, S.; Krausmann, J.; Dönges, I.; Pfleger, S.; Jörg, Engstlera, J.; Schneider, J. J. ZnS Nanoparticles have been Synthesized on Vertically Aligned Carbon Nano ZnS/ZnO@ CNT and ZnS@CNT Nanocomposites by Gas Phase Conversion of ZnO@CNT. A Systematic Study of Their Photocatalytic Properties. *Dalton Trans.* **2017,** *46*, 5189–5201.
42. Orge, C. A.; Faria, J. L.; Pereira, M. F. R. Photocatalytic Ozonation of Aniline with TiO_2–Carbon Composite Materials. *J. Environ. Manage.* **2017,** *195* (2), 208–215.
43. Pajootan, E.; Rahimdokht, M.; Arami, M. Carbon and CNT Fabricated Carbon Substrates for TiO_2 Nanoparticles Immobilization with Industrial Perspective of Continuous Photocatalytic Elimination of Dye Molecules. *J. Ind. Eng. Chem.* **2017,** *55*, 149–163.
44. Panahian, Y.; Arsalani, N. Synthesis of Hedgehoglike F-TiO_2(B)/CNT Nanocomposites for Sonophotocatalytic and Photocatalytic Degradation of Malachite Green (MG) Under Visible Light: Kinetic Study. *J. Phys. Chem. A* **2017**, *121* (30), 5614–5624.
45. Pyrgiotakis, G.; Lee, S. H.; Sigmund, W. Advanced Photocatalysis with Anatase Nano-coated Multi-walled Carbon Nanotubes. *MRS Online Proc. Lib. Arch.* **2005,** *876*. doi.org/10.1557/PROC-876-R5.7.
46. Radushkevich, L. V; Lukyanovich, V. M. The Structure of Carbon Forming in Thermal Decomposition of Carbon Monoxide on an Iron Catalyst. *Soviet J. Phys. Chem.* **1952,** *26*, 88–95.
47. Roozban, N.; Abbasi, S.; Ghazizadeh, M. The Experimental and Statistical Investigation of the Photo Degradation of Methyl Orange Using Modified MWCNTs with Different Amount of ZnO Nanoparticles. *J. Mater. Sci. Mater. Electron.* **2017,** *28*, (10), 7343–7352.
48. Saleh, T. A.; Gupta, V. K. Functionalization of Tungsten Oxide into MWCNT and Its Application for Sunlight-induced Degradation of Rhodamine B. *J. Colloid Interface Sci.* **2011,** *362* (2), 337–344.
49. Saleh, T. A.; Gupta, V. K. Photo-catalyzed Degradation of Hazardous Dye Methyl Orange by Use of a Composite Catalyst Consisting of Multi-walled Carbon Nanotubes and Titanium Dioxide. *J. Colloid Interface Sci.* **2012,** *371*, (1), 101–106.

50. Shakouri, A.; Heris, S. Z.; Etemad, S. G.; Mousavi, S. M. Photocatalytic Activity Performance of Novel Cross-linked PEBAX Copolymer Nanocomposite on Azo Dye Degradation. *J. Mol. Liq.* **2016,** *216*, 275–283.
51. Sobahi, T. R. A.; Abdelaal, M. Y.; Mohamed, R. M.; Mokhtar, M. Photocatalytic Degradation of Methylene Blue Dye in Water Using Pt/ZnO-MWCNT Under Visible Light. *Nanosci. Nanotechnol. Lett.* **2017,** *9* (2), 144–150.
52. Sun, T.; Cui, D.; Ma, Q.; Peng, X.; Yuan, L. Synthesis of $BiVO_4$/MWCNT/Ag@AgCl Composite with Enhanced Photocatalytic Performance. *J. Phys. Chem. Solids* **2017,** *111*, 190–198.
53. Tennent, H. G. Carbon fibrils, Method for Producing same and Compositions Containing same. US Patent 4,663,230, May 5, 1987.
54. Tettey, K. E.; Yee, M. Q.; Lee, D. Photocatalytic and Conductive MWCNT/TiO_2 Nanocomposite Thin Films. *ACS Appl. Mater. Interfaces* **2010,** *2* (9), 2646–2652.
55. Vu, T. H. T.; Au, H. T.; Nguyen, T. T. T.; Do, M. H.; Pham, M. T.; Bui, D. H.; Phan, T. S.; Nguyen, D. L. Synthesis of Carbon Nanotube/Titanate Nanotube Composites with Photocatalytic Activity for H_2S Oxidation. *J. Sulfur Chem.* **2017,** *38* (3), 264–278.
56. Wang, H.; Wang, H. L.; Jiang, W. F.; Li, Z. Q. Photocatalytic Degradation of 2,4-Dinitrophenol (DNP) by Multi-walled Carbon Nanotubes (MWCNTs)/TiO_2 Composite in Aqueous Solution Under Solar Irradiation. *Water Res.* **2009,** *43* (1), 204–210.
57. Wang, H.; Wang, H. W.; Jiang, W. F. Solar Photocatalytic Degradation of 2,6-dinitro-*p*-Cresol (DNPC) Using Multi-walled Carbon Nanotubes (MWCNTs)–TiO_2 Composite Photocatalysts. *Chemosphere* **2009,** *75* (8), 1105–1111.
58. Wang, X.; Chen, Y.; Zheng, B.; Jiarui, H.; Yu, B.; Zhang, W. Significant Enhancement of Photocatalytic Activity of Multi-walled Carbon Nanotubes Modified WSe_2 Composite. *Mater. Lett.* **2017,** *197*, 67–70.
59. Wang, Z.; Yin, L.; Zhang, M.; Zhou, G.; Fei, H.; Shi, H.; Dai, H. Synthesis and Characterization of Ag_3PO_4/Multiwalled Carbon Nanotube Composite Photocatalyst with Enhanced Photocatalytic Activity and Stability Under Visible Light. *J. Mater. Sci.* **2014,** *49* (4), 1585–1593.
60. Xu, Y. J.; Zhuang, Y.; Fu, X. New Insight for Enhanced Photocatalytic Activity of TiO_2 by Doping Carbon Nanotubes: A Case Study on Degradation of Benzene and Methyl Orange. *J. Phys. Chem. C.* **2010,** *114* (6), 2669–2676.
61. Yin, S.; Di, J.; Li, M.; Fan, W.; Xia, J.; Xu, H.; Sun, Y.; Li, H. Synthesis of Multiwalled Carbon Nanotube Modified BiOCl Microspheres with Enhanced Visible-light Response Photoactivity. *Clean Soil Air Water* **2016,** *44*, 781–787.
62. Yu, J.; Ma, T.; Liu, S. Enhanced Photocatalytic Activity of Mesoporous TiO_2 Aggregates by Embedding Carbon Nanotubes as Electron-transfer Channel. *Phys. Chem. Chem. Phys.* **2011,** *13*, 3491–3501.
63. Yu, Y.; Yu, J. C.; Yu, J. G.; Kwok, Y. C.; Che, Y. K.; Zhao, J. C.; Ding, L.; Ge, W. K.; Wong, P. K. Enhancement of photocatalytic Activity of Mesoporous TiO_2 by Using Carbon Nanotubes. *Appl. Catal. A Gen* **2005,** *289*, 186–196.
64. Yu, Y.; Yu, J. C.; Chan, C. Y.; Che, Y. K.; Zhao, J. C.; Ding, L.; Ge, W. K.; Wong, P. K. Enhancement of Adsorption and Photocatalytic Activity of TiO_2 by Using Carbon Nanotubes for the Treatment of Azo Dye. *Appl. Catal. B Environ.* **2005,** *61*, 1–11.
65. Yuan, C.; Hung, C. H.; Yuan, C. S.; Li, H. W. Preparation and Application of Immobilized Surfactant-modified PANi-CNT/TiO_2 Under Visible-light Irradiation, Material (Basel). **2017,** *10* (8). DOI: 10.3390/ma10080877.

66. Zhang, X.; Cao, S.; Wu, Z.; Zhao, S.; Piao, L. Enhanced Photocatalytic Activity Towards Degradation and H_2 Evolution Over One Dimensional TiO_2@MWCNTs Heterojunction. *Appl. Surf. Sci.* **2017,** *402*, 360–368.
67. Zhao, D.; Wang, W.; Sun, Y.; Fan, Z.; Du, M.; Zhang, Q.; Ji, F.; Xu, X. One-step Synthesis of Composite Material MWCNT@BiVO4 and Its Photocatalytic Activity. *RSC Adv.* **2017,** *7*, 33671–33679.
68. Zhao, D.; Yang, X.; Chen, C.; Wang, X.; Enhanced Photocatalytic Degradation of Methylene Blue on Multiwalled Carbon Nanotubes–TiO_2. *J. Colloid Interface Sci.* **2013,** *398*, 234–239.
69. Zhao, G.; Zhang, D.; Yu, J.; Xie, Y.; Hu, W.; Jiao, F. Multi-walled Carbon Nanotubes Modified Bi_2S_3 Microspheres for Enhanced Photocatalytic Decomposition Efficiency. *Ceram. Int.* **2017,** *43* (17), 15080–15088.
70. Zinadini, S.; Rostami, S.; Vatanpour, V.; Jalilian, E. Preparation of Antibiofouling Polyethersulfone Mixed Matrix NF Membrane Using Photocatalytic Activity of ZnO/MWCNTs Nanocomposite. *J. Membr. Sci.* **2017,** *529*, 133–141.

CHAPTER 4

ADSORPTION OF CHOLESTEROL BY CARBON NANOTUBES

ALEXANDER V. VAKHRUSHEV[1,2*], A. A. VAKHRUSHEV[3], N. N. CHUCKOVA[4], I. A. CHERENKOV[4], and N. V. CORMILINA[4]

[1]*Department of Mechanics of Nanostructures, Institute of Mechanics, Udmurt Federal Research Center, Ural Division, Russian Academy of Sciences, Izhevsk, Russia*

[2]*Department of Nanotechnology and Microsystems, Kalashnikov Izhevsk State Technical University, Izhevsk, Russia*

[3]*Department of Simulation and Modeling of Metallurgical Processes, Montan University, Leoben, Austria*

[4]*Department of Medical Biology, Izhevsk State Medical Academy, Izhevsk, Russia*

**Corresponding author. E-mail: vakhrushev-a@yandex.ru*

ABSTRACT

The aim of this work is devoted to the study of methods for extracting cholesterol from its solutions using sorbents based on carbon nanotubes. Numerical modeling of processes was carried out with the purpose of revealing mechanisms of interaction of nanotubes with cholesterol, as well as with high- and low-density lipoproteins. Numerical modeling was carried out using molecular dynamics. An experiment was performed to adsorb cholesterol from its solution using powdered nanotubes. This study qualitatively confirmed the fact of adsorption of cholesterol by carbon nanotubes from its solutions.

4.1 INTRODUCTION

Cholesterol is a natural lipophilic alcohol contained in the cell membranes of all the animal organisms. It is a hydrophobic substance. In the dissolved state, it can be due to fats and natural solvents. About 80% of cholesterol is produced by the body itself; the remaining 20% comes from food. Cholesterol provides stability of cell membranes in a wide range of temperatures.

It is important to monitor the blood levels of various cholesterols. Ideally, the level of "bad" low-molecular lipoproteins is below 100 mg/dL. For people at high risk of cardiovascular disease, this should be below 70 mg/dL. The higher the percentage of "good" high-molecular lipoproteins in the general level, the better. Elevated blood cholesterol levels can cause various diseases.

Cholesterol sorption was considered as one of the promising and effective methods for treatment and prevention of atherosclerosis and its consequences—coronary heart disease, heart attack, strokes, etc.[1] Widely used carbon sorbents, which have a pronounced lipophilicity, under certain conditions are able to form a large specific surface.[2]

The main disadvantage of coal sorbents is their aggressiveness toward blood cells. At sorption, the number of red blood cells decreases by 27–35% (depending on the brand of coal), leukocytes by 52–74%, and platelets by 84%. Moreover, the better the charcoal sorbs cholesterol, the stronger its negative effect on the uniform elements of blood.[3] This, in the final analysis, leads to the fact that hemosorption of cholesterol on coal sorbents is now practically not used.[4]

However, significant progress in the creation of new materials based on carbon, in particular nanotubes, allows us to reconsider the problem of sorption of cholesterol. The structure of carbon materials makes it possible to predict the processes of cholesterol binding at the molecular level. The possibility of integrating nanoparticles with biological molecules is a prerequisite for the creation of new highly effective and safe systems for biotechnological and medical applications.

This work is devoted to the theoretical (with the help of mathematical modeling methods) and experimental study of the filtration of cholesterol molecules from its solutions, as well as of lipoproteins of different densities based on sorbents consisting of carbon nanotubes (CNTs).

4.2 PROBLEM STATEMENTS

The problem was solved by molecular dynamics method (MD method).[5–7] MD method has been widely used when modeling the behavior of nanosystems due to the simplicity of implementation, satisfactory accuracy, and low costs of computational resources. A detailed description of the MD is not possible in this chapter. The reader can study in the book about the technique of modeling such complex, multiphase nanosystems.[8]

The chemical structure of cholesterol is shown in Figure 4.1. It consists of a nucleus, a tail part, and in the head part there is OH group, due to which a practically hydrophobic molecule of cholesterol can interact with an aqueous solution and form hydrogen bonds.

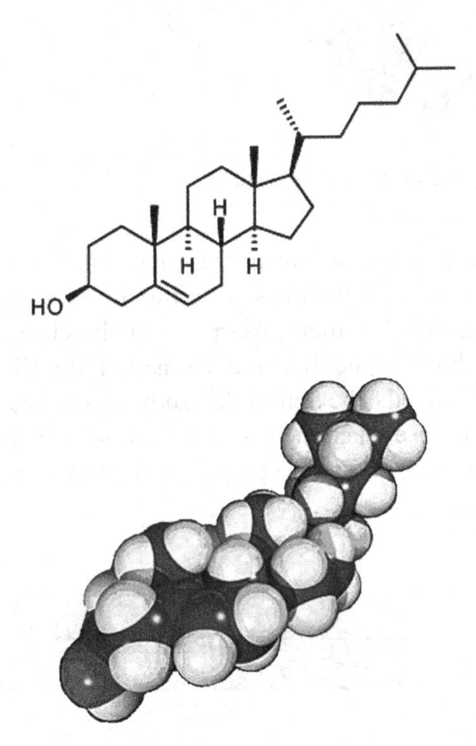

FIGURE 4.1 Structure of the cholesterol molecule.

At the first stage, the possibility of interaction between a molecule of cholesterol and a CNT was investigated. For this purpose, a mathematical model of the cholesterol-nanotube system was developed on the basis of the MD method. At the initial time, the cholesterol molecule was placed at a certain distance from the entrance to the CNT (Fig. 4.2). Calculations at the first stage were conducted without considering the influence of the environment.

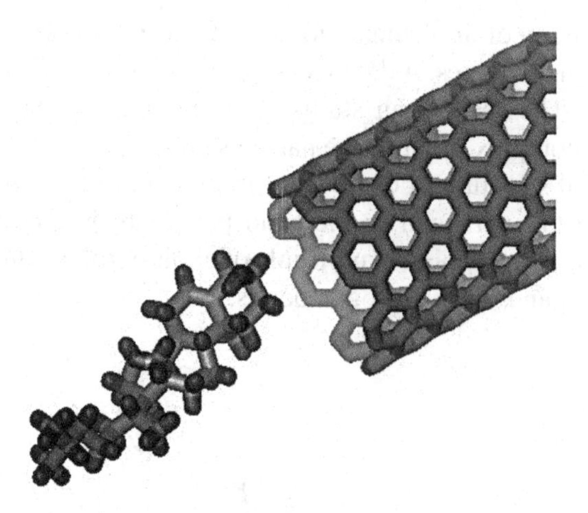

FIGURE 4.2 Cholesterol molecule and nanotube.

Further, the interaction between the elements of the system was considered; in particular, nanotubes of different diameters were investigated for the possibility of their absorption of the cholesterol molecule. In addition, the problem of the dynamic interaction of a CNT and a cholesterol molecule was solved. The aim of the study was to analyze the adsorption processes in the case when the cholesterol molecule is initially located not in the static position, but moves perpendicular to the axis of the carbon nanocannel (Fig. 4.3).

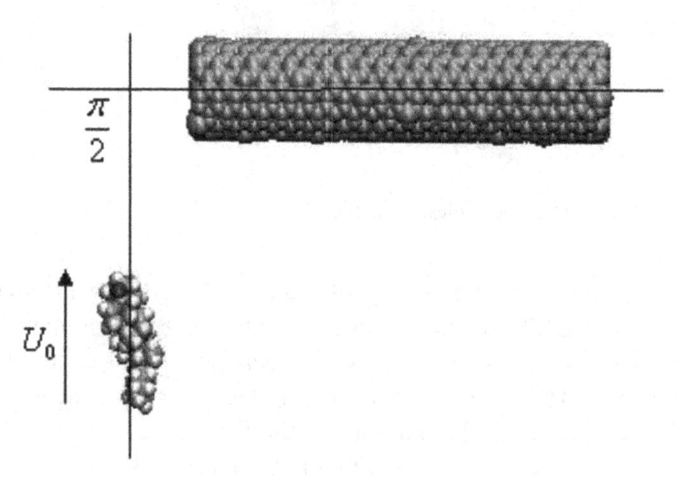

FIGURE 4.3 Dynamic interaction of cholesterol and nanotubes.

The solution of the problem in the above formulation can correspond to the model of the adsorption of cholesterol from the blood plasma by means of a filter built into the catheter connected to the circulatory system.

CNTs were chosen to solve the problem positively. Conventional sorbents are "crude" cleaning agents. They damage healthy cells in the process of purifying the blood from harmful constituents. In contrast, modern research in the field of bionanotechnology has shown that such refined structures as CNTs can be used even for intercellular interaction.[8]

The investigated mechanism of cholesterol adsorption is aimed at developing a method for purifying blood plasma from its free molecules. However, the percentage of free cholesterol in the blood plasma is low. More important is the task of filtering cholesterol esters along with low- and very low-molecular-weight lipoproteins.

Figure 4.4 shows the initial configuration of the modeled molecular system, which includes a nanotube placed next to a high-density lipoprotein (also called a nanodisk) based on the ApoA1 protein. All components are immersed in an aqueous solution.

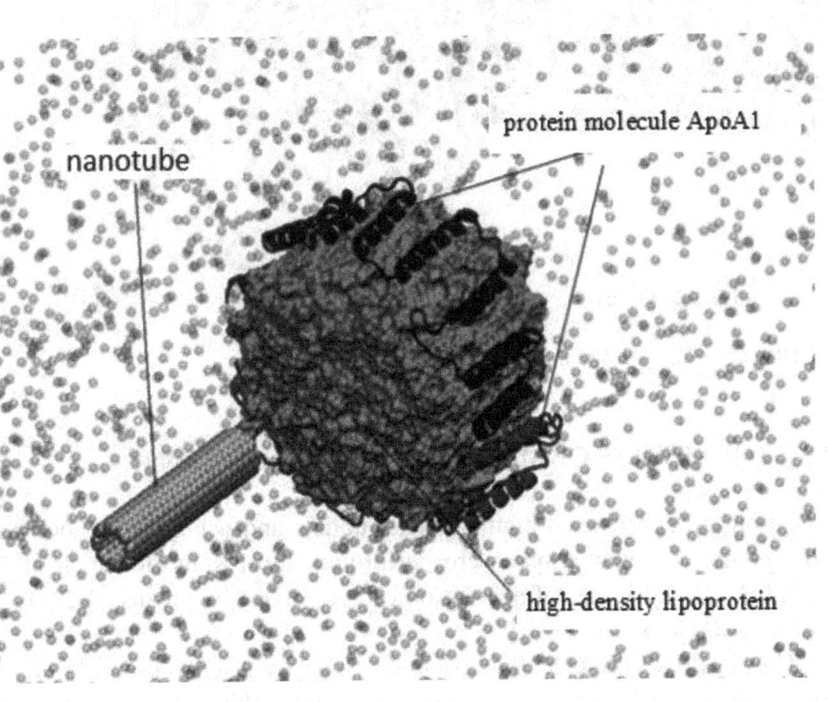

FIGURE 4.4 Nanodisk based on ApoA1 protein and a carbon nanotube in an aqueous medium.

The structure of the "harmful" low-density lipoprotein based on ApoB is shown in Figure 4.5. Compared to the ApoA1 nanodisk, it has a loose structure. These molecules play a decisive role in the formation of cholesterol plaques.

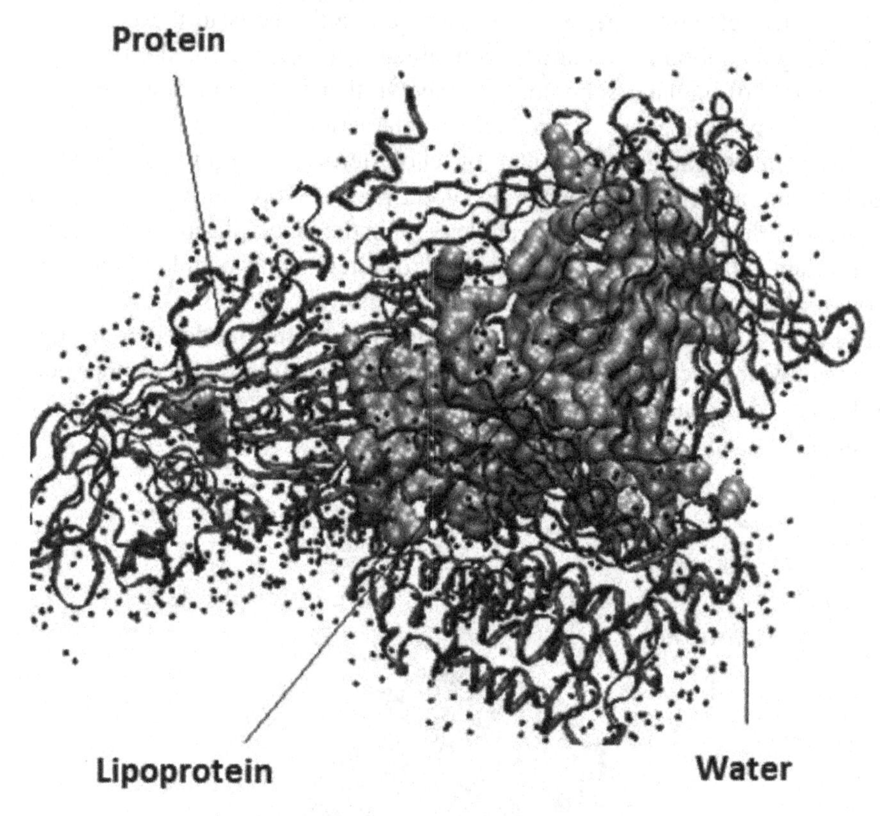

FIGURE 4.5 Low-density lipoprotein based on ApoB protein.

4.3 THE RESULTS OF MATHEMATICAL SIMULATION

Based on the constructed mathematical models and using the protocols of MD, numerical calculations were performed; the results of which allow analyzing the processes of adsorption of cholesterol by means of CNTs of various diameters.

The results of the calculations showed that small-enough CNTs do not pass a cholesterol molecule through themselves, although the head group enters the mouth of the canal (Fig. 4.6).

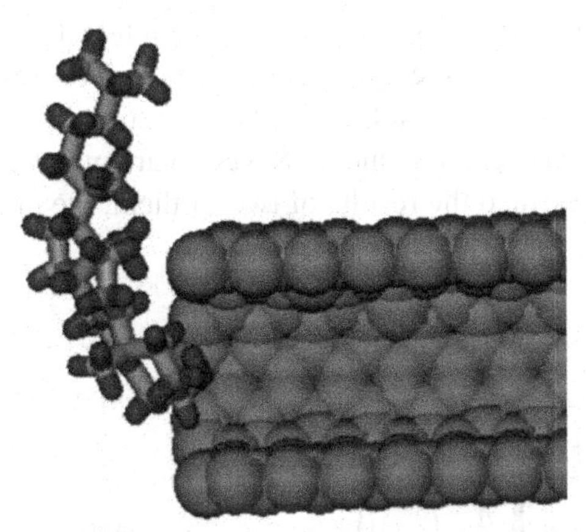

FIGURE 4.6 A tube of small diameter that filters out cholesterol molecule.

As the diameter of the nanotube increases, the adsorption process proceeds successfully, as the cholesterol molecule is drawn into the nanotube and remains inside it (Fig. 4.7).

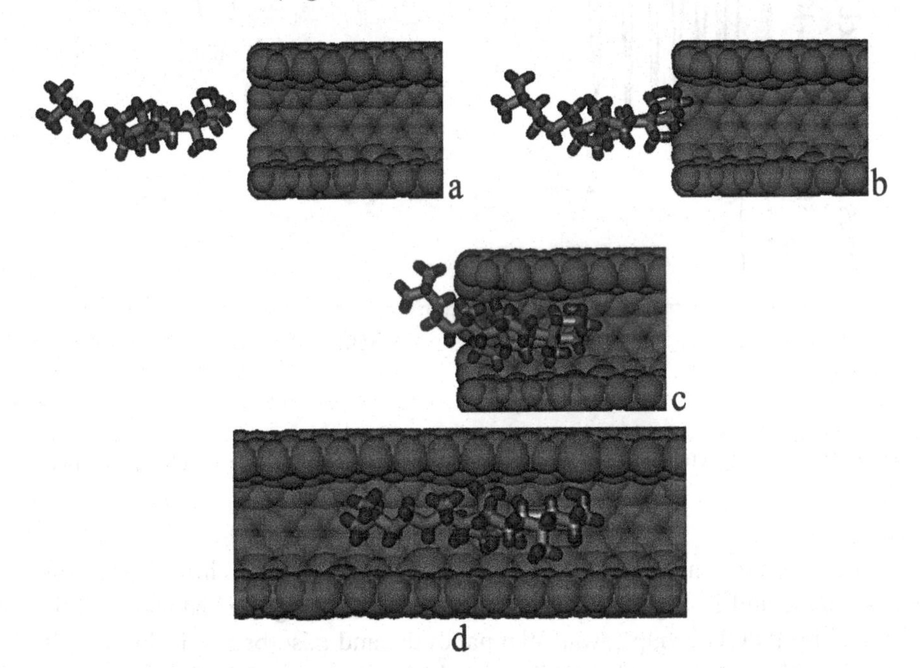

FIGURE 4.7 Stages of adsorption of a molecule of cholesterol by a nanotube: (a) the initial state and (b) cholesterol in the "mouth" region; (c) cholesterol is "absorbed" by a nanotube; and (d) final state.

Since at the first stage of the calculation, the interaction was studied in a vacuum; then, according to the law of conservation of energy, the cholesterol molecule continues to move inside the CNT under the action of stored potential energy, which has become kinetic. Several variants of calculations were made. We have presented the results of two of them, the most characteristic (Fig. 4.8).

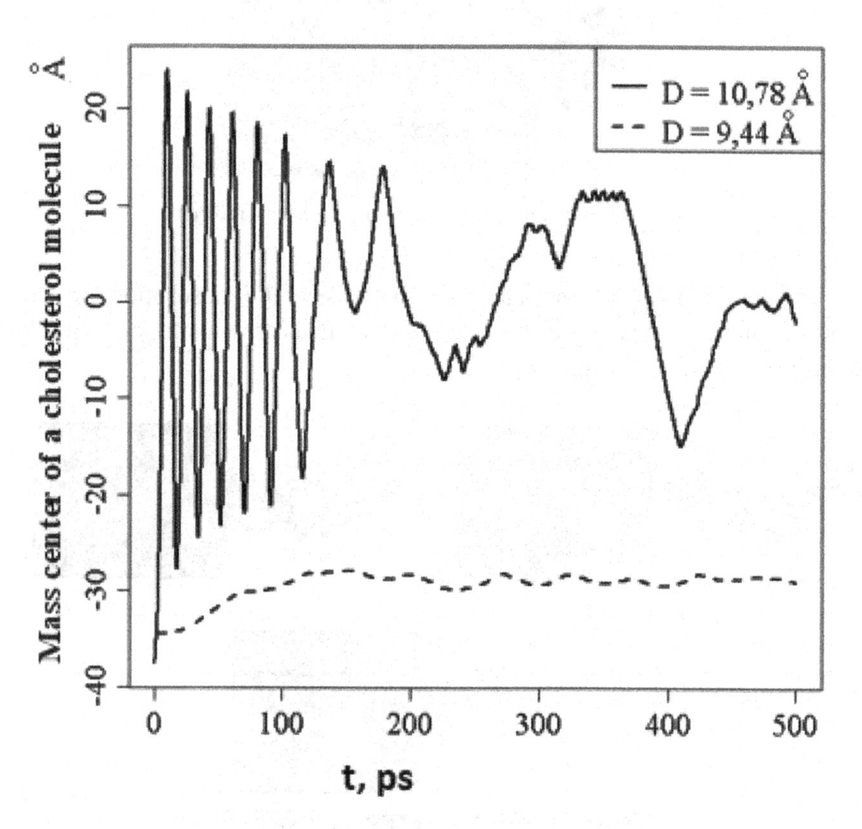

FIGURE 4.8 Trajectory of the mass center of a cholesterol molecule in tubes of various diameters.

Based on the calculations, the following results were obtained. In the first case, indicated in Figure 4.9, the cholesterol, reaching the channel axis, is turned by its OH group toward the nanotube and adsorbed to it. Further, it makes still movements along the channel (Fig. 4.9a), until the kinetic energy dissipates into the solid wall of the nanocannon. The position of the center of mass of the molecule remains constant in the radial direction (Fig. 4.9b).

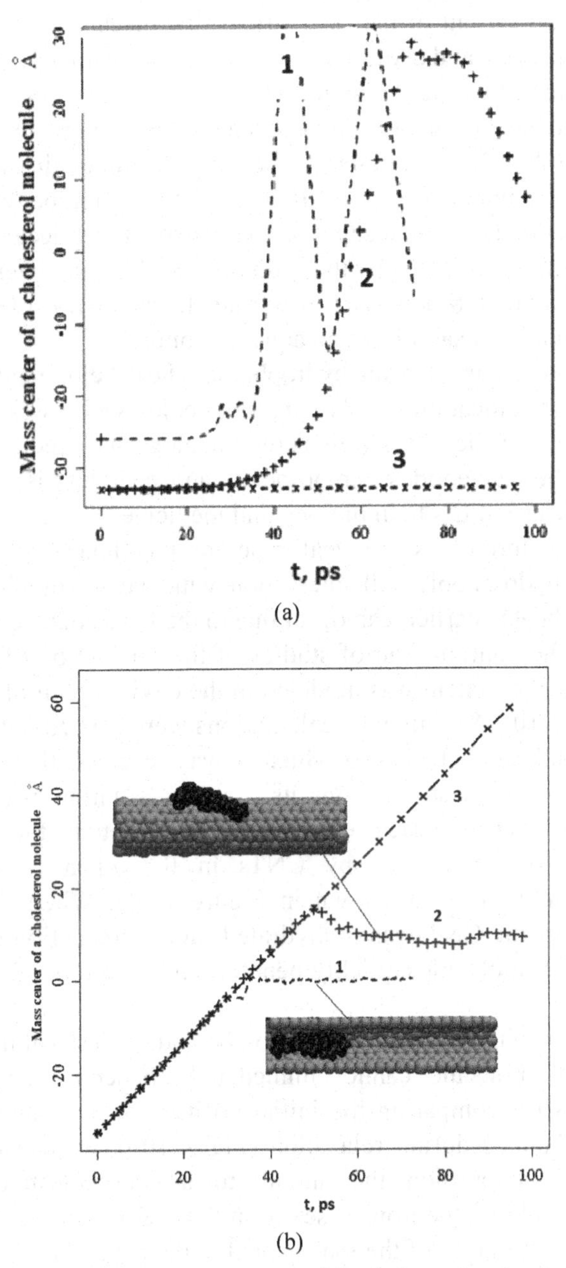

FIGURE 4.9 The trajectory of the movement of the cholesterol molecule for various types of interaction with a CNT: (a) trajectory along the axis of the nanotube and (b) trajectory along a radial direction (1—adsorption inside the nanotube, 2—adsorption on the surface of a nanotube, and 3—free state of the cholesterol molecule).

In the second variant of the calculations, the mechanism of interaction between the nanotube and cholesterol has a different character. In Figure 4.9, the curves indicated by the "+" sign correspond to these results. Analysis of the radial component of the trajectory of the molecule (Fig. 4.9b) shows that the cholesterol "flies" the entrance to the CNT; however, due to the forces of intermolecular interaction, it is attracted to the nanotube, changing the direction of motion to the opposite. For comparison, the trajectory of motion without interaction, indicated by the symbols "-x-," is also given.

The next important task was to investigate the interaction of components in a real medium, for example, in an aqueous solution.

CNTs in the ordinary state are hydrophobic. The development of methods for controlled modification of CNTs by molecules such as carbohydrates, proteins, and nucleic acids (as well as their analogs and precursors—oligosaccharides, oligonucleotides, amino acids, and peptides) is an important step toward the use of CNTs in biology and medicine.[9,10]

Due to the nature of its chemical structure, the cholesterol molecule is also virtually hydrophobic. All interaction with the surrounding aqueous solution, as indicated earlier, can occur due to the head charged OH group.

Hence, in the continuation of studies of the interaction of cholesterol and nanotubes, the system was modeled in the environment of an aqueous solution (Fig. 4.10). A number of calculations were carried out, confirmed by experimental data, thanks to which it was revealed that the process of adsorption of cholesterol from its aqueous solution was observed. The purpose of the numerical experiment was to study the mechanism of adsorption of cholesterol by CNTs in the aquatic environment. The initial configuration is shown in Figure 4.10a. When carrying out the calculations, the CNT was prehydrated: under the action of capillary forces, the water molecules quickly penetrate into the tube, filling it from the inside.

Unlike the results obtained for a nanotube and cholesterol in a vacuum, the cholesterol molecule cannot immediately penetrate the CNT. On the contrary, when comparing the initial position of the water molecules (Fig. 4.10a) obtained during relaxation (Fig. 4.10b), we can observe the occurrence of fluid flow along the nanotube toward the cholesterol molecule. It can be assumed that the flow arises from the OH group, the presence of which creates a similarity of the ion channel in the nanotube.

When analyzing trajectories, a rotation of the cholesterol molecule is observed (Fig. 4.10b), which, for a long period of time, cannot penetrate into the nanotube.

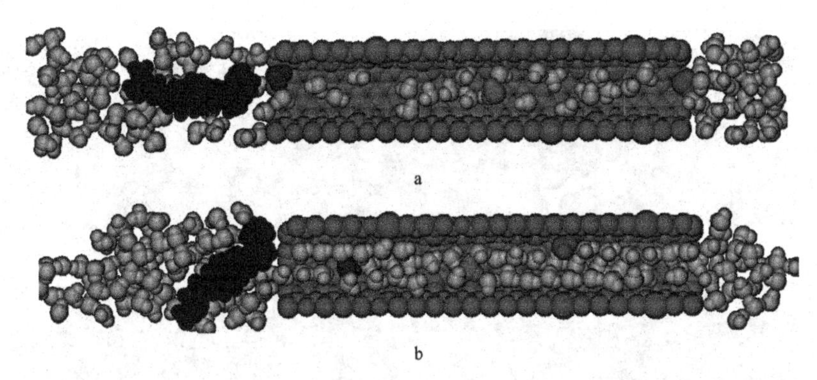

FIGURE 4.10 The interaction of nanotubes and cholesterol in water: (a) the initial state of the system and (b) rotation of the molecule in the "search" of an energetically favorable position.

Then, the molecule penetrates into the mouth of the nanotube and is "absorbed" in it completely (Fig. 4.11). The time of absorption of the cholesterol molecule upon its entry into the mouth of the nanotube was about 200 ps according to the calculations, which is confirmed by the work of other authors.[12] In addition to the above phenomena, it was found that initially, chaotic water molecules become homogeneous, aligned with an organized layered structure (Fig. 4.12).

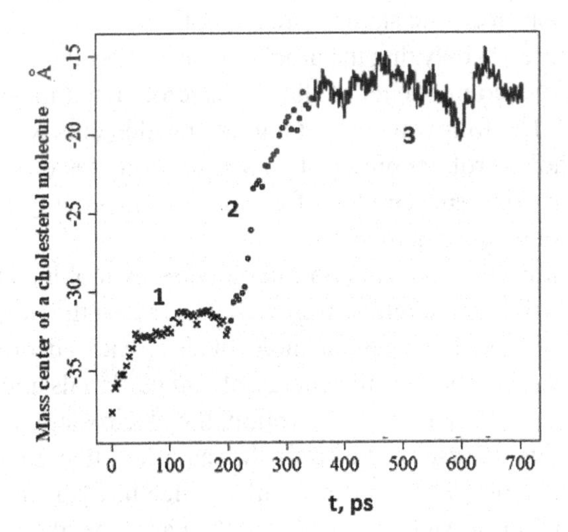

FIGURE 4.11 Trajectory of the movement of a cholesterol molecule in a single solution: 1—free state of the cholesterol molecule, 2—the cholesterol molecule enters the nanotube, and 3—the cholesterol molecule is adsorbed by a nanotube.

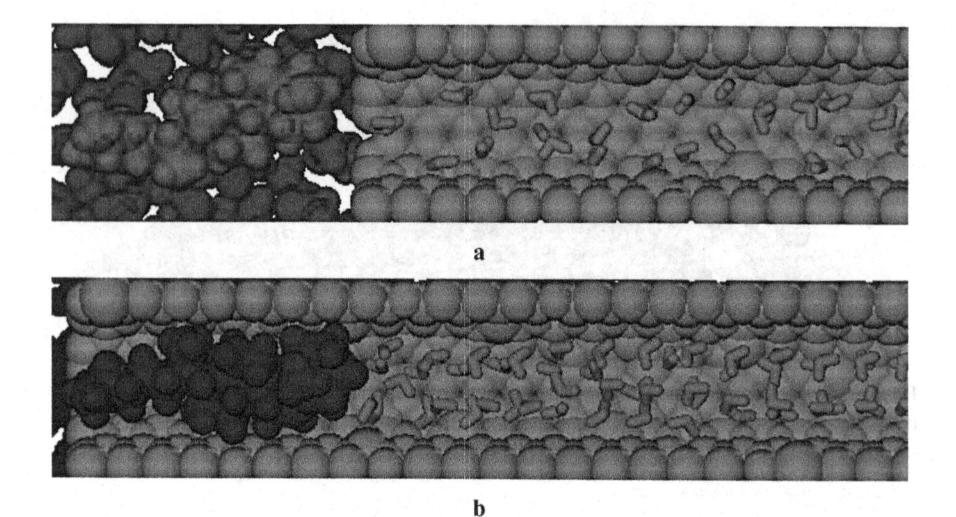

FIGURE 4.12 (See color insert.) The structure of molecular water inside a carbon nanotube: (a) in a free state and (b) after the adsorption of cholesterol.

While numerical simulation, there was considerable resistance from the aqueous solution to the progress of the cholesterol molecule inside the nanotube. If the cholesterol displacement curve in a vacuum is linear, then in the presence of a solution, the type of process becomes "activation," that is, it is a transition from one stable state to another.

For the purpose of analyzing the processes occurring, the interaction energies of the various components of the cholesterol–nanotube–water system were calculated. The following sets of interacting elements were considered: cholesterol—cholesterol, cholesterol—nanotube, cholesterol—water, and cholesterol—environment. Graphs of changes in the energy parameters of the system are shown in Figure 4.13.

It is easy to trace the onset of cholesterol entry by looking at its potential energy (Fig. 4.13d): its magnitude begins to decrease at the time of 200 ps at entry and reaches a plateau when the molecule is already adsorbed at 400 ps. The determining quantities are the energy of van der Waals and electrostatic interactions. After completion of adsorption, the electrostatic force becomes insignificant. In the course of the study, it was found that cholesterol loses favorable interactions with water molecules, since in CNT, it is surrounded by a smaller number of water molecules in the nanotube, the more energetically attractive attraction of the nanotube wall plays an important role and provides a common potential well.

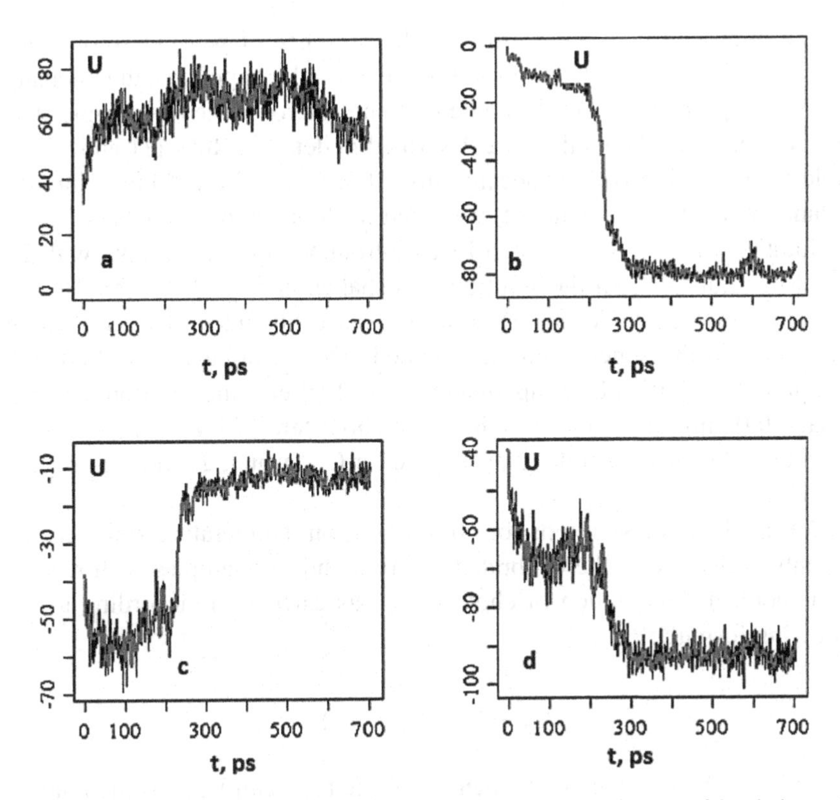

FIGURE 4.13 (See color insert.) The interaction energies U (kcal/mole) of the cholesterol–nanotube–water constituents in the system: (a) cholesterol–cholesterol; (b) cholesterol–a nanotube; (c) cholesterol–water; and (d) cholesterol–the surrounding system.

4.4 EXPERIMENT

The experimental procedure is as follows. To verify the calculated data, experiments were conducted on the binding of cholesterol by CNTs in the model system. The investigated material was a mixture of CNTs, including single-layer and multilayered nanotubes of various lengths and configurations. The task is to qualitatively evaluate the adsorption of CNT cholesterol. The level of cholesterol was determined by the enzymatic colorimetric method using the kit "Vital Diagnostics SPb."

Principle of the method—in determining cholesterol, its esters are hydrolyzed by cholesterol esterase to free fatty acids and cholesterol, which is oxidized by oxygen in the air to D 4-cholesterol under the influence of cholesterol oxidase.

The hydrogen peroxide formed in the presence of peroxidase oxidizes indicator substances with the formation of colored compounds, the intensity of which is proportional to the content of cholesterol in the test sample. The chemical bases of the method are described in detail in Refs. [11 and 12]. A cholesterol solution with a concentration of 5.17 mmol/L (200 mg/100 mL, included in the kit as a standard) was used for four samples of CNTs.

Solution of 0.1 mL cholesterol was introduced into a test tube with 25 mg of nanotubes. Then the solution was shaken and incubated for 10 min at room temperature. Then the solution was centrifuged. The content of cholesterol in the supernatant was studied. Three samples were prepared: experimental—0.02 mL of supernatant, 2.0 mL of enzyme solution; calibrational—0.02 mL of a standard solution of cholesterol, 2.0 mL of an enzyme solution and control sample (blank)—0.02 mL of water, 2.0 mL of enzyme solution.

We incubate the solutions for 10 min at room temperature and measure the optical density of the calibrated solution and test samples with respect to the control. Calculation of cholesterol was carried out according to the following formula:

$$Cch = \frac{E_{on}}{E_{\kappa}} \times 5.17 \text{ mol/L},$$

where E_{on} is the relative optical density of the test sample, E_{k} is the relative optical density of the calibration sample.

In the test sample, the cholesterol content was 2.44 mmol/L (94.44 mg/100 mL)—47.22% of the initial content. Thus, the studied drug CNT has a pronounced ability to bind cholesterol.

4.5 CONCLUSIONS

The mechanisms of interaction between a molecule of cholesterol and CNTs were studied. It was found that the adsorption process is observed both in a vacuum and in the presence of an environment, for example, an aqueous solution. In the latter case, adsorption has an activation character; the process of absorption of CNT cholesterol at a temperature of 300 K is about 200 ps.

It was found that CNTs can adsorb cholesterol to both internal and external surfaces, which increases their effectiveness as sorbents. The mechanism of interaction of the elements of the system is studied. Thus, for example, it was found that the molecule of cholesterol during the adsorption into the interior of CNT loses its bonds with water molecules, but it acquires an energetically more favorable interaction with the carbon wall.

It was found out that a tight packing of high-density nanodiscs keeps useful nutrients inside the molecule. As a result, CNTs interact only over the surface with ApoA1 in the case when the CNT is fixed on a substrate. In the free state, these compounds practically do not interact, since the nanotubes are carried away by the Brownian motion of the solution molecules.

An important result is that the conclusions obtained on the basis of MD calculations are confirmed by an experiment. This study qualitatively confirmed the fact of adsorption of cholesterol by the CNTs from its solutions.

KEYWORDS

- **carbon nanotubes**
- **adsorption**
- **cholesterol**
- **lipoproteins epitaxy**
- **molecular dynamics modeling**

REFERENCES

1. Gorchakov, V. D.; Sergienko, V. I.; Vladimirov, V. G. *Selective Hemosorbents*; Meditsina: Moscow, 1989; p 224.
2. Denty, E.; Walker, J. M. Active Carbon Materials: Properties, Selection Criteria and Evaluation Methods. In *Sorbents and Their Clinical Application;* Giorgiano, K. Ed.; Kiev High School: Kiev, Ukraine, 1989; pp 84–94.
3. Zemskova, L. A.; Sheveleva, I. V. Modified Sorption-active Carbon Fiber Materials. *Rus. Chem. Well.* (Zh. M., D. of Mendeleyev Chemical Society of the Russian Federation), **2004,** *XLVIII* (5), 53–57.
4. Lopukhin, Yu. M.; Archakov, A. I.; Vladimirov, Yu. A.; Kogan, E. M. *Cholesterol.* Medicine: Moscow, 1983; p 352.
5. Lennard-Jones, J. E. On the Determination of Molecular Fields. II. From the Equation of State of a Gas. *Proc. Roy. Soc. A* **1924,** *106*, 463–477.
6. Vakhrushev, A. V. *Computational Multiscale Modeling of Multiphase Nanosystems. Theory and Applications*. Apple Academic Press: Waretown, New Jersey, USA, 2017; p 402.

7. Turley, E. V. Molecular Dynamics and Diffusion in Biomembranes. *Diss. Cand. fiz.-mat. Sci.* **2006,** *119.*
8. Onfelt, B.; Davis, D. M. Can Membrane Nanotubes Facilitate Communication Between Immune Cells? *Biochem. Soc. Trans.* **2004,** *32* (5), 676–678.
9. Tzeng, Y. et al. Hydration Properties of Carbon Nanotubes and Their Effects on Electrical and Biosensor Applications. *New Diam. Front. Carbon Technol.* **2004,** *14* (3), 193–201.
10. Kouassi, G. K.; Irudayaraj, J.; McCarty, G. Examination of Cholesterol Oxidase Attachment to Magnetic Nanoparticles. *J. Nanobiotechnol.* **2005,** *3,* 1. DOI: 10.1186/1477-3155-3-1.
11. Pesce M. A.; Bodourian S. H. Enzymatic Rate Method for Measuring Cholesterol in Serum. *Clin. Chem.* **1976,** *22* (12), 2042–2045.
12. Shah, S.; Solanki, K.; Gupta, M. N. Enhancement of Lipase Activity in Non-aqueous Media upon Immobilization on Multi-walled Carbon Nanotubes. *Chem. Cent. J.* **2007,** *1*, 30. DOI: 10.1186/1752-153X-1-30.

CHAPTER 5

CARBON NANOTUBES: A CONCISE REVIEW OF THE SYNTHESIS TECHNIQUES, PROPERTIES, AND APPLICATIONS

SAUMYA SHALU, KAKOLI DASGUPTA, ANURADHA KUMARI, and BARNALI DASGUPTA GHOSH*

Department of Chemistry, Birla Institute of Technology, Mesra, Ranchi 835215, Jharkhand, India

**Corresponding author. E-mail: barnali.iitkgp@gmail.com*

ABSTRACT

Carbon nanotubes (CNTs) are the allotropes of carbon-containing continuous hexagonal network of atoms. They have attracted enormous interest over the last couple of decades due to their excellent mechanical and electrical properties as well as unique surface area. Nanotubes are classified into two categories: single walled nanotubes and multiwalled nanotubes. The synthesis of pure CNTs while controlling their properties, diameter, and wall number, remains a challenge. The most widely used synthesis methods, namely, arc discharge, laser ablation, and chemical vapor deposition have been discussed in this work. The functionalization of CNTs helps in overcoming the van der Waals interactions and greatly enhances their properties and application potential. Both covalent and noncovalent functionalization methods have been described. The thermal, mechanical, optical, and electrical properties of CNTs and their scope in energy storage devices have been highlighted. The properties of CNTs are still being explored by scientists for potential applications in nanomedicines, genetic engineering, solar cells, etc., and it continues to be an amazing field of research.

5.1 INTRODUCTION

Carbon nanotubes (CNTs) have generated enormous interest since their discovery by Iijima in 1991.[1] CNTs are the allotropes of carbon containing continuous unbroken hexagonal network of carbon that provide great strength to the structure.[2] They are long, thin hollow cylinders made of carbon atoms or graphite having diameter in the nanometer scale. They have some remarkable properties such as excellent electrical, thermal, magnetic, and mechanical properties[3] that make nanotubes (NTs) useful in a variety of applications such as nanoelectronic devices, electromechanical actuators, superconductors, electrochemical capacitors, nanocomposite materials, etc.[4] CNT has excellent elastic properties and it is a very strong material, almost 50 times stronger than steel. Heat transfer is another important property of CNTs. It can be used as both metallic and semiconducting material. These properties made it a very promising candidate in the field of electronics. Metallic CNTs can be used in high current carrying metal interconnects, whereas CNTs as semiconductor can be used in active devices, including transistors.[5,6] The outstanding properties of CNT make it a high-performance material.[7] A significant challenge, however, is the synthesis of CNTs while controlling their properties, diameter, chirality, and wall number. There are three main methods, namely, arc-discharge,[8] laser ablation,[9] and chemical vapor deposition (CVD)[10] that are used for the synthesis of CNTs.[11] Solid state precursors are used in the arc-discharge and laser ablation method, these facilitate the formation of large amounts of by-products. While in CVD method, hydrocarbon gases are the sources of carbon atoms, here catalyst plays an important role for the particular pattern of the NT structures.[6]

CNTs can be found in different forms that may differ in their structure, length, thickness, the type of helicity, and the number of layers (Fig. 5.1). Even though they are framed from basically a similar graphite sheet, their electrical properties differ depending on these variations.[12]

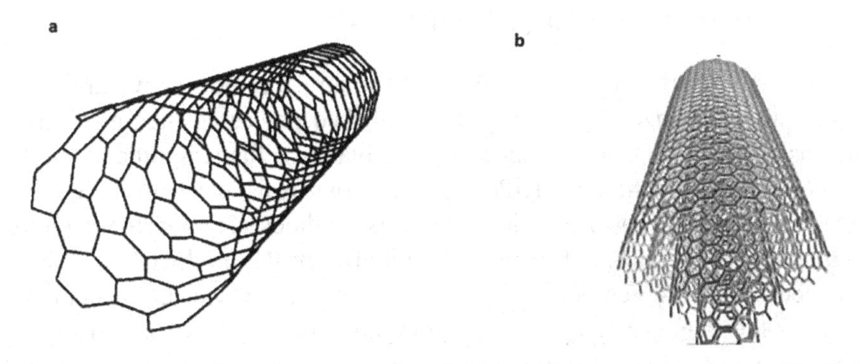

FIGURE 5.1 (a) A schematic representation of single-walled nanotube and (b) multiwalled carbon nanotube.

Source: Reproduced with permission (a) From Saether, E.; Frankland, S. J. V.; Pipes, R. B. *Compos. Sci. Technol.* **2003**, *63* (11), 1543–1550, © 2003 Elsevier and (b) From Siddique, R.; Mehta, A. *Constr. Build. Mater.* **2014**, *50*, 116–129, © 2014 Elsevier.

CNTs can be thought of as all sp^2 carbons arranged in graphene sheets, which have been moved up to shape a consistent empty tube. The tubes can be capped at the closures by a fullerene sort of structure containing pentagons, and can have lengths extending from many nanometers to a few microns. They can be subdivided into two classes—single-walled carbon nanotubes (SWNTs) and multiwalled carbon nanotubes (MWNTs). SWNTs, as the name suggests, consist of a solitary empty tube while MWNTs are made up of different concentric NTs. Iijima was first to recognize that NTs were concentrically rolled graphene sheets with a large number of potential helicities and chiralities rather than a graphene sheet rolled up like a scroll as originally proposed by Bacon. Iijima initially observed only MWNTs with between 2 and 20 layers, but in a subsequent publication in 1993, he confirmed the existence of SWNTs and elucidated their structure.[8,13]

CNTs may be used as filler particles and their addition into polymer matrices displays a new outlook for the development of composite materials. Since polymers are nonconducting in nature with low modulus, graphite, carbon black, or metal particles are used as conductive fillers for polymers to get properties for various applications. To enhance the electrical conductivity, conductive microfillers are added to polymer matrix.[14,15] The presence of CNTs helps in the improvement of tensile strength, toughness, thermal conductivity, glass transition temperature, electrical conductivity, solvent resistance, optical properties, etc., of the composite materials.[2]

5.2 SYNTHESIS TECHNIQUES OF CNTs

SWNT and MWNT can be produced by many techniques. A variety of CNTs are synthesized by some well-established processes such as electric arc discharge, CVD, laser vaporization (ablation), flame synthesis, high-pressure carbon monoxide (HiPco) process, pyrolysis, electrolysis, etc.[16,17] Arc discharge and laser ablation synthesis methods are high-temperature preparation techniques that were used initially for the production of CNTs but recently the low-temperature CVD technique has replaced these methods, because the diameter, purity, orientation, alignment, as well as density of CNTs can be easily controlled in the latter.[18]

The production methods of CNTs may be broadly classified into the following groups:

1. Physical processes
2. Chemical processes
3. Miscellaneous processes

5.2.1 *PHYSICAL PROCESSES*

Physical processes include two of the popular methods of CNT production which are arc discharge and laser ablation methods. These two are the widely used methods of CNT synthesis for experimental purpose.

5.2.1.1 ARC DISCHARGE

This method is the most common and the easiest process for producing CNTs. The arc-discharge method is known to be an established and convenient tool for producing high temperature (>3000°C) required to vaporize carbon atoms into plasma. A high current power supply and low voltage (~12–25 V) are employed in this technique, to produce an electric arc discharge between two graphite electrodes, namely, the cathode and the anode. The method is carried out under an inert atmosphere (helium or argon) at low pressure.[16,19,20] Helium is most preferred among other inert gases because of its high ionization potential.[21] Iijima first synthesized MWNTs using this method. He found the structure of MWNTs to consist of tubes (needlelike).[8] Fullerenes were also produced by the same process.[19] The carbon needles of diameter 4–30 nm and 1 mm in length were grown on cathode of the carbon electrode by direct current (DC) arc-discharge evaporation of carbon.[22] A pressurized chamber was used by Iijima in which a gas mixture of 10 Torr methane and 40 Torr argon was filled and the two graphite electrodes were placed at the center of the chamber (Fig. 5.2) maintained by a distance of 1–2 mm between the two rod tips.[17,23] The arc was produced between the two graphite electrodes by flowing a DC current of 200 A at 20 V.[17] As a result, NTs were formed along with fullerene and soot when DC arc stabilized. High-quality CNTs are obtained using this method.

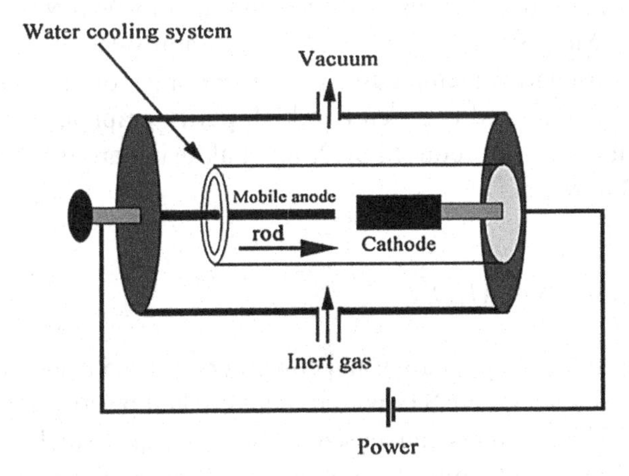

FIGURE 5.2 Schematic representation of arc-discharge method of CNT synthesis.

Source: Reproduced with permission from Journet, C.; Bernier, P. *Appl. Phys. A* **1998,** *67*, 1–9, © 2008 Springer.

On the other hand, conventional arc discharge process is unable to produce large quantities of CNTs. The process is unstable and discontinuous because of cathode spot phenomena. There is a nonuniformity in the temperature distribution and discontinuity in the current flow since the electrode spacing is not constant.[24] As a result, some impurities and carbon nanoparticles are always present with NTs.[16] So in view of this problem, many studies have been done for the growth mechanism of NTs. Lee et al. synthesized CNTs by plasma rotating arc method.[16,24] CNTs are synthesized in this method by rotating the anode at a high velocity which uniformly distributes the micro-discharges generating a stable plasma. As a result, the gas temperature of plasma, the discharge volume, and the carbon vapor perpendicular to the anode accelerate. The NTs are collected on the graphite collector and not stuck to the cathode surface due to the centrifugal force. The increase in the rotation speed of anode increases the yield of NTs. Since the NT does not stick to the surface of the cathode, the gap between the electrodes is easier to control. The plasma rotating electrode process produces massive high-quality NTs.

Arc-discharge method of MWNT production does not require any catalyst. High-yield production of MWNT was reported by Jung et al. in liquid nitrogen by arc-discharge method.[25] This technique can be a practical option for the large-scale synthesis of MWNTs of high purity. Sornsuwit and Maaithong used similar methods for high-quality MWNTs and SWNTs production.[26] Montoro et al. produced NTs in aqueous solution of H_3VO_4 through arc-discharge method using pure graphite electrodes.[27] DC arc discharge was generated between two high-purity graphite electrodes. The nanomaterials are also produced by Xing et al. using arc-discharge method under liquid nitrogen.[28]

5.2.1.2 LASER ABLATION

CNTs were produced by Smalley and coworkers in 1995 using laser ablation technique.[29] Single-walled NTs with top quality and purity are produced by this process. The carbon is condensed in this technique which deposits onto a substrate. In this technique, a continuous laser is irradiated at the graphite target under an inert medium (around 500 Torr of helium or argon) in a high-temperature reactor.[12] The reaction temperature is the most vital requirement since the yield and the quality of NTs is dependent on it. At 1200°C, the best quality NTs are obtained and when the temperature is lowered, the quality of the CNTs decreases producing many defects. The setup is depicted in Figure 5.3 consisting of a furnace, a target carbon composite doped with catalytic metals, a quartz tube with a window, a water-cooled trap, and flow systems for the buffer gas so that constant pressure as well as the rate flow is maintained. A laser beam is introduced through the window and focused on the target which is placed at the center of furnace. SWNTs are formed when the target is vaporized in high temperature Ar buffer gas. The Ar pressure and the rate of flow are 500 Torr and 1 cm s^{-1}, respectively. The SWNTs are carried to the trap by the buffer gas where they are collected. The focus point is changed or the target is moved so as to keep the vaporization surface as fresh as possible.[20]

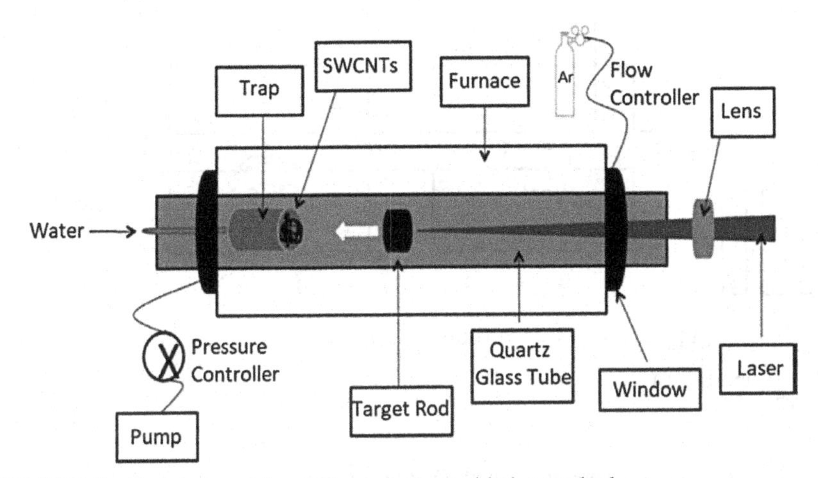

FIGURE 5.3 Carbon nanotube synthesis by laser ablation method.

Source: Reproduced with permission from Ref. [20]. © 2014 Elsevier.

5.2.2 CHEMICAL PROCESSES

5.2.2.1 CHEMICAL VAPOR DEPOSITION

This is an important process for the large-scale production of CNTs. By this technique, large quantities of NTs are produced and the direction of growth on substrate may be controlled to produce aligned vertical MWNTs.[17] It is basically a thermal dehydrogenation reaction where Fe, Ni, or Co (transition metal catalysts) lowers the temperature which is needed to "crack" a feed of gaseous hydrocarbon to hydrogen and carbon. It is a variable process to manufacture powders, coatings, fibers, and monolithic components. Most of the metals, many nonmetallic elements, and various compounds such as carbides, oxides, intermetallics, nitrides, etc., are produced by this method.[20] It is an efficient method for the manufacturing of semiconductors and is used in many applications, namely, optical, optoelectronic, and corrosion resistance. In this process, a solid is deposited on a heated surface in the vapor phase. CVD may be described as a vapor-transfer process which has atomic character in which the species deposited are atoms/molecules or a mixture of these two. The samples of CNTs are generated in the reactor as shown in Figure 5.4, which is composed of a quartz tube having 50 mm outer diameter, 40 mm inner diameter, and 1500 mm length which is extended through two furnaces.

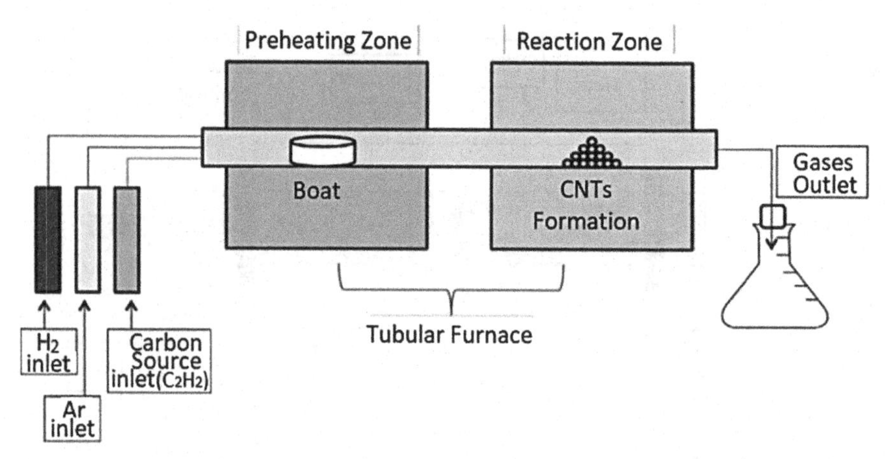

FIGURE 5.4 Pictorial representation of CVD process for CNT synthesis.
Source: Reproduced with permission from Ref. [20]. © 2014 Elsevier.

At the center of the first furnace, a ceramic boat was placed with the catalyst (ferrocene) inside the ceramic tube. The CNT growth occurred at the second furnace and was collected in ceramic boats which are placed at the center. The whole system was flushed with Ar to have an oxygen-free environment. The second furnace was then heated to the needed reaction temperature, and heating was continued till it reaches a steady state. Then the Ar flow was stopped and the first furnace was turned on until the temperature reaches 150°C. The C_2H_2 gas along with H_2 was released. The reaction was carried out for a required time. As a result, CNTs are produced in the inner walls and ceramic boats of the second furnace and are collected and weighed separately.[20]

5.2.2.2 HiPco REACTION

This method originated at the Rice University in 1991 for producing CNTs of remarkable purity. SWNTs of high quality are produced by this method.[30] In this method, the catalyst is infused in gaseous phase contrary to the other methods where the metal catalysts are accumulated before the carbon deposits on the surface.[17] The metal catalysts are developed in situ when the hydrocarbon gas [$Fe(CO)_5$ or $Ni(CO)_4$] is introduced into the reactor with carbon monoxide (CO) gas at high pressure of 30–50 atm and high temperature around 900–1110°C. This technique is designated as HiPco (high-pressure CO).[31] In this technique, the NTs are free from catalytic supports and may be functionalized continuously. Hence, this process synthesizes CNTs on a larger scale. This process can be tuned to control the diameters of the nanotubes by tuning the parameters such as CO pressure, temperature, and concentration of catalysts.[31]

5.2.3 MISCELLANEOUS PROCESSES

There are some miscellaneous processes for producing CNTs that are used comparatively less than the above mentioned techniques. These include methods such as helium arc discharge, electrolysis, and flame synthesis. These processes are not so effective in producing CNTs at a larger scale.[17]

The helium arc-discharge method was reported by the NASA's Goddard Space Flight Center scientists' in 2006. In this technique, helium arc welding process was used by the scientists to vaporize amorphous C-rods. And then the NTs were formed by settling the vapor into a water-cooled C-cathode. SWNTs were produced by this process with 70% yield at a lower cost. The electrolysis method was developed at University of Miskolc, Hungary. In this technique, the alkali metals were deposited on graphite cathode against high-temperature molten salt system. The NTs were formed due to the interfacial forces. The flame synthesis method, which uses hydrocarbon flames, is used to synthesize SWNTs in a controlled flame environment.[22] This environment produces the temperature and the C-atoms were formed from hydrocarbon fuels. This produces small islands of aerosol metal catalysts in which the SWNTs are grown.[17]

5.3 FUNCTIONALIZATION OF CNT

The functionalization of CNTs attracted appreciable interest in many fields since their discovery.[32] The stability of tubes by π–π electron interactions and their high surface energy makes CNTs prone to aggregate. As a result, they possess very limited applications due to their weak dispersions and insolubility into solutions and matrices. Therefore to enhance their properties, CNTs require an extended functionalization to make NTs highly processable.[33] The functionalization of CNTs prevents agglomeration of NT, which stabilizes the CNTs within the matrix and solution.[2] Two different methods, namely, chemical and mechanical techniques are used for the dispersion of CNTs. The mechanical technique may include grinding, high-shear mixing, ultrasonication, etc. They can separate NTs from one another but there is a possibility of breaking up of NTs which may lead to a decrease in their aspect ratio during process. These methods are, however, ineffective and time consuming. On the contrary, the chemical technique alters the surface energy of NTs which upgrades their adhesion and wetting properties and their stabilization to disperse. The surface chemistry of CNTs can be modified either noncovalently or covalently.[33] A functionalized NT may have some unique and different properties (optical, electrical, or mechanical) from that of the original one.[33] Thus, the functionalization of CNTs is very essential for all sorts of applications.

5.3.1 NONCOVALENT FUNCTIONALIZATION

Noncovalent functionalization of CNTs helps in improving the solubility and processability and does not destroy the final structural properties of CNTs.[2] NTs are functionalized noncovalently by surfactants, aromatic compounds, polymers, etc. These modifications of CNTs maintain their desired properties by improving their solubilities. Surfactants, biomacromolecules, or wrapping with polymers are mainly involved in this type of functionalization.[34] Cationic, anionic, and nonionic surfactants are employed to disperse CNTs in water.[2] For the dispersion of CNTs in water, the surfactant adsorption on the sidewalls of NTs is the broadly used method due to its ease of use, cheapness, and disposability. The surfactants are composed of a hydrophilic group (head) and a hydrophobic group (tail). The former facilitates the affinity of CNTs with aqueous bulk solvent and the latter assists adsorption to the hydrophobic NT walls. The interaction between CNTs and bulk medium can be modified by surface acting agents which provide electrostatic or steric repulsive forces as well as decreases the surface energy.[34] The two common surfactants that are used to decrease the accumulation of CNTs in water are sodium dodecyl sulfate (SDS—an ionic surfactant) and sodium dodecyl benzene sulfonate (NaDDBS).[2] CNTs were purified from graphite nanoparticles by sonicating them for a few minutes in 1% aqueous solution of SDS by Bonard et al. in 1997.[35] The adsorption (chemical) of SDS molecules on the surface of NTs affects the electrostatic repulsion between polar heads exposed in aqueous solutions, which prevents the aggregation of CNTs, forming stable aqueous black suspensions. The interaction of the surfactants and the CNTs is completely dependent on the surfactant nature, for example, its alkyl chain length, the size of the head group, as well as its charge. Since SDS does not have a benzene ring, its interaction is weaker as compared to NaDDBS and Triton-X100 ($C_{14}H_{22}O(C_2H_4O)_n$). NaDDBS on the other hand disperses better than Triton-X100 due to a slightly longer alkyl chain and its head group.[2]

There are countless polymers that assist CNT dispersion in water and many other organic solvents. The physical association of the polymers with NTs can be improved and that can be explained by the polymer wrapping mechanism. In this process, the polymers are wrapped around CNTs that form supramolecular complexes. The π-stacking interaction of the NT surface and the polymer accounts for the close association of the structures. A NT–polymer hybrid was prepared by suspending SWNTs in organic solvents poly(p-phenylenevinylene-co-2,5-dioctyloxy-m-phenylenevinylene) by Blau and coworkers by wrapping the copolymer around NTs.[36–38] As a result, the properties (electrical) of the hybrids were enhanced as compared to those of the individual components. For the modification of SWNTs, a noncovalent technique is used by the encapsulation of SWNTs within the amphiphilic and the cross-linked poly(styrene)-block-poly(acrylic acid) copolymer micelles which help in improving the dispersion of SWNTs into polar or nonpolar solvents as well as polymer matrices since the copolymer shell is fixed permanently. Hence, the encapsulated SWNTs' stabilization is possible with reference to other polymer synthesis. CNTs may also be dispersed into other different media by nonwrapping methods. In this method, the compositions and the copolymers of different structures act effectively as stabilizers, and it may also be suitable to disperse the tubes into different solvents. This mechanism also supports the adsorption of block copolymers to the NTs. It has been proposed by Nativ-Roth et al. that the solvophilic blocks behave as a steric barrier which leads to form stable dispersions of individual SWNTs as well as MWNTs superior to the threshold concentration of polymers.[39] The strong π–π interaction between the polymer and the surface of the NT gives soluble SWNTs. But the drawback of this noncovalent attachment is that there must be a weaker force between the wrapping molecule and the NTs. Hence, the effectiveness of the load transfer in a composite may be low as filler. The noncovalent functionalization also has one interesting feature, that is, reversibility. Bare CNTs can be regained from their dispersant-coated counterparts, both in aqueous and organic solvents. Reversibility is an effective tool which can be used in the enrichment, separation, or the purification of various types of NTs.

5.3.2 COVALENT FUNCTIONALIZATION

Since CNTs have π-orbitals of sp^2 hybridized C-atoms, they are more reactive than a flat graphene sheet. CNTs also possess an advanced tendency to get covalently attached with the chemical species. In this type of functionalization, the carbon atoms change from sp^2 to sp^3 and as a result, the translational symmetry of NTs is disturbed. Hence, it affects the transport and electronic properties of CNTs.[3] In covalent functionalization, functional groups –OH or –COOH are produced on the CNTs during the oxidation by air, oxygen, nitric acid, concentrated sulfuric acid, aqueous hydrogen peroxide, and acid mixture. The acid-treated MWNTs surface shows some defects in the carbon–carbon bonding due to the presence of the carboxylic acid group on the surface, whereas the raw MWNTs indicate uniform surfaces and a distinct diffraction pattern because of their perfect lattice structure of C–C bonds.[2] On the NT surface, the presence of carboxylic acid groups is advantageous because a number of reactions may be governed with this group. The presence of these groups on the NT surface assists the addition of organic as well as inorganic materials, which is essential for NT dispersion. CNTs are functionalized at their sidewalls or end caps to enhance their dispersion and their solubility in solvents and polymer matrices. The fluorination of SWNTs at the side walls occurs when elemental fluorine was passed at different temperatures, which shows improved dispersion while ultrasonicating in dimethyl formamide or isopropanol. The solubilization of SWNTs by direct functionalization on their sidewalls is brought about by carbenes, nitrenes, and arylation.

The functionalization of CNTs with polymers known as polymer grafting is also significant for the fabrication of polymer/CNT nanocomposites. The grafting of polymers covalently to the NTs, namely, "grafting to" and "grafting from" techniques have been reported. In the "grafting to" approach, polymer molecules are attached on the CNT surface by chemical reactions such as esterification, amidation, radical coupling, etc.[2] In this approach, for the preparation of composites, the polymers must have appropriate functional groups. The CNT functionalization using this method was reported by Fu et al where CNTs comprising carboxylic acid groups was refluxed with thionyl chloride for the conversion of acid groups into acylchlorides.[40] The fabrication of polymer grafted CNTs is done by adhering them to highly soluble linear polymers, like poly(vinyl acetate-co-vinyl alcohol) (PVA-VA) via ester linkages or poly (propionylethylenimine-co-ethylenimine) (PPEI-EI) via amide linkages. The derived CNTs (PVA grafted) were dissolvable in

PVA solution and the PVA-CNT nanocomposite films displayed high optical quality without any noticeable separation of phase. On the other hand, in "grafting from" technique, the polymer is fixed to the CNT surface by means of in-situ polymerization in the presence of reactive CNT or CNT supported initiators. By this method, the polymer CNT composites possessing high grafting density can be prepared.

Polymers such as poly(methyl methacrylate), polyamide 6, poly(acrylic acid), polystyrene, poly-(tert-butyl acrylate), poly(N-isopropylacrylamide), poly(N-vinylcarbazole), and poly(4-vinylpyridine) have been successfully used to graft onto CNT by cationic, anionic, radical, condensation, and ring-opening polymerizations. MWNT functionalization was reported by Gao et al. with a hydrophilic polymer, glycerol monomethacrylate (GMA) by this approach.[41] During this work, MWNT-Br macroinitiators were produced by treating oxidized MWNTs with thionyl chloride, glycol, and 2-bromo-2-methylpropionyl bromide for atom transfer radical polymerization. This functionalization is also known to reduce interfacial energy between polymers, monomers or CNT walls or solvent.[30]

5.4 PURIFICATION OF CNTs

The large-scale synthesis of CNTs leads to unique physical and chemical applications. But in all CNT preparation methods, the as-prepared CNTs comprise a number of impurities including amorphous carbon, metal particles or multishell carbon nanocapsules. The type and the amount of the impurities depend on the method used for synthesizing CNTs.[42,43] Therefore, the NTs need to be purified so as to recognize their applications.[44] The CNT soot produced during the synthesis of NTs comprise several impurities. Amorphous carbon, graphite sheets, metal catalysts, etc., are the major unwanted elements in the soot which disrupt the application of CNTs to obtain desirable properties.[42] So, in view of this problem, CNTs are purified thoroughly to eliminate the impurities while holding the structure of CNT intact.[44] Some common techniques for the purification of CNTs include oxidation, acid treatment, annealing and thermal treatment, ultrasonication, etc.

The oxidative treatment of the NTs eliminates the carbonaceous impurities such as amorphous carbon, carbon nanoparticles, and fullerenes. Amorphous carbon is easier to eliminate since it has high density of defects due to which it effectively oxidizes under gentle conditions. Fullerenes can also be removed easily due to their solubility into organic solvents. The disadvantage of the oxidative treatment of CNT is that CNTs too are oxidized with the impurities. However, the damage caused to CNTs is lesser compared to the adverse effect of the impurities. The yield and the efficiency of this process depend on the oxidation time, temperature, metal content, oxidizing agent, and environment. Another purification method is the acid treatment which removes the metal catalysts. Here, the metal surfaces are activated by either sonication or oxidation and then exposed to the acid. The NTs stayed in suspended form and when treated with nitric acid, the metal catalyst was affected and not the CNTs and other particles.[42,43] The refluxing of the sample in the presence of strong acid helps in the reduction of the extent of the metal particles as well as the amorphous carbon. The CNTs were purer after acid refluxing but the tubes were tangled so many of the impurities (carbon particles and catalyst) were trapped. These impurities were hard to remove with filtration.[45] So, the sonication method was carried out to untangle CNTs. The surfactant-assisted sonication with organic solvent was favored since it took longer for CNTs to settle down, showing an even state of suspension. The stabilization of the dispersed tubes is affected by the nature of the solvent.[42, 45] Annealing and thermal treatment is another technique which is useful in optimizing the structure of CNTs. Hence, to obtain CNTs of high quality, purification is necessary. The purification techniques vary as per the features of raw CNT and the techniques used for the synthesis.[44]

5.5 PROPERTIES OF CNTs

5.5.1 MECHANICAL PROPERTIES

The unprecedented interest in CNTs ever since their discovery is largely due to their remarkable properties. Iijima was the first to visualize CNTs as graphene sheets rolled concentrically. However, the high flexibility and strength of CNTs combined with its high stiffness make CNTs quite mechanically advanced to graphite fibers.[46] The high mechanical strength of CNTs can be attributed to the strongly bonded sp^2 C=C and their high aspect ratio.[47] The Young's modulus and yield strength of CNTs have been measured as ~0.64 TPa and 37 GPa respectively, making them the strongest known materials.[3,48] The yield strength of steel corresponds to ~300 MPa, which gives us a perspective of the strength of CNTs, at about one-sixth the weight of steel.[30] SWNTs are also highly flexible and continue to retain their shape under large strains.

There have been numerous theoretical studies on the mechanical properties of CNTs but the experimental measurements have been few owing to the technological challenges involved.[46,49,50] Temperature is an important factor that influences the strength of a material as the defects and dislocations in the structure of materials are activated with increase in temperature. At room temperature, the ductile-type bond flip and the covalent bond brittleness are the major shortcomings of CNTs. However, the hexagonal network has the unique ability to bear stress and impart flexibility to the CNT chains. Huang et al. reported that even at high temperatures, SWNTs become 15 times narrower and 280% elongated before breaking. The superplasticity of the NTs can be explained on account of nucleation and motion of kinks in the NT structure. They argued that such strain resistant behavior could be employed to reinforce ceramics and nanocomposites even at elevated temperatures. [51]

In addition to high Young's modulus along their axes, CNTs also exhibit buckling instability and fracture strength and withstand high strains without any atomic rearrangements. Georgakilas et al. explained the exceptional behavior of CNTs beyond small strains with the sudden release of energy accompanied by a reversible structural switch.[47] CNTs thus far have proved to be ideal carbon fibers and their impressive mechanical properties can be exploited for many potential applications.

CNTs are currently being investigated to fabricate mechanically robust composites based on polymer matrices. However, the properties of CNT–polymer composites are influenced by several other factors. CNT-based transparent electrodes for applications as touch screens and OLEDs (organic light-emitting diode) are in demonstration in major labs worldwide. CNT-based research is still in its early stages and there is a long way to go in terms of material stability, reproducibility, and scalability.

5.5.2 ELECTRICAL PROPERTIES

CNTs derive their electrical properties from graphene which is not surprising as CNTs themselves are cylindrically fixed graphene monolayers (SWNTs) or multilayers (MWNTs).[52–55] The electrical conductivity of CNTs, such as metals or semiconductors, depends on the exact way the graphene monolayer is wrapped into the cylindrical form. To put precisely, the electrical properties of CNTs are determined by the diameter and chirality of the NTs. Armchair nanotubes (when $n = m$; Fig. 5.5) show metallic behavior as they have a finite density of states at the Fermi level.[56] However, they may become semiconducting if their diameter is less than a threshold value as the energy gaps in the semiconducting CNTs is proportional to 1/d (d = NT diameter). Chiral NTs, for which m ≠ n and m − n is a multiple of 3 (Fig. 5.5), are semiconducting.[56] Other carbon nanotubes are described as moderate semiconductors. Hence, it can be said that the electronic states of an infinitely long nanotube are parallel lines in k-space, continuous along the tube axis and quantized along the circumference.[47]

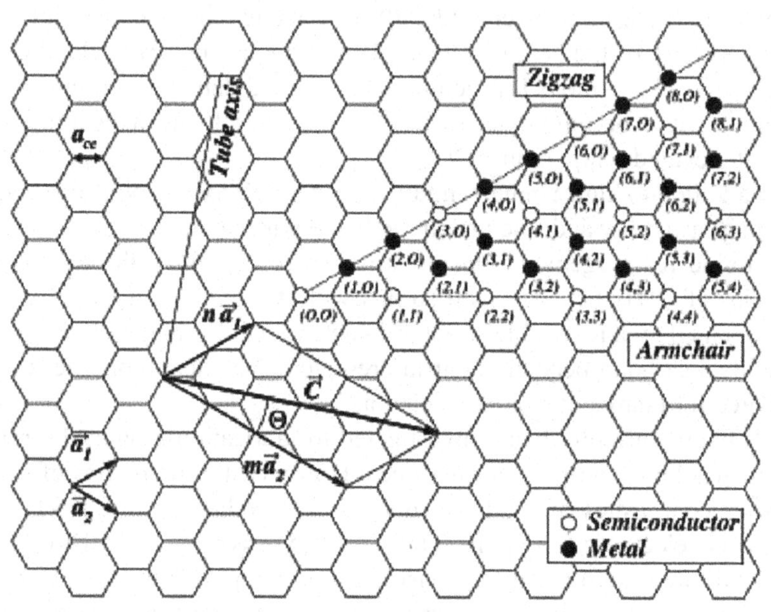

FIGURE 5.5 Representation of the chiral vector $C = na_1 + ma_2$ and chiral angle θ for a nanotube on a graphene sheet. Here n and m are integers and a_1 and a_2 are the unit cell vectors of the 2D-sheet.

Source: Reproduced with permission from Ref. [56]. © 2014 Elsevier.

On doping CNTs with metals, the resulting materials may be p-type, n-type, or intrinsic semiconductors depending on the type and concentration of the metal.[57] SWNTs have exhibited very high carrier mobility of up to 104 $cm^2/(V\ s)$ with an electric current density of about 4 × 109 A/cm^2, surpassing traditional conductors such as copper or aluminum by more than three times.[58] In fact, metallic CNTs have excellent conductivities (i.e., low resistivities) and can carry very high current densities but practically these high conductivities are not realized due to impurities and defects in the nanotubes that increase the resistivity.[59,60] CNTs show higher conductivity when isolated as compared to when they are organized into superstructures due to the contact resistance that arises at the points of connection between individual CNTs and to the presence of impurities that causes scattering in the latter case.[61,62]

The electrical conductivity of CNT functionalized polymer composites are determined by the degree of nanocomponent dispersion in the polymer and by the quality of the tube–tube connections. MWCNTs exhibit superconductivity below 12 K when they are entirely band-ended, the superconducting temperature being 30 times higher than that reported for SWCNTs and MWCNTs with noninterconnected shells.[63] The strong dependence of the conductivity on the electrochemical potential in the CNT can be utilized to design field effect transistor (FET) which is of immense technological value. It conducts current when the potential on a capacitively coupled gate moves the Fermi level into the valence or conduction band and acts as an insulator when the Fermi level is in the band gap. CNTs to be relevant in electronic applications face big challenges such as self-assembly of tubes into large-scale arrays, chirality control, and diameter specificity.

5.5.3 *OPTICAL PROPERTIES*

CNT is an important material for nanoscale light emitters and other optical applications. Optical properties of the SWNTs are related to their one-dimensional nature. It comprises absorption, photoluminescence, and Raman scattering spectroscopy of the CNTs. Diameter of the CNTs plays an important role to determine their optical properties. The energy of the dipole-allowed optical transitions depends directly on the diameter of NTs and the transition occurs between valence and conduction bands.[29,64,65] It was reported that the optical activity of the material is lost if the NTs become larger.[12] The semiconducting nature of the CNTs is responsible for the photonic device applications, whereas their metallic characteristic make them suitable for the electronic applications. CNT is an excellent material for the optoelectronic applications.[65] According to the p-tight-binding model, the one-third of the NTs are metallic and the rest parts are semiconducting (depending on their indices, n or m).[29] Kataura et. al. observed the optical absorption spectrum of the SWNT. They reported three peaks at 0.68, 1.2, and 1.7 eV, two for the semiconducting and one for the metallic tubes. A large absorption band at 4.5 eV was also observed for the π-plasmon of the SWNTs.[66] Similar optical absorption spectra were also reported in other paper.[64] Many researchers also reported the optical properties of the CNTs and their related compounds.[66–68] The optical properties can be described by using a Gaussian distribution of NT diameters.

5.5.4 THERMAL PROPERTIES

The potential applications of CNTs in integrated circuits and/or as thermal interface materials require an acute understanding of the thermal properties of CNTs. CNTs show good thermal conductivity along the tube axis, owing to the property of "ballistic conduction."[12] However, laterally to the NT axis, they behave as insulators. Theoretical investigations again exceed the number of experimental studies on the thermal conductivity of CNTs. Pop et al. studied the thermal conductance of SWNTs in the temperature range 300–800 K. At room temperature the reported thermal conductivity of SWNTs with diameter 1.7 nm reached values up to 3500 W/m K.[69] Similar values were obtained for MWNTs (~3000 W/m K) by Kim and coworkers.[70]

CNTs have high thermal stability and are stable up to 750°C in air and 2800°C in vacuum.[12] The thermal stability decreases with increase in the defects in the CNT structure. The thermal properties of CNTs are similar to graphite but the thermal expansion of CNTs is strongly isotropic whereas for graphite fibers the expansion is anisotropic. For major electronic device applications, CNTs need to be thermally stable as well as resistant to chemicals and moisture.[71,72]

5.6 APPLICATIONS OF CNTs (Fig. 5.6)

5.6.1 CNTs FOR ENERGY STORAGE

Over the last few decades the energy problem has emerged as the greatest question surrounding the scientific community. As such *batteries* and *electrochemical capacitors* have been developed extensively for better energy storage applications. CNT-based nanoporous electrodes have been explored for fabricating high-performance batteries and supercapacitors with improved life cycle and power delivery ability.[71]

Ever since SONY commercialized Li-ion batteries in 1991, they have become the backbone of the electronics industry.[73] Compared to the Pb-acid and Ni–Cd batteries, Li-ion batteries have exhibited better performance and are environmentally safe. However, to improve the performance of the

Li-ion battery (Fig. 5.7), to satisfy the ever-increasing demands of modern electronics, better electrode options are being investigated. One-dimensional CNTs are favored materials for use in Li-ion batteries owing to their high surface area, high electrical conductivity, and high mesoporosity (pores 2–5 nm).[72] Both SWNTs and MWNTs have been investigated as pristine anode materials and also as host materials for anode applications. CNTs have also been studied as conductive additives that improve the transport of Li ions and electrons. The excellent electrical and thermal conductivity of CNTs make them safe and viable option for use in batteries. The hollow porous structure of CNTs allows Li^+ intercalation within the core and at the walls of CNTs.[74] A number of factors such as chirality, length, diameter, and defects influence the capacity of CNT electrodes.[72]

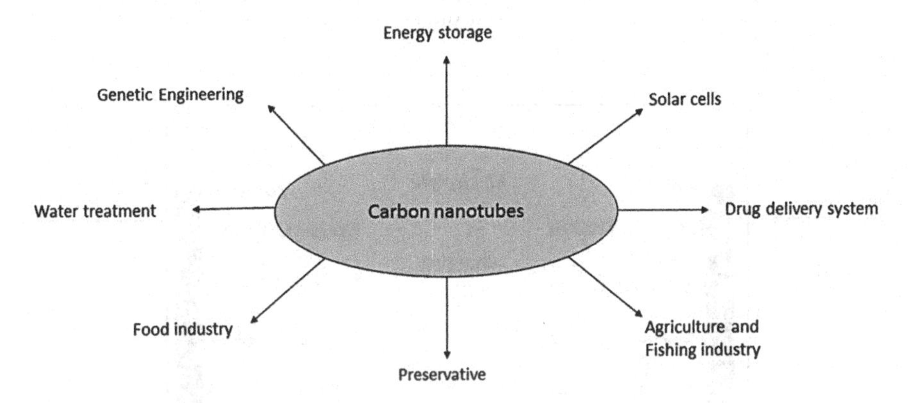

FIGURE 5.6 Recent advances in the applications of carbon nanotubes.

However, by controlling these factors SWCNTs have been reported to achieve a reversible capacity of 1000 mAh/g.[75] A major shortcoming of CNT-based electrodes is the formation of the solid electrolyte interface which greatly compromises the reversible capacity. Efforts are ongoing to overcome these shortcomings and realize the full benefits of CNTs.

Electrochemical capacitors or supercapacitors as they are commonly known, were first conceptualized by Becker in 1957 and later commercialized by Panasonic in 1978.[74, 76] Supercapacitors combine the characters of batteries and capacitors and have the benefits of longer life cycles, faster charging and better power delivery. The capacitance of a supercapacitor is directly dependant on the factor A/d where A is the surface area and d is the distance between the two electrodes.[71] As already discussed, the high surface area of CNTs makes them an attractive option for use in supercapacitors. Both SWNTs and MWNTs have been widely investigated as electrodes for supercapacitors with excellent results, the specific capacitances being 102 F g^{-1} (for MWNTs) and 180 G g^{-1} (for SWNTs).[77,78] The electrolyte used also influences the capacitance of CNT supercapacitors. Research is ongoing to enhance the performance of CNT-based supercapacitors for better stability and lifetimes by using functionalized CNTs, incorporating the CNTs with metal oxides and conducting polymers and using other carbon nanoallotropes.[72]

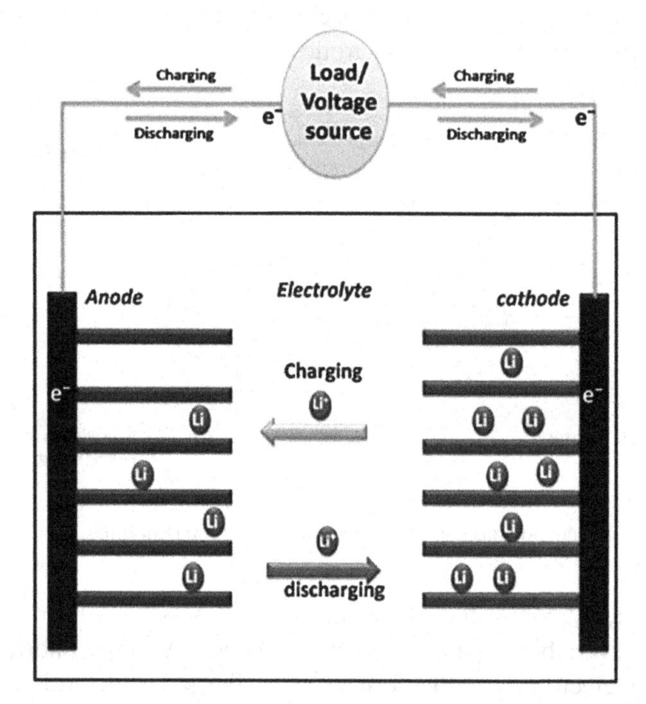

FIGURE 5.7 A basic representation of the working of a Li-ion battery.

Source: Reproduced with permission from Sehrawat, P.; Julien, C.; Islam, S. S. *Mater. Sci. Eng. B* **2016,** *213*, 12–40. © 2014 Elsevier.

5.6.2 IN GENETIC ENGINEERING

CNTs are used in the improvement of tissue engineering, proteomics, and bioimaging genomes. Because of its unique cylindrical structure and properties it can be used in the treatment of bronchitis, asthma, genetic disorders, and cancer. It helps in the gene therapy and to overcome the problems such as cytotoxic effects. CNTs can be used as a carrier for genes. DNA analysis can be done with the help of CNTs, in this process single-stranded DNA wraps the CNTs (a sequence dependent process) which changes their electrostatic property. It has some transfection importance because of the tendency of CNTs to release DNA after getting attached with DNA before it was eradicated by the cell defense system.[79]

5.6.3 IN DRUG DELIVERY SYSTEMS

There are three main factors that make CNT a very effective substance in the medical field: (1) having small size, (2) have high surface area to volume ratio, and (3) their capacity to contain chemical. It can be very small for passing through the holes in tumors. The large surface area to volume ratio gives a good base for the transporting chemicals and helps in the reaction for ultra-sensitive glucose detection.[80] It may also be used as a carrier for the drug delivery system.[81] CNTs and carbon nanohorns (CNHs) can hold a variety of antigens on their surface; therefore they can be used as a source of antigen in vaccines.[82] CNT shows higher efficiency as compared to classical drug delivery systems such as peptides or liposomes.[83] The poor solubility of medicines can be solved by CNTs and also improve the contact and penetration into the target cells. NTs help in the reduction of side effects of medicines.

5.6.4 IN SOLAR CELLS

CNTs increase the effectiveness of the solar cells. The bandgaps of CNT match with the solar spectrum having strong photo absorption and high carrier mobility which make CNT as an ideal photovoltaic material. SWNT can form ideal p–n junction diodes which is the main requirement of many electronic materials. That means SWNT can show enhanced properties of diodes. Because of low processing cost, simple fabrication process and high-efficiency CNTs can be used in dye-sensitized solar cells.[84]

5.6.5 IN WATER TREATMENT

There are some anthropogenic activities, urbanization and industrialization which have been continuously causing water pollution. It affects the fresh water availability in earth's surface. Several methods are available for the treatment of water. An innovative technology involves CNT for the purification of water, due to its lower cost, greater chemical reactivity, large surface area, and less chemical mass. CNT is very effective for the decontamination processes—desalination, adsorption, disinfection and hybrid catalysis, etc. It increases efficiency of water purification technology and reduces efforts and time.[85] It also helps in the treatment of inorganic, organic, and biological water pollutants.[86]

5.6.6 IN PRESERVATIVES

The antioxidants such as CNHs and CNTs are used to prevent oxidation of certain compounds in the medical field. It can be used in the sunscreen to avoid oxidation of some skin components, that is, to protect skin against free radicals formed by ultra-violet sunlight or by the body.[87,88] In future, CNTs may be promising antioxidants for disease prevention and health protective effect.[89]

5.6.7 IN THE AGRICULTURE AND FISHING INDUSTRIES

CNT is used in carbon dioxide sensor which is an effective device to supervise the CO_2 concentration inside the greenhouse or to obtain an optimal condition for the growth of plants. CNT can also be used in the pressure sensor for the uniform spraying of liquid pesticides, herbicides, and insecticides. In the fishing industries, pH balance sensors containing CNT are very useful for maintaining the pH of water, which directly affects the growth of the fishes.[90]

5.6.8 IN THE FOOD INDUSTRY

NTs are becoming increasingly significant for the food industry. CNT-based biosensors and chemical sensors are used to detect the quality of the food. Biosensors are used to identify the contamination of foods caused by bacterial pathogens and the chemical sensor are used to identify the unwanted chemical residues caused by food additives, pesticides, herbicides, and other environmental contaminants in uncooked and cooked foods.[91]

5.7 CONCLUSIONS

Carbon has emerged as the one of the most important elements in the last century and the nanoallotropes of carbon are the way to the future. One-dimensional SWNT and MWNT have opened the road to electronic miniaturization. SWNTs were successfully incorporated into integrated circuits at IBM, Watson Research Center, New York.[92] CNT-based field emission transistors (CNT-FET) continue to be a major area of active research. CNTs have many prospective applications as outlined earlier; however, two major areas continue to be in focus—electronics and green energy.

As the world consumption of energy continues to shoot up, the conventional energy sources need to be compensated with alternative options. With countries all around the globe signing up for the Paris Agreement to cut down CO_2 emissions, researchers now more than ever are driven to find clean energy alternatives. CNTs could be our potential answer to the growing energy needs. CNTs have been used as electrodes in lithium-ion batteries and electrochemical capacitors. However, much more efforts are required to improve their synthesis, design, and performance. The electrical conductivities of CNT fibers need to be enhanced greatly and the problem of solid electrolyte interphase has to be overcome.

For viable applications of CNTs in devices, the factors that drive commercialization should be addressed. There is still a lot more to be done in terms of purification and doping of CNTs, cost control, and scalability of CNT-based devices. In spite of numerous hurdles, the progress in research in the CNT spectrum continues to be promising. Suffice to say that CNTs are here to stay and revolutionize the future, and may as well propel us into a space elevator![93]

KEYWORDS

- **carbon nanotubes**
- **laser ablation**
- **chemical vapor deposition**
- **functionalization of CNTs**
- **mechanical properties**
- **energy storage**

REFERENCES

1. Iijima, S. *Nature* **1991,** *354* (6348), 56–58.
2. Sahoo, N. G.; Rana, S.; Cho, J. W.; Li, L.; Chan, S. H. *Prog. Polym. Sci.* **2010,** *35* (7), 837–867.
3. Treacy, M. J.; Ebbesen, T.; Gibson, J. *Nature* **1996,** *381* (6584), 678–680.
4. Gooding, J. J. *Electrochim. Acta* **2005,** *50* (15), 3049–3060.
5. Compano, R.; Molenkamp, L.; Paul, D. *Technology Roadmap for Nanoelectronics*; European Commission: Brussels, 2000.
6. Fan, S.; Chapline, M. G.; Franklin, N. R.; Tombler, T. W.; Cassell, A. M.; Dai, H. *Science* **1999,** *283* (5401), 512–514.
7. Vilatela, J. J.; Khare, R.; Windle, A. H. *Carbon* **2012,** *50* (3), 1227–1234.
8. Iijima, S.; Ichihashi, T. *Nature* **1993,** *363* (6430), 603–605.
9. Guo, T.; Nikolaev, P.; Thess, A.; Colbert, D. T.; Smalley, R. E. *Chem. Phys. Lett.* **1995,** *243* (1), 49–54.
10. Huang, S.; Woodson, M.; Smalley, R.; Liu, J. *Nano Lett.* **2004,** *4* (6), 1025–1028.
11. Dresselhaus, M. S.; Dresselhaus, G.; Avouris, P., Eds. *Carbon Nanotubes;* Topics Applied Physics; Springer Verlag: Berlin, 2001; Vol. 80.
12. Varshney, K. *Int. J. Eng. Res.* **2014,** *2* (4), 660–677.
13. Wang, L.; Davids, P.; Saxena, A.; Bishop, A. *Phys. Rev. B* **1992,** *46* (11), 7175.
14. Goh, P.; Ismail, A.; Ng, B. *Compos. Part A Appl. Sci. Manuf.* **2014,** *56,* 103–126.
15. Amrin, S.; Deshpande, V. *Physica E Low-dimens. Syst. Nanostruct.* **2017,** *87,* 317–326.
16. Jawahar, N.; Surendra, E.; Radha, K. *J. Pharm. Sci. Res.* **2015,** *7,* 141–154.
17. Rafique, M. M. A.; Iqbal, J. *J. Encapsul. Adsorpt. Sci.* **2011,** *1* (02), 29.
18. Prasek, J.; Drbohlavova, J.; Chomoucka, J.; Hubalek, J.; Jasek, O.; Adam, V.; Kizek, R. *J. Mater. Chem.* **2011,** *21* (40), 15872–15884.
19. Paradise, M.; Goswami, T. *Mater. Des.* **2007,** *28* (5), 1477–1489.
20. Mubarak, N.; Abdullah, E.; Jayakumar, N.; Sahu, J. *J. Ind. Eng. Chem.* **2014,** *20* (4), 1186–1197.
21. Ebbesen, T. W. *Carbon Nanotubes: Preparation and Properties;* CRC Press: New York, 1996.
22. Ibrahim, K. S. *Carbon Lett.* **2013,** *14* (3), 131–144.
23. Ando, Y.; Zhao, X. *New Diam. Front. Carbon Technol.* **2006,** *16* (3), 123–138.
24. Lee, S. J.; Baik, H. K.; Yoo, J.-e.; Han, J. H. *Diam. Relat. Mater.* **2002,** *11* (3), 914–917.
25. Jung, S.-H.; Kim, M.-R.; Jeong, S.-H.; Kim, S.-U.; Lee, O.-J.; Lee, K.-H.; Suh, J.-H.; Park, C.-K. *Appl. Phys. A* **2003,** *76* (2), 285–286.
26. Sornsuwit, N.; Maaithong, W. *Int. J. Precis. Eng. Manuf.* **2008,** *9* (3), 18–21.
27. Montoro, L. A.; Lofrano, R. C.; Rosolen, J. M. *Carbon* **2005,** *43* (1), 200–203.
28. Gang, X.; Jia, S.-L.; Shi, Z.-Q. *New Carbon Mater.* **2007,** *22* (4), 337–341.
29. Popov, V. N. *Mater. Sci. Eng. R Rep.* **2004,** *43* (3), 61–102.
30. O'Connell, M. J. *Carbon Nanotubes: Properties and Applications*; CRC Press: Boca Raton, 2006.
31. Bronikowski, M. J.; Willis, P. A.; Colbert, D. T.; Smith, K.; Smalley, R. E. *J. Vac. Sci. Technol. A Vac. Surf. Films* **2001,** *19* (4), 1800–1805.
32. Balasubramanian, K.; Burghard, M. *Small* **2005,** *1* (2), 180–192.
33. Mallakpour, S.; Soltanian, S. *RSC Adv.* **2016,** *6* (111), 109916–109935.
34. Vaisman, L.; Wagner, H. D.; Marom, G. *Adv. Colloid Interface Sci.* **2006,** *128,* 37–46.

35. Di Crescenzo, A.; Ettorre, V.; Fontana, A. *Beilstein J. Nanotechnol.* **2014,** *5*, 1675.
36. Curran, S. A.; Ajayan, P. M.; Blau, W. J.; Carroll, D. L.; Coleman, J. N.; Dalton, A. B.; Davey, A. P.; Drury, A.; McCarthy, B.; Maier, S. *Adv. Mater.* **1998,** *10* (14), 1091–1093.
37. Fujigaya, T.; Nakashima, N. *Sci. Technol. Adv. Mater.* **2015,** *16* (2), 024802.
38. Mc Carthy, B.; Coleman, J.; Czerw, R.; Dalton, A.; Carroll, D.; Blau, W. *Synth. Met.* **2001,** *121* (1–3), 1225–1226.
39. Nativ-Roth, E.; Shvartzman-Cohen, R.; Bounioux, C.; Florent, M.; Zhang, D.; Szleifer, I.; Yerushalmi-Rozen, R. *Macromolecules* **2007,** *40* (10), 3676–3685.
40. Fu, K.; Huang, W.; Lin, Y.; Riddle, L. A.; Carroll, D. L.; Sun, Y.-P. *Nano Lett.* **2001,** *1* (8), 439–441.
41. Gao, C.; Vo, C. D.; Jin, Y. Z.; Li, W.; Armes, S. P. *Macromolecules* **2005,** *38* (21), 8634–8648.
42. Aqel, A.; El-Nour, K. M. A.; Ammar, R. A.; Al-Warthan, A. *Arab. J. Chem.* **2012,** *5* (1), 1–23.
43. Hou, P.; Bai, S.; Yang, Q.; Liu, C.; Cheng, H. *Carbon* **2002,** *40* (1), 81–85.
44. Hou, P.-X.; Liu, C.; Cheng, H.-M. *Carbon* **2008,** *46* (15), 2003–2025.
45. Hirlekar, R.; Yamagar, M.; Garse, H.; Vij, M.; Kadam, V. *Asian J. Pharm. Clin. Res.* **2009,** *2* (4), 17–27.
46. Salvetat, J.-P.; Bonard, J.-M.; Thomson, N.; Kulik, A.; Forro, L.; Benoit, W.; Zuppiroli, L. *Appl. Phys. A* **1999,** *69* (3), 255–260.
47. Georgakilas, V.; Perman, J. A.; Tucek, J.; Zboril, R. *Chem. Rev.* **2015,** *115* (11), 4744–4822.
48. Nardelli, M. B.; Bernholc, J. *Phys. Rev. B* **1999,** *60* (24), R16338.
49. Nardelli, M. B.; Fattebert, J.-L.; Orlikowski, D.; Roland, C.; Zhao, Q.; Bernholc, J. *Carbon* **2000,** *38* (11), 1703–1711.
50. Ruoff, R. S.; Lorents, D. C. *Carbon* **1995,** *33* (7), 925–930.
51. Huang, J.; Chen, S.; Wang, Z.; Kempa, K.; Wang, Y.; Jo, S.; Chen, G.; Dresselhaus, M.; Ren, Z. *Nature* **2006,** *439* (7074), 281–281.
52. Dai, H. *Acc. Chem. Res.* **2002,** *35* (12), 1035–1044.
53. Charlier, J.-C.; Blase, X.; Roche, S. *Rev. Mod. Phys.* **2007,** *79* (2), 677.
54. Andrews, R.; Jacques, D.; Qian, D.; Rantell, T. *Acc. Chem. Res.* **2002,** *35* (12), 1008–1017.
55. Ajayan, P. *Chem. Rev.* **1999,** *99* (7), 1787–1800.
56. Belin, T.; Epron, F. *Mater. Sci. Eng. B* **2005,** *119* (2), 105–118.
57. Dai, H. *Surf. Sci.* **2002,** *500* (1), 218–241.
58. Hong, S.; Myung, S. *Nat. Nanotechnol.* **2007,** *2* (4), 207–208.
59. Dai, H.; Wong, E. W.; Lieber, C. M. *Nature* **1996,** *384*, 147–150.
60. Song, S. N.; Wang, X.; Chang, R.; Ketterson, J. *Phys. Rev. Lett.* **1994,** *72* (5), 697.
61. Zhu, H.; Xu, C.; Wu, D.; Wei, B.; Vajtai, R.; Ajayan, P. *Science* **2002,** *296* (5569), 884–886.
62. Li, Q.; Li, Y.; Zhang, X.; Chikkannanavar, S. B.; Zhao, Y.; Dangelewicz, A. M.; Zheng, L.; Doorn, S. K.; Jia, Q.; Peterson, D. E. *Adv. Mater.* **2007,** *19* (20), 3358–3363.
63. Takesue, I.; Haruyama, J.; Kobayashi, N.; Chiashi, S.; Maruyama, S.; Sugai, T.; Shinohara, H. *Phys. Rev. Lett.* **2006,** *96* (5), 057001.
64. Hartschuh, A.; Pedrosa, H. N.; Peterson, J.; Huang, L.; Anger, P.; Qian, H.; Meixner, A. J.; Steiner, M.; Novotny, L.; Krauss, T. D. *ChemPhysChem* **2005,** *6* (4), 577–582.
65. Vajtai, R. *Springer Handbook of Nanomaterials;* Springer Science & Business Media: Heilderberg, 2013.
66. Kataura, H.; Kumazawa, Y.; Maniwa, Y.; Umezu, I.; Suzuki, S.; Ohtsuka, Y.; Achiba, Y. *Synth. Met.* **1999,** *103* (1–3), 2555–2558.

67. Zhao, G.; Bagayoko, D.; Yang, L. *J. Appl. Phys.* **2006,** *99* (11), 114311.
68. Zhao, J.; Chen, X.; Xie, J. R. *Anal. Chim. Acta* **2006,** *568* (1), 161–170.
69. Pop, E.; Mann, D.; Wang, Q.; Goodson, K.; Dai, H. *Nano Lett.* **2006,** *6* (1), 96–100.
70. Kim, P.; Shi, L.; Majumdar, A.; McEuen, P. *Phys. Rev. Lett.* **2001,** *87* (21), 215502.
71. Hu, L.; Hecht, D. S.; Gruner, G. *Chem. Rev.* **2010,** *110* (10), 5790–5844.
72. Yang, Z.; Ren, J.; Zhang, Z.; Chen, X.; Guan, G.; Qiu, L.; Zhang, Y.; Peng, H. *Chem. Rev.* **2015,** *115* (11), 5159–5223.
73. Nagaura, T.; Tozawa, K. *JEC Press* **1990,** *9*, 209.
74. Dai, L.; Chang, D. W.; Baek, J. B.; Lu, W. *Small* **2012,** *8* (8), 1130–1166.
75. Li, X.; Kang, F.; Bai, X.; Shen, W. *Electrochem. Commun.* **2007,** *9* (4), 663–666.
76. Kötz, R.; Carlen, M. *Electrochim. Acta* **2000,** *45* (15), 2483–2498.
77. Niu, C.; Sichel, E. K.; Hoch, R.; Moy, D.; Tennent, H. *Appl. Phys. Lett.* **1997,** *70* (11), 1480–1482.
78. Zilli, D.; Bonelli, P.; Cukierman, A. *Nanotechnology* **2006,** *17* (20), 5136.
79. Duesberg, G.; Burghard, M.; Muster, J.; Philipp, G. *Chem. Commun.* **1998,** (3), 435–436.
80. Kushwaha, S. K. S.; Ghoshal, S.; Rai, A. K.; Singh, S. *Braz. J. Pharm. Sci.* **2013,** *49* (4), 629–643.
81. Wu, W.; Wieckowski, S.; Pastorin, G.; Benincasa, M.; Klumpp, C.; Briand, J. P.; Gennaro, R.; Prato, M.; Bianco, A. *Angew. Chem. Int. Ed.* **2005,** *44* (39), 6358–6362.
82. Pantarotto, D.; Partidos, C. D.; Hoebeke, J.; Brown, F.; Kramer, E.; Briand, J.-P.; Muller, S.; Prato, M.; Bianco, A. *Chem. Biol.* **2003,** *10* (10), 961–966.
83. Kartel, M.; Ivanov, L.; Kovalenko, S.; Tereschenko, V. Carbon Nanotubes: Biorisks and Biodefence. In *Biodefence*; Springer: Dordrecht, 2011; pp 11–22.
84. Chappel, S.; Chen, S.-G.; Zaban, A. *Langmuir* **2002,** *18* (8), 3336–3342.
85. Savage, N.; Diallo, M. S. *J. Nanopart. Res.* **2005,** *7* (4), 331–342.
86. Kar, S.; Subramanian, M.; Pal, A.; Ghosh, A.; Bindal, R.; Prabhakar, S.; Nuwad, J.; Pillai, C.; Chattopadhyay, S.; Tewari, P. In *Preparation, Characterisation and Performance Evaluation of Anti-biofouling Property of Carbon Nanotube-polysulfone Nanocomposite Membranes*, AIP Conference Proceedings, Mumbai, India, 2013; pp 181–185.
87. Pai, P.; Nair, K.; Jamade, S.; Shah, R.; Ekshinge, V.; Jadhav, N. *Curr. Pharma. Res. J.* **2006,** *1*, 11–15.
88. Galano, A. *J. Phys. Chem.* **2008,** *112* (24), 8922–8927.
89. Galano, A. *Nanoscale* **2010,** *2* (3), 373–380.
90. Luong, J. H.; Hrapovic, S.; Wang, D. *Electroanalysis* **2005,** *17* (1), 47–53.
91. Sinha, N.; Ma, J.; Yeow, J. T. *J. Nanosci. Nanotechnol.* **2006,** *6* (3), 573–590.
92. Collins, P. G.; Arnold, M. S.; Avouris, P. *Science* **2001,** *292* (5517), 706–709.
93. Yakobson, B. I.; Smalley, R. E. *Am. Sci.* **1997,** *85* (4), 324–337.

CHAPTER 6

CONDUCTING POLYMER/CNT-BASED NANOCOMPOSITES AS SMART EMERGING MATERIALS

NEHA KANWAR RAWAT[1*], P. K. PANDA[1], and ANUJIT GHOSAL[2,3]

[1]*Materials Science Division, CSIR-National Aerospace Laboratory, Bengaluru 560017, India*

[2]*Plant NanoBiotechnology Laboratory, School of Biotechnology, Jawaharlal Nehru University, New Delhi 110067, India*

[3]*School of Life Science, Beijing Institute of Technology, Beijing, China*

Corresponding author. E-mail: neharawatjmi@gmail.com

ABSTRACT

Conducting polymers/carbon nanotube (CNT) nanocomposites (NCs) have emerged as one of the most fascinating polymers in the field of material science and engineering. Their marvelous resourceful properties make them useful in every field of research. They find application in varying arenas such as supercapacitors, sensors, electronic materials, drug release, biosensors, tissue engineering, as well as stimuli-responsive and biomimetic polymeric materials. The chapter discusses role of conducting polymers and CNTs with their vast applications as NCs and their significant contribution in materials science and engineering as well as their prospects.

6.1 INTRODUCTION

The emergence of carbon nanotubes (CNTs) since their discovery two decades back, has created new prospects for the creation of polymer composites having potential for a wide spectrum of applications. Copious noteworthy advances have been found up to date, and more technological challenges await the optimization of a system to fill the gap between prospects and practical performance. Despite this tremendous progress, challenging issues related to directional alignment and CNT assembly within a polymer matrix remain and many researches are in progress.

Therefore, the formation of CNTs/functional polymer composites would be a meaningful way to improve or extend the properties of polymers or CNTs. Some shortcomings like poor processibility of CNTs can be modified by combining these carbon materials with other things, making their composites, hybrids or interpenetrating polymer networks, etc.[1–4]

Conducting polymers (CPs) since long has been considered as prominent carbon fillers in almost all material science areas, owing to their easy synthesis.[3,5] The features like processibility and numerous marvelous properties make them strong prospective candidate. CPs find vast applications in the field of fuel cells, computer displays, super capacitors, corrosion-resistant coatings, and many other applications. These versatile polymers have many ways to be synthesized, which includes alone as polymers, interpenetrating polymer networks, composites, hybrids, and nanostructured materials.[6]

In this chapter, we focus on the preparation, characterization of fascinating CNT–CP NCs and their novel properties and applications. This chapter presents the development of existing approaches in the arena of CNTs and their composites where fillers are CPs, with emphasis being placed on the recent progress, exciting advances, and active pursuit in this growing field of material science and engineering.

6.2 BRIEF INTRODUCTION OF CNTs AND CPs

6.2.1 CARBON NANOTUBES

In 1991, discovery of CNTs by Iijima et al. paved a way in a history of materials science and engineering.[7,8] After that, there is no looking back for this class of carbon materials, in way of their preparation and applications. Apart from having various attributes such as high mechanical strength and modulus, good electrical conductivity, and excellent chemical stability, they also have characteristics of other nanomaterials having low dimensions, which make them promising in the area of nanodevices.[9]

However, they have certain drawbacks like poor processability which leads to limit their applications. Therefore, many researchers have actively engaged in their chemical modification with other counterparts such as organic molecules or polymers to enhance their solubility.[1,10]

The most prominent method for organic molecules or polymers to increase their functionalization on the surface of CNTs is the process known as "grafting to" and "grafting from" techniques using reactions, such as esterification reaction, amide reaction, atom transfer radical polymerization (ATRP), radical addition-fragmentation chain transfer polymerization (RAFT), ring-opening, etc. Moreover, due to the inert chemical properties of CNTs in some respects, they cause a great hindrance in the applications mainly in sensors.[11,12] Therefore, the importance of CNTs/functional polymer composites would be a most convenient way to improve or extend the properties of polymers or CNTs. The aligned array of CNTs have proved to be an excellent substrate for the electrodeposition of CPs as they grow/respond conveniently on the surface of each CNT, smoothing the creation of uniform coaxial CNTs/CP nanocomposites.

6.2.2 CONDUCTING POLYMERS

CPs are considered as "synthetic metals," having highly conjugated polymeric chain.[3] For their discovery, 2000 Noble prize in chemistry was given to Alan J. Heeger, Alan G. MacDiarmid, and Hideki Shirakawa.[6,13,14] Typical CPs (Fig. 6.1) include polyacetylene (PA), polyaniline (PANI), polypyrrole (PPy), polythiophene (PTh), poly(para-phenylene) (PPP), poly(phenylenevinylene) (PPV), polyfuran (PF), etc. The chemical structures of these polymers are illustrated in Figure 6.1. They can be regarded as electrical insulators, semiconductors, or conductors, depending on the level of doping and nature of the dopants. Upon adding dopants and/or subjecting to chemical or electrochemical redox reactions, the electrical conductivity of these conjugated polymers can increase by some orders of magnitude. For example, the emeraldine base PANI (undoped form) is an insulator having conductivity of only 10^{-10} to 10^{-8} S/cm; when doping with some agents such as acids, carbon fillers, etc., it can increase to 10^{2} to 10^{3} S/cm or higher. The high conductivity upon doping makes these polymers promising materials for applications ranging from electro-optic, molecular, and nanoelectronic devices to microwave absorbing and corrosion protection coatings. Additionally, they have the ability of the reversible doping–dedoping which provides them advantage in applications areas such as sensors, actuators, and separation membranes (Table 6.1; Figs. 6.2 and 6.3).[9,15]

6.2.2.1 CLASSIFICATION OF CPs

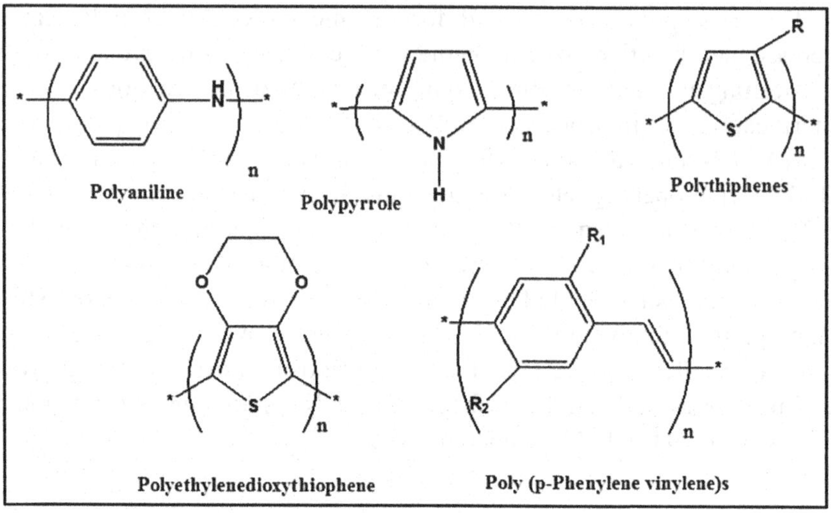

Polyaniline
Polypyrrole
Polythiphenes
Polyethylenedioxythiophene
Poly (p-Phenylene vinylene)s

FIGURE 6.1 Various conducting polymers with their structures.

Conducting polymers (CPs)
Intrinsically CPs
Extrinsically CPs
Conjugated CPs
Doped CPs
Polymers with conducting blends
Polymers with conducting elements

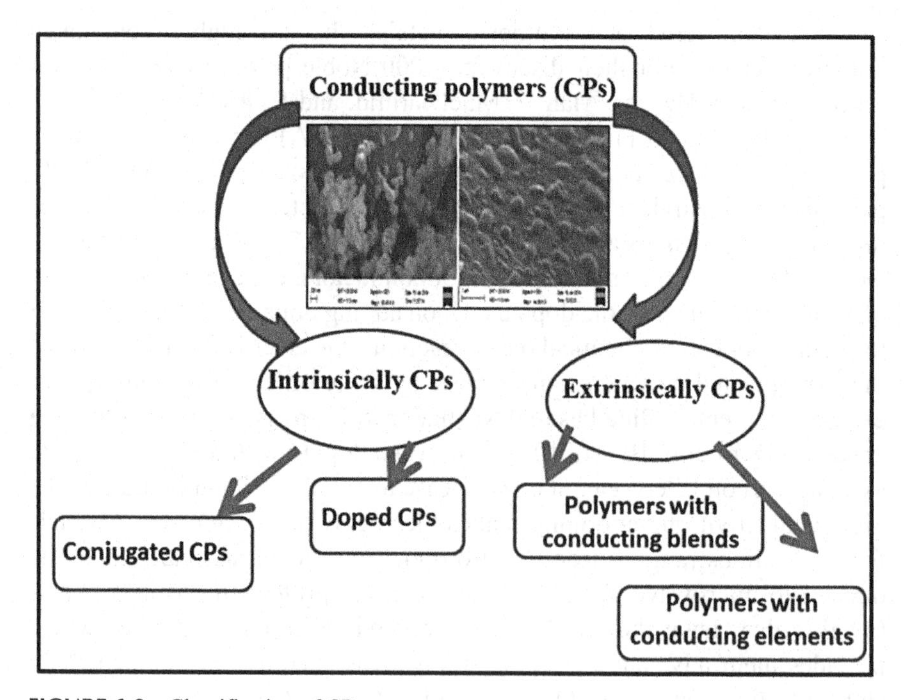

FIGURE 6.2 Classification of CPs.

TABLE 6.1 CPs and Their Properties Including Prominent Works in Varying Field.

Polymer	Abbreviation	Family	Conductivity (in order)	Color	Advantages	Shortcomings	Applications
Polyaniline	PANI	Amine	10^2	Green, black	Cost effective, easy synthesis, processibility	Less biodegradability	Biosensors Corrosion Electromagnetic interference shielding
Polypyrrole	PPY	Hetroaro-matic	$10–10^3$	Green, blue	Varying forms, cost effective, easy modification	Health hazards, low processibility	Hydrogels Corrosion Electromagnetic interference shielding
Polythiophene	PTh	Hetroaro-matic	$10^2–10^3$	Brown, brick	Cost effective, bio-degradable	High cost, nonbiodegradable	Biocompatibility Sensors Supercapacitors
Poly(o-ansidine)	PoA	Amine	10^2	Dark green	Easy synthesis, processibility	Less processible, bulky side group	Biosensors
Poly(3,4-ethylene-dioxythiophene)	PEDOT	Hetroaro-matic	$10^2–10^3$	Yellow, blue	Highly conducting	High cost, nonbiodegradable	Bioelectronics Sensors FETs
Polyparavinylene	PPV	Amine	0.1–10	Brown, brick	Easy synthesis, processibility	Easy fusible rings	Neural probes
Polyfuran	PFu	Hetroaro-matic	$1–10^2$	Green	Small chains, easy fusible	Small ring size easy to modify	Biodegradable polymers Limited application

FET, field emission transistors.

6.3 CNTs-BASED CP NCs

6.3.1 SYNTHESIS OF CNTs–CPs NCs

The formulation of CNTs/functional polymer composites would be a convenient method to improve or extend the properties of polymers or CNTs. The principle of interaction between CNTs and CPs are phenomenon of electron/hole transport. CNTs being 1D nanostructures, the easiest way is to form a layer of CPs on CNTs by coating onto their surface. Therefore, it is important to investigate effective methods to make CPs grow on the surface of CNTs (Fig. 6.3). The approach for this is a method known as electrochemical polymerization, in which the morphology and properties of the NCs can be controlled by controlling the applied potential or current density. Ajayan et al. have reported the electrochemical oxidation of aniline in H_2SO_4 on the CNTs electrode to fabricate CNT/PANI composites.[16,17] Another work by Chen et al. demonstrated the formation of CNT/PPy nanocomposites, the first example of anionic CNTs acting as the dopant of a CP.[18] Their results showed that even at low potential PPy was uniformly coated on the surface of individual CNTs results in prospective applications in nanoelectronics devices.

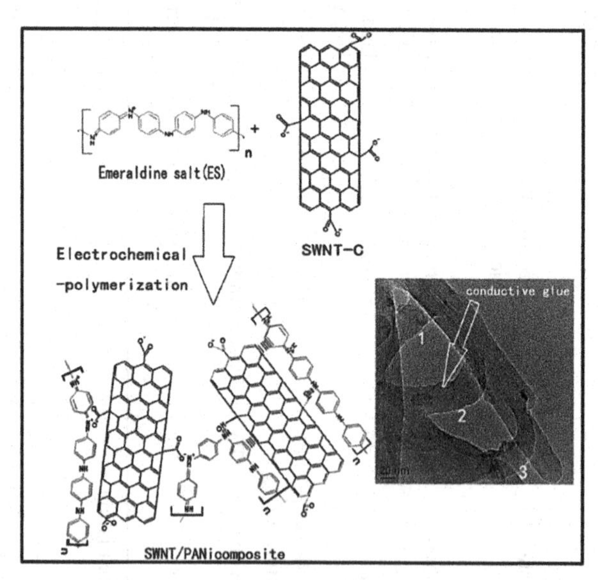

FIGURE 6.3 Schematic illustration of the in situ electrochemical polymerization of the SWNT–PANI composites (inset is the TEM image of SWNT–PANI composites).

Source: Reproduced with permission from Ref. [61]. © 2010 American Chemical Society.

The other method in which growth of films and their thickness can be controlled is a template-directed electropolymerization method. On comparing the results obtained from, "ex-situ" polymerization method (i.e., combining the CPs with CNTs directly), can have better site-selective interactions between the CPs rings and CNTs like in case of PANI and CNTs, improving their charge-transfer processes. However, the morphology of the as-synthesized CNTs/CP NCs (Fig. 6.4) is not very uniform. To enhance the interactions between CPs and CNTs, introduction of covalent bonding between Ps and CNTs has been explored. Philip et al. functionalized multi-walled carbon nanotubes (MWCNTs) with p-phenylenediamine, giving their phenylamine functional groups on the surface of MWCNTs. Zhang et al. have prepared a series of CNTs/CP nanocables by such an in situ chemical polymerization directed by the cationic surfactant CTAB.[20]

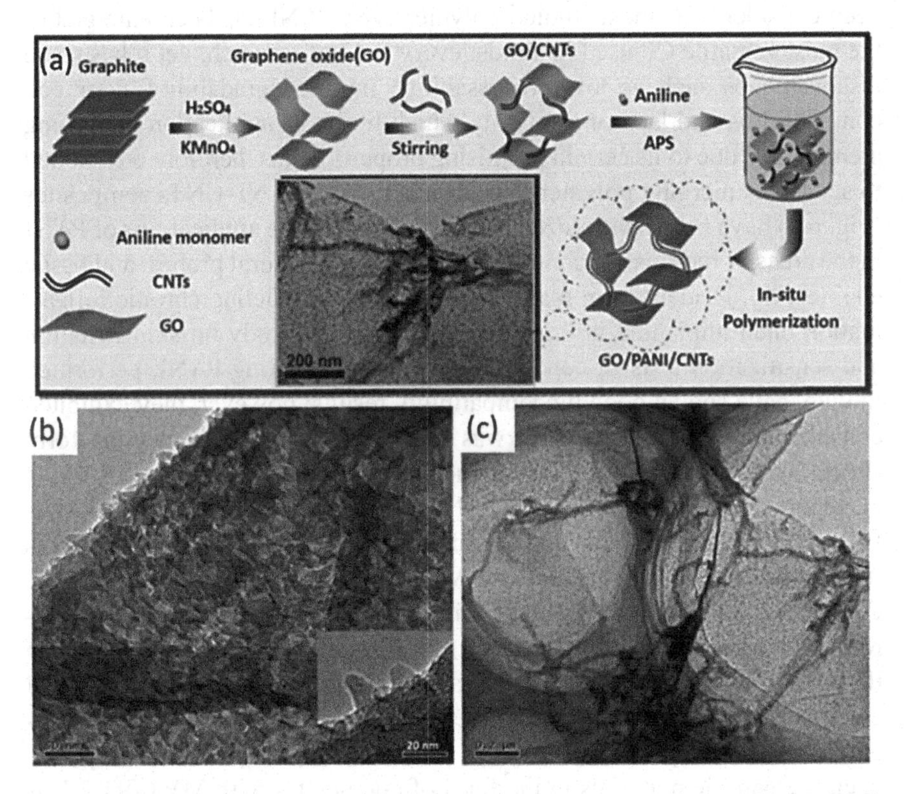

FIGURE 6.4 (a) Schematic diagram of the preparation of PANI–GO–CNT (PGC) composites. (b) and (c): SEM images of PGC.

Source: Reproduced with permission from Ref. [62]. © 2013 Elsevier.

Vapor deposition polymerization is very important approach explored for the preparation of core–shell or core–sheath nanostructures. Jang and Bae have fabricated carbon nanofibers/PPy nanocables using this method.[21] Microscopic studies revealed the formation of uniform PPy layer on the surface of carbon nanofibers. The thickness of PPy layer can be well-controlled by varying the amount of the monomer in the feed. It is noted that not all the core–sheath-shaped CNTs/CP NCs are necessary to have structures with CNTs component as core and CPs components as shell layer.

6.3.2 CNTs–PANI COMPOSITES

One of the most investigated CPs is PANI or aniline black,[3,22] which has been considered as most studied polymer ever. PANI has been emerged as the most versatile CP used in almost every field of research, yet it has some disadvantages such as low processability, nonbiodegradability, poor cell compatibility, and lack of flexibility which limits its application in varying arena; PANI due to its certain promising properties was therefore envisioned to act as an emerging polymer in biomedical field. PANI–CNTs composites (Fig. 6.5) have been synthesized and carried out on the applications of PANI in the areas of biosensors, controlled drug delivery, neural probes, and tissue engineering. Evident from many reports present, depicting chronic inflammation once implanted in human body.[23] A recent study on skin irritation and sensitization tests of conducting and nonconducting PANI, performed in vivo, have shown good biocompatibility results; however, they exhibited considerable cytotoxicity which was found more in their conducting form (PANI hydrochloride) than nonconducting form (emeraldine base).[24] Wu et al. report the synthesis of doped PANI in its emeraldine salt form (PANI-ES) with carboxylic acid and acylchloride groups contained MWCNTs (designated as c-MWCNTs and a-MWCNTs) by in situ polymerization. The conductivities of 0.5 wt% functionalized MWCNT containing PANI-ES/c-MWCNT and PANI-ES/a-MWCNT composites are 60–70% higher than that of PANI without MWCNT.[25] Frackowiak et al. studied three types of electrically conducting polymers (ECPs), that is, PANI, PPy, and poly(3,4-ethylenedioxythiophene) (PEDOT), which have been tested as supercapacitor electrode materials in the form of composites with MWCNT.[25] It is also noteworthy that such a type of ECP–CNTs composite did not need any binding substance that is an important practical advantage.

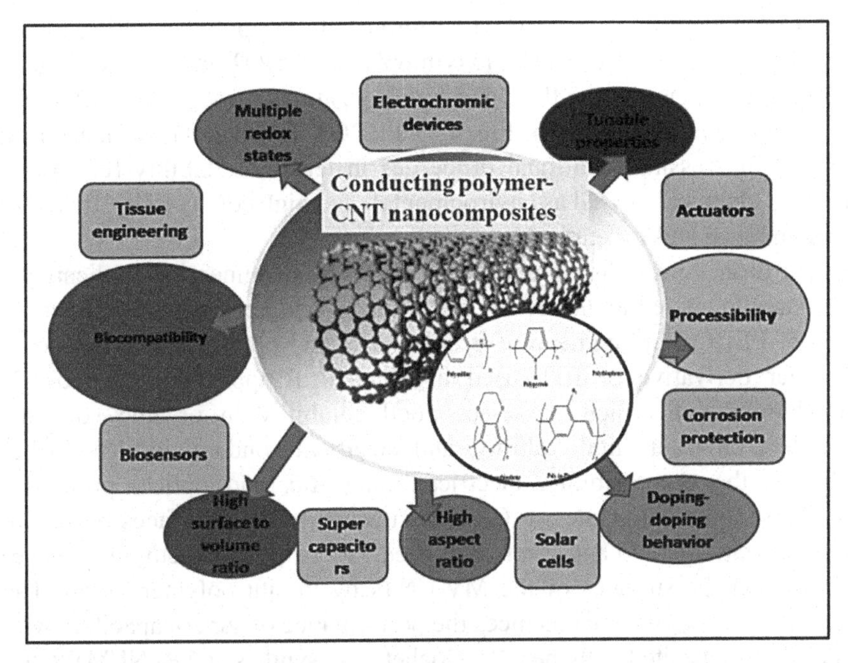

FIGURE 6.5 Various properties and applications of CNT–CP nanocomposites.

6.3.3 CNTs–PPY COMPOSITES

PPy-based CNT NCs have advanced features, which arise only after certain modifications, as the pure PPy is crystalline, brittle, insoluble, and not suitable for most applications, for example, in biomedical field, a carboxyethyl derivative of PPy is modified easily at N-position with biological moieties; this modification promotes tissue response and enhances the biomaterial–tissue interface. PPy nanoparticles demonstrated cytotoxic effects, at higher concentrations, in a recent in vitro study; however, at lower concentrations no negative effect was found on cell proliferation.[26] PPy is used in numerous applications such as biosensors, drug delivery systems, implants, scaffolds, neural probes, tissue engineering, and others.[27,28]

6.3.4 CNTs–PTh COMPOSITES

PTh is heteroatomic member of CP family. Poly(3,4-ethylenedioxythiophene) generally known as PEDOT[29] is a fascinating conjugated polymer and a derivative of PTh; PEDOT is formed by the polymerization of the bicyclic monomer 3,4-ethylenedioxythiophene. PEDOT has a dioxyalkylene bridging group across the third and fourth positions respective to heterocyclic ring as compared to PTh (Fig. 6.4). PEDOT has emerged as an aspiring new CP possessing additional properties in terms of stability (electrical, thermal, chemical as well as environmental) and high conductivity. It is used in biomedical and for engineering applications.

The prominent areas include biosensing and bioengineering applications, for example, in neural electrodes, nerve grafts, and heart patches. The main area of PEDOT in engineering is solar cells and opto-elctronic devices.[30] Another derivative of PTh used in biomedical applications is poly(3-hexylthiophene), which possesses good solubility in organic solvents, excellent environmental stability, and electrical conductivity. To further improve the processibility, electrical, magnetic, and optical properties of CNTs, some conjugates or CPs are attached to their surfaces by in situ polymerization. Xiao and Zhou deposited PPy or poly(3-methylthiophene) (PMeT) on the surfaces of the MWCNTs by in situ polymerization. The Faraday effect of the CP enhances the performance of super-capacitors with MWCNTs deposited with the CP.[31] Cochet et al. synthesized PANI/MWCNT composites by means of in situ polymerization in the presence of MWCNTs.[32] Their results state, the site-selective interaction between the quinoid ring of PANI and MWCNTs, thus confirms how far charge transfer processes lead to improve the electric properties of PANI/MWCNT composite.

6.4 APPLICATION OF CNT/CPs IN VARYING ARENA

6.4.1 SENSORS

Literature reports that the applications of CNT/CPs composites have high potential in area of sensors (Fig. 6.6).

Highly sensitive ammonia gas sensors were fabricated using MWCNTs reinforced ECP composites following solution casting method. Two types of CPs such as PEDOT–polystyrene sulfonic acid (PEDOT–PSS) and PANI were used and compared for their ammonia gas sensing properties at room temperature. Both the sensors were found to exhibit excellent sensitivity and poor recovery for ammonia gas at room temperature, but as compared to PANI, PEDOT–PSS polymer composite was found to be more sensitive (with sensitivity of ~16%) with less response time (~15 min).

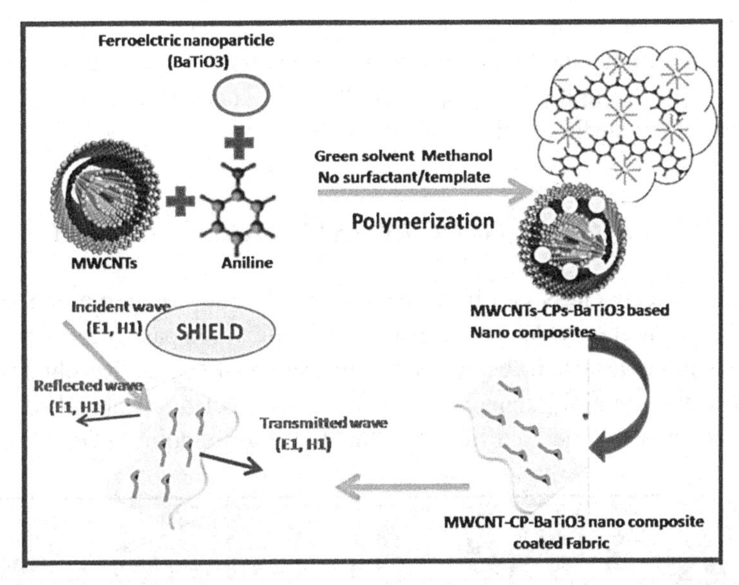

FIGURE 6.6 Mechanism of CPs–CNT NCs shielding for electromagnetic radiations.

In one of the few studies that utilized synthetic chemistry of CPs in a hybrid setting, Wang et al. demonstrated that calixarene-substituted PTh-coated SWNT sensors were used to detect differentiation of xylene isomers. In this work, the calixarene groups were attached to the hexane tails in P3HT and then chemically polymerized to create the isomer-selective CP. The CP was then mixed with SWNTs and spin coated on electrodes to create hybrid sensor device. These sensors were shown to have increased sensitivity to p-xylene by virtue of the calixarene shape selective hydrophobic pocket, while unmodified SWNT/P3HT devices displayed no difference in response to individual xylene isomers.[33] Rahman et al. formulated POAS–Ag/MWCNT NCs which were prepared by an adsorption process. A selective 3-methoxyphenolic sensor was developed by a current–voltage (*I–V*) technique for the first time. It displays the highest sensitivity (~3.829 μA cm^{-2} mM^{-1}) ever published in the literature.

Generally, literature shows that PANI NCs with CNTs and/or graphene are also extensively used for cholesterol detection.[34–36]

PPy/CNT composite nanowires were prepared by a template-directed electrochemical synthetic route, involving plating of PPy into the pores of a host membrane in the presence of shortened and carboxylated CNT dopants (without added electrolyte). Cyclic voltammetric growth profiles indicate that the CNT is incorporated within the growing nanowire and serves as the sole charge-balancing "counterion". Transmission electron microscopy images indicate high-quality straight PPy/CNT nanowires with a smooth and featureless surface and a uniform diameter.[37]

6.4.2 ELECTROMAGNETIC INTERFERENCE SHIELDING

Shielding effectiveness can be considered as the ratio of impinging energy to the residual energy. When an electromagnetic wave passes through some shielding materials used, absorption and reflection take place. Residual energy is part of the remaining energy that is neither reflected nor absorbed by the shield, but it is emerged out from the shield into surroundings (Fig. 6.7).

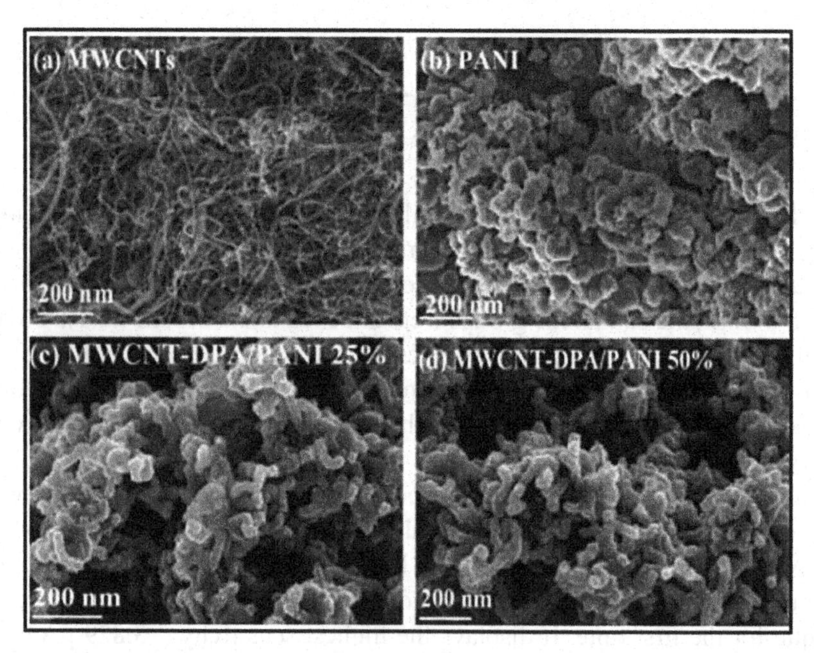

FIGURE 6.7 SEM images of (a) MWCNTs, (b) PANI, (c) MWCNT-DPA/PANI 25%, and (d) MWCNT-DPA/PANI 50%.

Source: Reproduced with permission from Ref. [54]. © 2014 Elsevier.

Recently, Gupta and colleagues studied enhanced field emission properties of MWCNT/PPy nanocomposites. Structural properties of these CNT/polymer NCs are very important in aerospace field as paints, anti-radar protectors, antistatic, etc.[38] The group at UoAlberta, Canada is actively engaged in finding the role of these nanocarbon-based fillers. The electromagnetic interference (EMI) shielding effects of MWCNT/polymer were studied by Al-Saleh and Sundararaj et al.[39]

The results indicated that shielding was done by absorption and reflection. Due to heat-absorbing capability, these NCs can be used in aerospace industries as electromagnetic wave absorption materials.[40]

Maiti et al. prepared a kind of electrically conducting polystyrene (PS)/MWCNTs/graphite nanoplate (GN) NCs to get high and constant electromagnetic wave interference shielding effectiveness at X-band (8~12 GHz), which has high application in both commercial and military applications. Besides more multiple reflection, the electrical conducting network among graphene sheets formed by the inside CNTs could also strengthen the conducting absorption. They proved that these composites when applied as shielding materials, the electromagnetic wave was attenuated greatly, and such attenuation had slightly been affected by frequency, even after the multiple reflection and absorption, as a result constant electromagnetic shielding effectiveness was produced.[41]

Sobha et al. prepared a system (PANI/FMWCNT/TPU) by in situ polymerization of aniline in thermoplastic polyurethane (TPU) solution assisted by ultrasonication of TPU composites based on PANI and functionalized MWCNT (FMWCNT). Their results proved a conductivity value of 28.6 S/m and EMI-shielding efficiency (SE) of 31.35 dB in X-band region at 8% filler concentration. On the contrary, for neat FMWCNT/TPU composites, the values were 1.52 S/m and 19.65 dB, respectively.[42]

6.4.3 BIOMEDICAL AREA

The controlled delivery of chemical compounds has been a great challenge, but now the use of CPs and their composites has facilitated this issue. The advantage of CPs are that suitable for drug release applications as they are porous and have delocalized charge carriers that aid in the diffusion of the bound molecules.[23,24,43–48]

6.4.3.1 TISSUE ENGINEERING

This field uses the combination of cells, engineering and materials methods, and suitable biochemical and physicochemical factors to improve or replace biological tissues. Tissue engineering involves the use of a tissue scaffold for the formation of new viable tissue for a medical purpose. The CNT–CPs NCs have become pertinent part of this application in biomedical field.

Luo et al. reported the preparation of composite of nylon resin and CNTs by melt compounding using a twin-screw extruder. The composites with loading of CNTs exhibited marvelous mechanical properties like strength modulus, elongation modulus, and hardness than that of pure nylon. The rheological behavior of nylon/CNTs composites showed that composites possessing high nanotube loadings exhibited a large decrease in viscosity with increasing shear frequency.[49] Furthermore, there are various reports found in literature on use of CNTs as conductive filler to make conductive polymer composites for various biomedical applications such as scaffolds for bone regeneration, tissue engineering, and nerve regeneration.[50]

Farid et al. have reported that CNTs and their derivatives as promising nanomaterials for stem cell research (i.e., culture, maintenance, and differentiation) and tissue engineering, as well as for regenerative, translational, and personalized medicine (e.g., bone reconstruction, neural regeneration). Also, from scarce nanotoxicological data, they also highlighted the importance of functionalizing graphene-based nanomaterials to minimize the cytotoxic effects, as well as other critical safety parameters that remain important to take into consideration when developing nanobiomaterials.[45,51]

6.4.3.2 DRUG DELIVERY

CPs and their nanocomposites are promising suitable matrices for biomolecules as they possess high electronic and ionic conductivities, which are significant for stability, speed, and sensitivity required for device applications in biomedical arena.

Elisa et al. have demonstrated CNT–CP-based composite applications in electrocorticography (ECoG) a tool for clinical applications. They successfully demonstrated their role in precise neural activity localization and novel applications, such as neural prosthetics.[52]

The devices were characterized for both electrochemically and by recording from rat somatosensory cortex in vivo. In this work, they proposed an ultra-flexible and conformable polyimide-based micro-ECoG array. The reason that these composites have high surface area, conductive polymer-CNT composites leads to improve their brain-electrical coupling capabilities.[53]

6.4.4 CORROSION PROTECTION

In area of corrosion, nanocomposite coatings have played a significant role, as literature depicts their application as successful candidates. Meeki et al. reported that functionalized carbon nanotubes (*f*-CNTs) are used as reinforcement to enhance the corrosion and mechanical behavior of PANI coatings for the protection of mild steel (MS) structures. They concluded that PANI/*f*-CNTs (Fig. 6.8) nanocomposite coatings are potentially employed as anticorrosive coatings for MS against corrosive environment.[54]

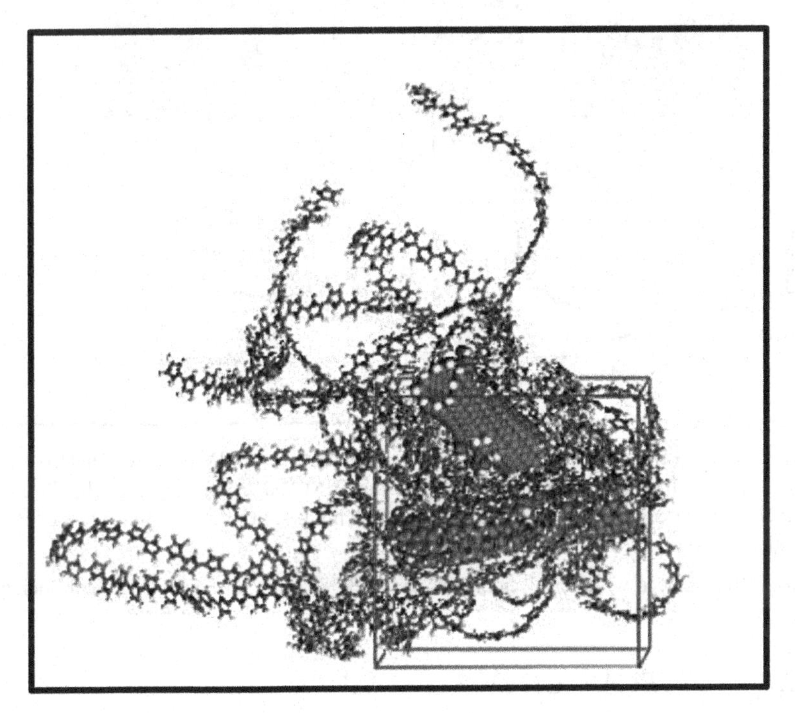

FIGURE 6.8 (See color insert.) PPy/CNT-CA 90:10 (w/w) bulk model after refinement stage.

Source: Reproduced from permission from Ref. [55]. © 2011 Elsevier.

Mariana et al. presented a computational method based on molecular mechanics and dynamics, to predict mechanical properties of PPy/polyaminobenzene sulfonic acid-functionalized single-walled CNTs (CNT-PABS) and PPy/carboxylic acid-functionalized single-walled CNTs (CNT-CA) composite. Their findings clearly confirmed that the CNT-PABS and

CNT-CA are properly dispersed in the composite coatings and have beneficial effect on mechanical integrity (Fig. 6.9). Moreover, the anticorrosion protecting ability of the composite coatings is significantly higher than the one characteristic to pure PPy.[55]

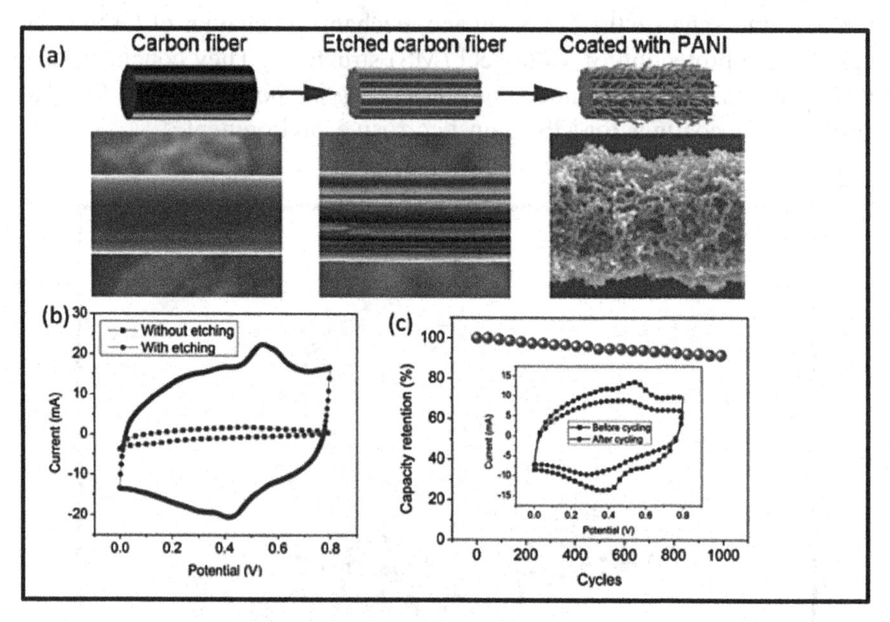

FIGURE 6.9 (a) Schematic diagram of the preparation of PANI-coated electro-etched carbon fiber cloth electrodes and the corresponding SEM images of every step. (b) CV curves of PANI coating carbon fibers with and without etching. (c) Cycling stability test of PANI-coated electrode.

Source: Reproduced with permission from Ref. [63]. © 2011 American Chemical Society.

6.4.5 SUPERCAPACITORS

The most challenging area for CNTs/CPs NCs are supercapacitors and batteries.[21,56–60] The methods from where they can be prepared via chemical synthesis, electrochemical deposition on preformed CNT electrodes, or by electrochemical co-deposition. The main attributes of CPs for supercapacitors formulations is their fast charging/discharging double-layer capacitance and excellent mechanical properties of the CNTs. The homogeneous nature of electrochemically co-deposited composites and their unusual interaction between the polymer and nanotubes, makes electron delocalization and conjugation along the polymer chains. As a result, they exhibit outstanding electrochemical charge storage properties along with fast charge/discharge switching, significant for high charge/discharge capacitors or supercapacitors. The prominent works includes those by Peng et al.[58]

6.5 CONCLUSION AND OUTLOOK

Novel functional polymer composites of CNTs–CPs show promising performance and applications, but more work is still necessary in this regard. One of the most crucial factors, to be assessed for their application is to enhance their properties, together with the ability to use them in controllable mode, which would be complimentary for varying area of applications. Therefore, future improvements should focus on refining synthetic methods and developing novel assembly ways for better structure, morphology, control of the size and composition of these nanocomposites.

ACKNOWLEDGMENTS

Dr. Neha Rawat and Dr. Anujit Ghosal are thankful to Government of India, Science and Engineering Research Board (SERB) for financial support in the form of National Postdoctoral Fellowship (PDF/2017/002907) and (PDF/2016/003866). The authors are also thankful to the Materials Science Division, CSIR-NAL Bangalore 560017, India and School of Biotechnology, Jawaharlal Nehru University, New Delhi 110067, India for implementation of fellowship.

KEYWORDS

- **conducting polymers**
- **carbon nanotubes**
- **nanocomposites**
- **material applications**
- **polyaniline**

REFERENCES

1. Thostenson, E. T.; Ren, Z.; Chou, T.-W. *Compos. Sci. Technol.* **2001,** *61*, 1899–1912.
2. Gospodinova, N.; Terlemezyan, L. *Prog. Polym. Sci.* **1998,** *23*, 1443–1484.
3. Bhadra, S.; Khastgir, D.; Singha, N. K.; Lee, J. H. *Prog. Polym. Sci.* **2009**, *34*, 783–810.
4. Bredas, J. L.; Street, G. B. *Account. Chem. Res.* **1985,** *18*, 309–315.
5. Fattahi, P.; Yang, G.; Kim, G.; Abidian, M. R. *Adv. Mater.* **2014,** *26*, 1846–1885.
6. MacDiarmid, A. G. *Synth. Metals* **2001,** *125*, 11–22.
7. Iijima, S. *Nature* **1991,** *354*, 56–58.
8. Iijima, S.; Ichihashi, T. *Nature* **1993,** *363*, 603.
9. Lorenzo, M.; Zhu, B.; Srinivasan, G. *Green Chem.* **2016,** *18*, 3513–3517.
10. Esposito, L.; Ramos, J.; Kortaberria, G. *Prog. Org. Coat.* **2014,** *77*, 1452–1458.
11. Do, T. N.; Visell, Y. *Sci. Rep.* **2017,** *7*, 1753.
12. Li, C.; Thostenson, E. T.; Chou, T.-W. *Compos. Sci. Technol.* **2008,** *68*, 1227–1249.
13. Heeger, A. J. *Synth. Metals* **2001,** *125*, 23–42.
14. Huang, W.-S.; Humphrey, B. D.; MacDiarmid, A. G. *Faraday Transactions 1: Physical Chemistry in Condensed Phases. J. Chem. Soc.* **1986,** *82*, 2385–2400.
15. Jaymand, M.; Hatamzadeh, M.; Omidi, Y. *Prog. Polym. Sci.* **2015,** *47*, 26–69.
16. Ajayan, P. M.; Schadler, L. S.; Giannaris, C.; Rubio, A. *Adv. Mater.* **2000,** *12*, 750–753.
17. Gao, W.; Alemany, L. B.; Ci, L.; Ajayan, P. M. *Nat. Chem.* **2009,** *1*, 403–408.
18. Guo, D.-j.; Li, H.-l. *J. Solid State Electrochem.* **2005,** *9*, 445–449.
19. Wang, H.; Lin, J.; Shen, Z. X. *J. Sci. Adv. Mater. Devices* **2016,** *1*, 225–255.
20. Zhang, X.; Zhang, J.; Wang, R. Zhu, T.; Liu, Z. *ChemPhysChem*, **2004,** *5*, 998–1002.
21. Feng, L.; Xie, N.; Zhong, J. *Materials* **2014,** *7*, 3919–3945.
22. Belaabed, B.; Lamouri, S.; Naar, N.; Bourson, P.; Ould Saad Hamady, S. *Polym. J.* **2010,** *42*, 546–554.
23. Nambiar, S.; Yeow, J. T. W. *Biosens. Bioelectron.* **2011,** *26*, 1825–1832.
24. Huang, Z.-B.; Yin, G.-F.; Liao, X.-M.; Gu, J.-W. *Front. Mater. Sci.* **2014,** *8*, 39–45.
25. Xu, J.; Yao, P.; Wang, Y.; He, F.; Wu, Y. *J. Mater. Sci. Mater. Electron.* **2009,** *20*, 517–527.
26. Yildirimer, L.; Thanh, N. T. K.; Loizidou, M.; Seifalian, A. M. *Nano Today* **2011,** *6*, 585–607.
27. Davis, F.; Higson, S. P. J. *Biosens. Bioelectron.* **2005,** *21*, 1–20.
28. Yang, G.; Kampstra, K. L.; Abidian, M. R. *Adv. Mater.* **2014,** *26*, 4954–4960.
29. Kirchmeyer, S.; Reuter, K. *J. Mater. Chem.* **2005,** *15*, 2077–2088.
30. Yin, Y.; Li, Z.; Jin, J.; Tusy, C.; Xia, J. *Synth. Metals* **2013,** *175*, 97–102.
31. Zhang, J.; Zhao, X. S. *J. Phys. Chem. C* **2012,** *116*, 5420–5426.
32. Ginic-Markovic, M.; Matisons, J. G.; Cervini, R; Simon, G. P.; Fredericks, P. M. *Chem. Mater.* **2006,** *18*, 6258–6265.
33. Hangarter, C. M.; Chartuprayoon, N.; Hernández, S. C.; Choa, Y.; Myung, N. V. *Nano Today* **2013,** *8*, 39–55.
34. Ahmad, S.; Riaz, U.; Kaushik, A.; Alam, J. *J. Inorg. Organomet. Polym. Mater.* **2009,** *19*, 355–360.
35. Kaushik, A.; Khan, R.; Solanki, P. R.; Pandey, P.; Alam, J.; Ahmad, S.; Malhotra, B. *Biosens. Bioelectron.* **2008,** *24*, 676–683.
36. Solanki, P. R.; Kaushik, A.; Agrawal, V. V.; Malhotra, B. D. *NPG Asia Mater.* **2011,** *3*, 17–24.

37. Wang, J.; Dai, J.; Yarlagadda, T. *Langmuir* **2005,** *21*, 9–12.
38. Gupta, B. K.; Kedawat, G.; Kumar, P.; Singh, S.; Suryawanshi, S. R.; Agrawal, N.; Gupta, G.; Kim, A. R.; Gupta, R. K.; More, M. A.; Late, D. J.; Hahm, M. G. *RSC Adv.* **2016,** *6*, 9932–9939.
39. Arjmand, M.; Chizari, K.; Krause, B.; Pötschke, P.; Sundararaj, U. *Carbon* **2016,** *98*, 358–372.
40. Pawar, S. P.; Arjmand, M.; Gandi, M.; Bose, S.; Sundararaj, U. *RSC Adv.* **2016,** *6*, 63224–63234.
41. Maiti, S.; Shrivastava, N. K.; Suin, S.; Khatua, B. B. *ACS Appl. Mater. Interfaces* **2013,** *5*, 4712–4724.
42. Sobha, A. P.; Narayanankutty, S. K. *Sens. Actuators A Phys.* **2015,** *233*, 98–107.
43. Khandhar, A. P.; Ferguson, R. M.; Simon, J. A.; Krishnan, K. M. *J. Biomed. Mater. Res. Part A* **2012,** *100A*, 728–737.
44. Lee, J. W.; Serna, F.; Nickels, J.; Schmidt, C. E. *Biomacromolecules* **2006,** *7*, 1692–1695.
45. Shadjou, N.; Hasanzadeh, M. *J. Biomed. Mater. Res. A* **2016,** *104*, 1250–1275.
46. Vogl, T. J.; Then, C.; Naguib, N. N. N.; Nour-Eldin, N.-E. A.; Larson, M.; Zangos, S.; Silber, G. *Acad. Radiol.* **2010,** *17*, 1486–1491.
47. Wang, D. W.; Li, F.; Lu, G. Q.; Cheng, H. M. *Carbon* **2008,** *46*, 1593–1599.
48. Zare, Y.; Shabani, I. *Mater. Sci. Eng. C* **2016,** *60*, 195–203.
49. Zhang, W. D.; Shen, L.; Phang, I. Y.; Liu, T. *Macromolecules* **2004,** *37*, 256–259.
50. Aljohani, S.; Alrawashdeh, A. I.; Khan, M. Z. H.; Zhao, Y.; Lagowski, J. B. *J. Phys. Chem. C* **2017,** *121*, 4692–4702.
51. Menaa, F.; Abdelghani, A.; Menaa, B. *J. Tissue Eng. Regen. Med.* **2015,** *9*, 1321–1338.
52. Castagnola, E.; Ansaldo, A.; Maggiolini, E.; Ius, T.; Skrap, M.; Ricci, D.; Fadiga, L. *Front. Neuroeng.* **2014,** *7*, 8.
53. Castagnola, E.; Maiolo, L.; Maggiolini, E.; Minotti, A.; Marrani, M.; Maita, F.; Pecora, A.; Angotzi, G. N.; Ansaldo, A.; Boffini, M.; Fadiga, L.; Fortunato, G.; Ricci, D. *IEEE Trans. Neural. Syst. Rehabil. Eng.* **2015,** *23*, 342–350.
54. Mekki, A.; Samanta, S.; Singh, A.; Salmi, Z.; Mahmoud, R.; Chehimi, M. M.; Aswal, D. K. *J. Colloid Interface Sci.* **2014,** *418*, 185–192.
55. Ioniţă, M.; Prună, A. *Prog. Org. Coat.* **2011,** 72, 647–652.
56. de Oliveira, H. P.; Sydlik, S. A.; Swager, T. M. *J. Phys. Chem. C* **2013,** *117*, 10270–10276.
57. Lang, X.; Hirata, A.; Fujita, T.; Chen, M. *Nat. Nanotechnol.* **2011,** *6*, 232–236.
58. Peng, C.; Zhang, S.; Jewell, D.; Chen, G. Z. *Prog. Nat. Sci.* **2008,** *18*, 777–788.
59. Sun, L.; Tian, C.; Li, M.; Meng, X.; Wang, L.; Wang, R.; Yin, J.; Fu, H. *J. Mater. Chem. A* **2013,** *1*, 6462–6470.
60. Wang, M.; Zhang, H.; Wang, C.; Wang, G. *Electrochimica. Acta* **2013,** *106*, 301–306.
61. Liu, J.; Sun, J.; Gao, L. *J. Phys. Chem. C* **2010,** *114*, 19614–19620.
62. Ning, G.; Li, T.; Yan, J.; Xu, C.; Wei, T.; Fan, Z. *Carbon* **2013,** *54*, 241–248.
63. Cheng, Q.; Tang, J.; Ma, J.; Zhang, H.; Shinya, N.; Qin, L.-C. *J. Phys. Chem. C* **2011,** *115*, 23584 -23590.

CHAPTER 7

CARBON NANOTUBES AND THEIR APPLICATIONS IN CHEMICAL ENGINEERING SCIENCE: A FAR-REACHING REVIEW

SUKANCHAN PALIT[1,2*]

[1]*Department of Chemical Engineering, University of Petroleum and Energy Studies, Energy Acres, Post Office Bidholi via Premnagar, Dehradun 248007, Uttarakhand, India*

[2]*43, Judges Bagan, Post Office Haridevpur, Kolkata 700082, India*

[*]*E-mail: sukanchan68@gmail.com; sukanchan92@gmail.com*

ABSTRACT

Human civilization and human scientific endeavor are today moving from one scientific paradigm to another. Engineering science and technology are in the path of newer scientific regeneration. Environmental engineering science, petroleum engineering, and chemical engineering are today the branches of engineering which are highly challenged as well as surpassing vast and versatile scientific frontiers. Global climate change, loss of ecological diversity, and depletion of fossil fuel resources has challenged the scientific firmament of human civilization and human scientific progress today. Nanotechnology and the world of carbon nanotubes are the utmost scientific needs of today. Science and engineering are two huge colossus with a definite and path-breaking vision of its own. In this chapter, the author deeply comprehends the scientific success, the deep scientific ingenuity, and the visionary world of carbon nanotubes application in chemical engineering and environmental engineering science. Carbon nanotubes are types of nanomaterials. Nanotechnology and nanomaterials applications are changing the vision of future scientific research pursuit today. Human scientific vision and the vast scientific fortitude of carbon nanotubes in diverse areas of science and engineering are the pillars of this well-researched chapter. Technological vision and scientific validation are the opposite sides of the visionary scientific coin today. In this chapter, the author reiterates the scientific success and the vast scientific potential in the application of nanotechnology in varied avenues of science and engineering. The vast technological and scientific advancements in the field of carbon nanotubes application are deeply elucidated in this chapter.

7.1 INTRODUCTION

Nanoscience and nanotechnology are today in the path of newer scientific rejuvenation and vast scientific vision. Nanomaterials and carbon nanotubes (CNTs) are the smart materials of tomorrow and are in the path toward a newer era in the field of science and engineering. Technology and engineering science are highly challenged today with the ever-growing concerns of global energy and environmental crises. Global climate change, the frequent environmental catastrophes and the depletion of fossil fuel resources are some of the global issues of immense scientific concern. Here comes the need of nanotechnology, nanomaterials and CNTs. There are immense applications of CNTs in chemical engineering science and environmental engineering. Human civilization and human scientific endeavor today stands in the midst of devil and the deep sea. Man's immense scientific prowess, mankind's vast scientific grit, and determination and the futuristic vision of energy sustainability will all lead a long and effective way in the true realization of energy security and the true mitigation of global climate change and environmental calamities. In this chapter, the author deeply elucidates the success of the science of nanotechnology and the application of nanomaterials in diverse areas of engineering and science such as chemical process engineering and environmental engineering. Human society today stands in the midst of vast scientific introspection and scientific determination. Technological challenges, the scientific motivation, and the vast world of scientific validation will all evolve toward a newer era in the field of chemical process engineering, petroleum engineering, and environmental engineering science. The vision and the scientific imagination of research pursuit in nanomaterials and CNTs are immense and path-breaking today. The world of science and engineering today stands in the midst of deep scientific dilemma and scientific fortitude. This study opens up a newer chapter in the research endeavor in nanoscience and nanotechnology.

7.2 THE VISION OF THE STUDY

Scientific research pursuit and vast scientific regeneration in the field of nanotechnology are surpassing vast and versatile scientific boundaries. Technology and engineering science has few answers to the scientific intricacies and scientific difficulties in the field of energy engineering and nanotechnology. The vision of this study is to target the needs of nanotechnology to human society. The primary aim and objective of this study is to delineate the scientific relevance and the scientific provenance in the field of nanotechnology applications and nanomaterials applications to human scientific progress. Today the scientific progress in nanotechnology is replete with scientific and intense academic rigor. This well-researched treatise goes beyond scientific imagination and targets the immense importance of scientific vision in nanoscience and nanotechnology applications in human scientific advancements. Global warming and global climate change are of immense concern to modern-day human civilization. Thus technology and engineering need to be reenvisioned and reenvisaged with the passage of scientific history and visionary timeframe. The scientific success of human civilization and the vision of science of modern-day human civilization are today groundbreaking. Thus, the need of science of sustainability whether it is environmental, energy, social, or economic. This chapter unfolds and unravels the success of science of nanotechnology and nanomaterials over the years. Human science, the vast need of nanotechnology, and the scientific fortitude of research pursuit will all today evolve into a newer dimension in the field of CNT applications in human society. The imminent needs of nanoscience and nanotechnology are vastly explored in this chapter.

7.3 ENERGY AND ENVIRONMENTAL SUSTAINABILITY AND THE VISION FOR THE FUTURE

The world of science and engineering are today witnessing immense challenges. The challenge and the vision of science today stand in the midst of deep scientific introspection. Environmental catastrophes are today reframing the scientific research pursuit in engineering science and technology. Grave concerns for energy security and the visionary world of energy and environmental sustainability are today challenging the scientific fabric today. Breach of environmental sustainability, water purification, and industrial wastewater treatment are the torchbearers toward a newer knowledge dimension in environmental engineering and energy engineering. The necessities and the vision of science are the pallbearers toward a newer global environmental engineering order.

7.4 WHAT DO YOU MEAN BY CNTs AND NANOMATERIALS?

CNTs are allotropes of carbon with cylindrical nanostructures. These cylindrical carbon molecules have unusual and unique properties, which are valuable for nanotechnology, electronics, optics, and other domains of material science and technology. In addition, owing to their extraordinary thermal conductivity, mechanical, and thermal properties, CNTs find applications as additives to various structural materials. Today, the world of material science and nanotechnology are surpassing scientific and technological boundaries. Nanomaterials and engineered nanomaterials (ENPs) are the veritable challenges of science today. Nanomaterials in principle, materials of which a single unit is sized (in at least one dimension) between 1 and 1000 nm (10^{-9} m) but usually is 1–100 nm (the usual definition of nanoscale). Today nanomaterials research takes a material science-based approach to nanotechnology. Materials with structure at the nanoscale often have unique optical, electronic, and mechanical properties.

7.4.1 CNTs IN INDUSTRIAL WASTEWATER TREATMENT AND ENVIRONMENTAL ENGINEERING

CNTs and its application in environmental engineering are today surpassing vast and versatile scientific frontiers. Waste water discharge from domestic, industrial, or agricultural sources involves a wide range of contaminants and is a matter of grave concern globally since they adversely affect the quality of drinking water. Drinking water treatment, the needs for industrial wastewater treatment and the futuristic vision of nanotechnology, will all lead a long and visionary way in the true emancipation of engineering science and technology today. The contaminants found in water such as heavy metal ions, 1,2-dichlorobenzene, and dioxin are nondegradable, highly toxic, and carcinogenic and results in cancer and nervous system damage. Here comes the importance of CNTs applications. CNTs with their high surface active site to volume ratio and controlled pore size distribution have an excellent and improved sorption capability and high-sorption efficiency. Environmental engineering science today is in the path of newer scientific regeneration and is today replete with vision of scientific fortitude. This treatise unfolds and unravels the deep and vast scientific truth in CNTs applications in human scientific progress.

7.5 THE SCIENTIFIC VISION, THE VAST SCIENTIFIC COGNIZANCE, AND THE SCIENTIFIC DOCTRINE OF NANOTECHNOLOGY

Scientific research pursuit in the field of nanoscience and nanotechnology are crossing vast and versatile scientific boundaries. Mankind's immense scientific prowess, man's unending scientific struggle and the futuristic vision of interdisciplinary research will all lead a long and effective way toward the true emancipation of nanotechnology today. The vast scientific vision, the wide scientific cognizance, and the scientific doctrine of nanotechnology are ever-growing and groundbreaking. The scientific world today stands in the midst of deep introspection and scientific fortitude. Nanotechnology is today linked to diverse areas of engineering science endeavor such as chemical process engineering and environmental engineering. Water purification, drinking water treatment, and industrial wastewater treatment are also veritably linked with the vast domain of nanotechnology. In this chapter, the author also deeply elucidates with cogent insight the contribution of nanomaterials and CNTs in the furtherance of the science and engineering of chemical processes and environmental engineering.

7.6 CHEMICAL ENGINEERING SCIENCE AND THE APPLICATIONS OF CNTs

Chemical engineering science, petroleum engineering, and environmental engineering are today in the path of newer scientific regeneration and scientific vision. Today, drinking water treatment and industrial wastewater treatment are in the midst of deep scientific comprehension and vast scientific forbearance. Chemical engineering and environmental engineering have few answers to the burning issue of groundwater remediation and provision of pure drinking water. Scientific profundity and scientific ingenuity are the pillars of science and engineering today. Chemical process engineering is today branching to diverse areas of scientific endeavor. Technological and scientific validations are the pillars of today's scientific research pursuit. Water process engineering, water purification innovations, and industrial wastewater treatment techniques will all lead a long and visionary way in the true emancipation of scientific endeavor in water science and technology. Today, water science and technology and chemical process engineering are two opposite sides of the visionary scientific coin. Membrane science and membrane separation processes are the scientific marvels of today which involve application of chemical engineering and environmental engineering concepts and fundamentals. Chemical process engineering is veritably linked with the science of nanofiltration, a membrane-separation technique.

7.7 RECENT SCIENTIFIC RESEARCH PURSUIT IN THE FIELD OF NANOTECHNOLOGY

Nanotechnology today stands in the midst of scientific fortitude and vast scientific comprehension. Technology and engineering science needs to be reenvisioned and streamlined with respect to the need of water purification technologies in human society. Human civilization today stands in the midst of vast scientific comprehension and deep scientific intricacies. Nanotechnology is the branch of science which is surpassing vast scientific frontiers. In this section, the author deeply comprehends the necessities of science and engineering of nanotechnology in diverse branches of science and engineering such as chemical engineering, environmental engineering, and petroleum engineering science.

The Royal Society and Royal Academy of Engineering Report[1] discussed with lucid and cogent insight the vast opportunities and uncertainties in nanoscience and nanotechnologies. The Report discussed science and applications of nanoscience and nanotechnologies. The report widely discussed nanomaterials, nanometrology, biotechnology, and nanomedicine.[1] The other areas of this research endeavor are nanomanufacturing and the industrial applications of nanotechnologies, and the intricate world of possible health risks and environmental and safety impacts.[1] Social and ethical issues in nanotechnologies applications are deeply discussed in this report. Today, deep scientific profundity and vast scientific acuity are the needs of endeavor in technology and engineering science today. In the similar manner, nanoscience and nanotechnology has vast and versatile scientific vision. This report deeply pronounces the scientific far-sightedness as well as the scientific foresight in the furtherance of science and engineering of nanotechnology. This report presents an overview of some key developments in nanoscience and nanotechnologies and also delineates some possible future applications and future trends in research pursuit.[1] The chapter is widely informed by evidence from scientists, engineers, and technologists in the academia and industry. Validation of nanoscience and nanotechnology stands as a major scientific imperative toward the march of human civilization today. This report also vastly illustrates the wide-ranging interest in these major areas and provides a strong background to the research and development forays addressing health, environmental, social, ethical, and regulatory application of nanotechnologies.[1] Human civilization's vast scientific determination, the immense technological prowess of nanotechnology, and the futuristic vision of nanomaterials and ENPs will all today lead a long and visionary way in the true realization of global scientific efforts today. The entire treatise encompasses areas which are considered into six broad categories: (1) nanomaterials, (2) nanometrology, (3) electronics, (4) optoelectronics, (5) information and communication technology, and (6) bio-nanotechnology and nanomedicine.[1] The report also elucidates on the development of future applications of nanotechnology as short (under 5 years), medium (5–15 years), and long term (over 20 years).[1] The authors also delineated potential in environmental, health and safety, ethical, societal implications or uncertainties in the vast vision of nanotechnology. Human scientific pursuit in nanoscience and nanotechnology are surpassing wide visionary boundaries. The challenge and the vision of nanotechnology in the 21st century are vastly elucidated in this report.[1]

Lloyd's Emerging Risks Team Report[2] discussed with deep and cogent insight recent developments, risks, and opportunities in the domain of nanotechnology. The report focuses on nanotechnology; a visionary class of products containing the materials built on the atomic scale. Today, nanotechnology has immense scientific potential and immense opportunities.[2] Nanotechnology represents an entire scientific and technological field, and not just a single product or even group of products. This branch of human scientific endeavor is visionary and far reaching. Nanotechnology is veritably linked with diverse areas of engineering and science in today's modern human civilization. The chemical reactivity of a material is related to its surface area when it is compared to its volume. Scientific far-sightedness, scientific acuity, and deep scientific discernment are the pillars of research pursuit in nanotechnology today. Dissecting a 1 cm^3 of any material into 1 nm cubes increases the total combined surface area some 10 million times.[2] The entire paper elucidates on the hazards to human, the effects on environment, nanoremediation, regulation, and risk assessment.[2] Impact of nanotechnology and nanomaterials on human health stands as a major scientific research pursuit in today's scientific world. This report focuses on nanotechnology, a class of products containing materials built on the atomic scale.[2] Nanotechnology today is already in use and has an immense scientific potential to become as commonly used as plastic.[2] Technological advancements, scientific profundity, and vast scientific ingenuity will all today lead a long and visionary way in the true emancipation of nanotechnology.[2] Due to the potential impact to the insurance industry, Lloyd's report enhances the deep scientific understanding and the vast scientific wisdom in nanotechnology and nanomaterials applications to human society.[2]

Chaudhry et al.[3] deeply discussed with lucid and cogent insight new opportunities, new questions, and newer concerns in the applications of nanotechnologies in the food arena. Scientific adjudication, the vast world of scientific validation, and the scientific truth behind nanotechnology are all the torchbearers toward a newer visionary era in the field of technology and engineering science. Nanotechnology is a broad terminology used to represent an assemblage of processes, materials, and applications that span physical, chemical, biological, and electronic engineering domains.[3] A nanomaterial has been defined as a "material having one or more external dimensions in the nanoscale or which is nanostructured" where the nanoscale size range is between 1 and 100 nm.[3] It has been widely suggested that for some time, materials and substances may be manipulated at the nanometer level through atom-by-atom assembly. The march of engineering and the advent of nanotechnology in recent decades have provided a systematic approach for the study and fine-tuning of material properties in the nanometer range. Food engineering and food technology today stands in the midst of deep scientific comprehension and scientific fortitude. Materials with all three external dimensions in the nanoscale are classified as nanomaterials.[3] Nanomaterials also exist in other forms, such as nanorods and nanotubes with two dimensions in the nanoscale range, or nanolayers, coatings, or sheets with just one dimension in the nanoscale. Human scientific foresight and deep technological insight are the torchbearers toward a newer visionary eon in the field of nanotechnology, material science, and chemical process engineering.[3] ENPs are the smart materials of today and the visionary materials of tomorrow. Of immense interest today to most nanotechnological applications are ENPs that are manufactured specifically to achieve a certain property or composition.[3] The current applications of nanotechnology span a wide arena and vast sectors, predominantly cosmetics and personal care, health care, paints and coatings, and electronics. These areas are comprehended in this chapter with immense scientific determination and scientific truth. As in these sectors, nanotechnology is also promising to revolutionize the food industry—from food production, processing, packaging, transportation, and storage to the new developments and innovations in food tastes and textures and vastly innovative food packaging applications. The authors in this chapter deeply enumerated an impartial view of the potential prospects, benefits, and risks that nanotechnology can bring to the food sector and its customers.[3] This well-researched paper also discussed the questions and grave concerns that the newer technological advancements will raise in the furtherance of food engineering and food science. Food technology and food engineering are today in the midst of deep scientific introspection and vast scientific wisdom. In the similar manner, nanotechnology and chemical process engineering are in the path toward newer scientific regeneration. The authors in this chapter deeply discussed evolution of new technologies in the food sector, public perception of nanotechnology food products, natural nanostructures in food, potential benefits and market drivers, current and projected applications of nanotechnology in the food sector, nanoingredients and additives, potential health risks and governance of health risks, and a vast introspection into regulatory frameworks.[3] Human scientific vision, the futuristic vision of scientific validation and the needs of nanotechnology in food engineering will all lead a long and visionary way toward the true realization of engineering and science in human society. Food and water are the parameters toward human civilization's scientific progress. Nanotechnology, water purification, and food engineering are the needs of human today, and research and development forays should be targeted in that direction. This treatise unfolds the world of application of nanotechnology in food science and food engineering.[3]

The Royal Society Report[4] discussed insight and scientific conscience knowledge, networks, and nations as a visionary path toward global scientific collaborations in the 21st century. Science and technology today are huge colossus with a definite vision of its own.[4] The success of technology and engineering science, the vast scientific profundity of engineering, and the needs of science to human society will all lead a long and visionary way in the true realization of scientific truth. The report delineates scientific landscape in 2011, international collaboration, global approaches to global problem, and the deep scientific understanding of global scientific landscape. Science is a veritable global enterprise.[4] Today, in modern civilization, there are over 7 million researchers around the world drawing on a combined international research and development spend of over US $1000 billion (a 45% increase since 2002).[4] Technology and engineering science are undergoing immense transformations and a vast metamorphosis globally. Science globally today stands in the midst of deep comprehension and unending vision. The world of scientific challenges, the scientific advancements, and the vast world of scientific validation are the pallbearers toward a newer era in the domain of engineering and science. Knowledge, networks, and nations paper reviews, based on available data, the changing pattern of science and scientific research pursuit, and scientific collaboration, in order to provide a strong foothold for understanding such visionary changes. Cross-disciplinary research endeavor today stands as a major pillar of vast scientific wisdom and an intricate scientific discernment. Thus, scientific collaboration across nations stands as a major scientific imperative. This paper enumerates the following stances which are:

- Support for international science should be enhanced and strengthened.[4]
- Internationally collaborative science should gain immense importance.[4]
- National and international strategies for science are required to address global disasters and global challenges.
- International capacity building should be the pillar of science and engineering today.
- Evaluation of better assessment tools in science are of immense need.[4]

Human civilization's immense scientific prowess and cross-boundary research are the cornerstones of any scientific achievement today. The vast scientific triumph of nanoscience and nanotechnology should be more readdressed with the passage of scientific history and time. This report widely addresses the vast potential and the imminent needs of global scientific collaboration.[4]

UNESCO Report[5] discussed with vast scientific insight and scientific conscience the ethics and politics of nanotechnology. The report delineates the present research status of nanotechnology, and ethical, legal, and political implications of nanotechnology. Nanotechnology research today is crossing vast and versatile scientific boundaries.[5] Nanotechnology could become the most important force in modern civilization today since the rise and march of internet and computer science. Scientific regeneration and vast scientific vision are the necessities of research pursuit in nanotechnology today. The challenges and the futuristic vision of nanoscience and nanotechnology are immense and groundbreaking as regards ethics and social implications.[5] Mankind's vast scientific hope and determination, the needs of human scientific progress, and the grave concerns of global energy and environmental scenario will all lead a long and effective way in the true realization of nanoscience and nanotechnology. Global scientific scenario today stands in the midst of introspection and deep vision. Nanotechnology in the similar vein is in the process of newer scientific regeneration. Nanotechnology could enhance the speed of memory chips, remove pollutants in water and air, and find cancer cells quicker.[5] Technological splendor, scientific excellence, and the achievements of engineering science are the torchbearers toward a newer eon in science and technology. Nanotechnology should be reframed and readdressed with immense scientific might and scientific ingenuity globally. The development of science and technology is rapidly transforming human existence significantly. Scientific research pursuits in the international platform are in the process of newer scientific rejuvenation. Technology has transformed the human society. Medical science has greatly and veritably improved the health of citizens. Medical technology and biomedical engineering are changing the face of human civilization and human scientific progress. The state of public health globally has changed drastically. Information technology has increased the possibilities and scientific potential of human communication and the domain of information and communication technology.[5] The science of ecology and the vast world of environmental engineering science have developed more sustainable ways of production and consumption. Life sciences and biological engineering are inventing new products and new innovations in medication. Nanotechnology encompasses all these fields and raises some valid ethical questions. In order to address a global scientific order and a global research question in nanotechnology, ethics, and social stances are the pillars of research pursuit in nanoscience. This report vastly addresses the scientific needs of ethics, the social and the economic sustainability behind scientific advancements and the futuristic vision of nanotechnology. This report is an ethical mandate of UNESCO and its policies in science and technology. Since nanotechnology is advancing at a rapid pace, ethics in society and science needs to be addressed vehemently. UNESCO is today contributing vastly from a global perspective and at an international level to promote the communication between various stakeholders and decision makers with the sole vision of furtherance of the science of nanotechnology.

Royal Commission on Environmental Pollution Report[6] deeply comprehended and elucidated novel materials in the environment as a veritable case of nanotechnology. Environmental engineering science and nanotechnology are two opposite sides of the visionary coin. This report elucidates novel materials, application of novel materials, the concerns for functionality, purpose, production, and properties of novel materials, environmental and health effects of nanomaterials, and finally, the challenges of designing an effective governance framework.[6] The success of science, the subtleties of technology and engineering science, and the vast futuristic vision of nanotechnology are the veritable pillars of this well-researched report. Nanotechnology is today integrated with diverse areas of engineering and science. Nanomaterials or novel materials are the visionary innovations of tomorrow. ENPs are the necessities of human civilization today. Scientific judgment, scientific truth, and vast scientific understanding in the field of nanotechnology are the torchbearers toward a newer era in the field of CNTs and nanomaterials applications today. Technology and engineering science are today gearing forward toward a newer scientific eon as nanoscience and nanotechnology moves forward.[6] The discovery, the vast development, and deployment of novel materials have always been a significant issue in the development of human civilization and the scientific progress. Prehistoric and historical epochs are even named according to new materials that were successively introduced and entered into common use during what we know as Stone Age, Bronze Age, and Iron Age.[6] Mankind's immense technological prowess, the futuristic vision of nanomaterials applications, and the vast scientific adroitness of nanoscience and nanotechnology will all lead a long and visionary way in the true emancipation of engineering science and technology in global scientific scenario today. The main upshot of this paper is to classify novel materials that take into account their functionality and the visionary applications of novel engineered materials. The classifications are: (1) evolutionary materials, (2) revolutionary materials, (3) combination materials, and (4) materials with the potential for unusual biological impact. This entire report unfolds and uncovers the scientific success and the deep scientific potential behind nanomaterials application and the futuristic vision of nanotechnology.

Borm et al.[7] presented with vast scientific conscience uncertainties and risks in nanotechnology applications in human society today. Nanoscience and nanotechnologies are vastly expected to change industrial production and the economics over the decades to come.[7] Technological advancements, scientific profundity and deep vision are the salient features of this entire treatise. Nanomaterials are being introduced into the global market on the basis of claimed benefits and their chemical identity is targeted toward already existing legislation, regardless of some unique and significant characteristics. The authors in this paper deeply comprehend uncertainties and risks, real life examples, and life cycle assessment and the vast world of scientific perceptions and scientific insight. Titanium dioxide, amorphous silica, and iron oxides are bulk nanoproducts that are available in the global market readily and are present in many consumer products, including food additives, pigments, paints, and cosmetics. The health and the environmental impact of nanomaterials and ENPs assumes immense importance today as nanotechnology surpasses vast and versatile scientific boundaries. The authors in this paper reiterated the needs of regulations and scientific restrictions today in the research endeavor in the field of nanomaterials applications.[7]

Nanoscience and nanotechnology are today in the midst of immense scientific regeneration and vast scientific vision. Technology and engineering science are today advancing at a rapid pace. Scientific adroitness, the vast scientific rejuvenation, and the scientific adjudication today stand tall in the midst of scientific advancements. In this paper, the author pointedly focuses on the deep scientific understanding and the scientific vision behind nanomaterials applications and the vast scientific emancipation of nanotechnology.

7.8 SIGNIFICANT RESEARCH ENDEAVOR IN THE FIELD OF CNTs APPLICATIONS

CNTs are the smart materials of future. The wide vision of material science, the vast needs of human civilization, and the futuristic vision of nanoscience and nanotechnology will all lead a long and effective way in the true emancipation of nanotechnology today. Technology has few answers to the ever-growing issue of groundwater arsenic contamination and water process engineering. Nanoscience and nanotechnology are the visionary avenues of science and engineering today. Nanomedicine and nanofiltration in the similar vein are today surpassing vast and versatile scientific frontiers.

Ong et al.[8] reviewed with deep insight and scientific fortitude CNTs in environmental protection with a vision toward green engineering perspective. Recent research and development forays in nanotechnologies have vastly helped to benchmark CNTs as one of the visionary nanomaterials.[8] In this short review, the authors deeply discussed the contribution of CNTs in terms of sustainability and green technology perspective. The areas of applications are waste water treatment, air pollution monitoring, biotechnologies, renewable energy technologies, supercapacitors, and green nanocomposites. Scientific knowledge and scientific vision in terms of large-scale applications of CNTs from the aspect of economics and potential environmental hazards are deeply envisioned and readdressed in this paper.[8]

Rafique et al.[9] discussed with immense scientific vision production of CNTs by different routes. The need for scientific vision and the scientific understanding in the field of production of nanotubes are immense, visionary, and groundbreaking. CNTs are one of the most important smart materials of the future. They possess remarkable and splendid electrical, mechanical, optical, thermal, and chemical properties which make them a perfect material for engineering applications.[9] In this paper, various methods of production of CNTs are deeply discussed outlining their capabilities, efficiencies and possible avenues in full-scale industrial production. Chemical vapor deposition (CVD) is the success of human scientific endeavor in nanoscience today. The production of CNTs involves: (1) physical processes, (2) chemical processes, and (3) miscellaneous processes. Technological advancements, scientific profundity, and scientific ingenuity are the pillars of the science of nanotechnology today. The physical processes include: (1) arc discharge and (2) laser ablation processes. Chemical processes include: (1) CVD and (2) high pressure carbon monoxide reaction technique. Miscellaneous processes encompasses: (1) helium arc discharge method, (2) electrolysis, and (3) flame synthesis. The entire paper deeply elucidates the scientific vision and the scientific forbearance in the field of CNTs production techniques.[9]

Kusworo et al.[10] discussed and deeply elucidated the use of CNTs mixed matrix membranes for biogas purification. A new innovation of mixed matrix membrane consisting of polyether sulfone (PES) and CNTs is prepared for biogas purification application. Scientific vision in the field of membrane science applications are highly advanced today and are groundbreaking. PES mixed matrix membrane with and without modification of CNTs were prepared by a dry/wet phase-inversion technique using a pneumatically membrane-casting machine system.[10] The scientific prowess of environmental engineering science, the vast vision, and scientific fortitude of chemical engineering and environmental protection and the needs of mankind will all lead a long and visionary way in the true emancipation of CNTs and the wide world of nanomaterials. Nanomaterials and ENPs are the visionary avenues of scientific ingenuity and scientific forbearance in today's world of scientific advancement. Membrane preparation and membrane production are the challenges of engineering and science today. This treatise opens up new visionary avenues in nanotechnology and CNTs in years to come.[10]

The world of science and technology are today gearing toward one scientific paradigm toward another. Mankind's immense scientific prowess and fortitude are changing the vision and challenges of nanotechnology, nanomaterials, and ENPs today. Chemical engineering applications and the wide vision of environmental engineering are the necessities of human civilization today. This entire paper goes beyond vast scientific imagination and vision, and veritably targets the imminent needs of human society.

7.9 SIGNIFICANT RESEARCH PURSUIT IN THE FIELD OF CHEMICAL ENGINEERING SCIENCE

Chemical process engineering and chemical technology are the challenging areas of human scientific endeavor today. Chemical engineering today encompasses nanotechnology, petroleum engineering, and other diverse areas of engineering and science today. Chemical engineering science and chemical technology are connected to diverse areas of science and engineering such as environmental engineering and novel separation processes. Human civilization's immense scientific grit and determination are the over-arching goals of human scientific progress today. Environmental protection and environmental pollution control today stands in the midst of scientific forbearance and vast hope. The ingenuity and scientific profundity of environmental protection science are deeply discussed in this section. The largest applications today in the field of chemical engineering science are environmental pollution control and nanotechnology. The state of environment is highly dismal globally.

Shannon et al.[11] discussed with immense insight and scientific conscience science and technology for water purification in the coming decades and in this 21st century. Human scientific progress, the futuristic vision of nanotechnology, chemical process engineering, and the vast world of environmental protection will all lead a long and visionary way toward the true realization of environmental sustainability. Global scientific order in environmental protection and the world of environmental sustainability are the two opposite sides of the visionary coin. This paper opens new windows of scientific knowledge in the field of water purification, drinking water treatment, and industrial wastewater treatment. The over-arching goals of this paper are: (1) disinfection of water bodies, (2) decontamination, (3) reuse and reclamation of water, and (4) desalination of seawater and salt water, and (5) a vast scientific perspective and a wide scientific deliberation on drinking water treatment.[11] Arsenic and heavy metal groundwater contamination are changing the vast scientific fabric of environmental engineering science.[11] The concerns of provision of clean drinking water to human society are deeply discussed in this treatise.[11]

Houmann[12] discussed in a Senior Thesis water purification device for a developing country constructed from local materials.[12] Water purification science, the immense scientific prowess, and scientific discernment of drinking water treatment and the needs of industrial wastewater treatment will all lead a long and visionary way in the true realization of environmental sustainability today globally.[12] Provision of safe drinking water stands as a major scientific imperative and a major scientific vision in the true pursuit of science and engineering today. In developing and underdeveloped countries around the world, thousands of people die for lack of clean drinking water, and diarrheal diseases are an ever-growing epidemic in human civilization.[12] The World Health Organization (WHO) and United Nations Children's Fund (UNICEF) estimate that 1.5 million children die annually because of diarrhea and water-related disease, and that nearly 1 billion people lack safe drinking water.[12] Human civilization thus stands in the midst of deep scientific ingenuity and scientific revelation. Here comes the need of modern science and engineering. Several water purification devices have been implemented by health organizations with varying degrees of scientific success and vision. The author in this research endeavor devised an innovative water purification kit. Two 2 ft-long and 2 in. diameter polyvinyl chloride (PVC) pipes were collected from a hardware store to be the column for the filter.[12] The author in this research pursuit constructed a filter that will be easy to use and composed of items that are readily available to villagers in the developing world. The experiment was a pioneer project to construct and test a water filter which would be efficient and effective. The time limitations and the small scale of this study enhanced the difficulty of obtaining encompassing and effective results. Technology and engineering science are today in the path of newer scientific regeneration. Innovative devices for water purification are the need of the day and will go a long and effective way in the true realization of environmental protection and environmental engineering science. This research pursuit is an effective eye-opener to that scientific and technological vision.[12]

Shon et al.[13] discussed with deep scientific fortitude and insight nanofiltration for water and wastewater treatment. The application of membrane technology in drinking water treatment and industrial wastewater treatment is increasing due to environmental engineering regulations. Human civilization today stands in the midst of deep insight, determination, and vast scientific ingenuity. Nanofiltration is one of the widely used membrane processes for water and industrial wastewater treatment in addition to other visionary applications such as desalination. Developed and developing countries around the world are witnessing immense scientific challenges and scientific forbearance. This paper vastly reviews the application of nanofiltration for drinking water and wastewater treatment including fundamentals, mechanisms, fouling challenges, and their controls.[13] Membrane filtration is a pressure-driven process in which the membrane acts as a selective medium to restrict the passage of pollutants such as organics, nutrients, turbidity, microorganisms, inorganic metal ions, and other oxygen depleting pollutants.[13] This is a comprehensive treatise toward scientific emancipation in the field of membrane science today. Water quality and water reuse are the scientific imperatives and the scientific vision of modern human civilization today. Membrane science and novel separation processes are today in the path of scientific rejuvenation. The authors in this paper discussed: (1) fundamentals of membrane processes, (2) nanofiltration, (3) separation mechanisms of nanofiltration, (3) mathematical modeling of nanofiltration processes, (4) membrane fouling in the nanofiltration process, and (5) membrane fouling control.[13] Membrane fouling is a vexing issue in the furtherance of science and engineering of membrane science today. In this paper, the authors deeply comprehend the vast scientific profundity, the scientific adroitness, and the scientific fortitude in nanofiltration applications in water treatment and novel separation techniques.

Chaudhary et al.[14] discussed with immense scientific insight biofilter in water and wastewater treatment. The needs for environmental protection and environmental engineering science are immense and scientifically inspiring. Biofilter is one of the important separation techniques that can be employed to remove organic pollutants from air, water, and industrial wastewater.[14] Mankind's immense scientific grit and determination, the vast scientific articulation, and the academic rigor of nanotechnology will all lead a long and visionary way for the true emancipation of water treatment globally. In this paper, the fundamentals of biological processes involved in the biofilter are critically reviewed together with mathematical modeling approach. The laboratory and full-scale applications of biofilter in water and wastewater treatment are also presented in minute details.[14]

Thiruvenkatachari et al.[15] discussed with cogent insight and deep scientific conscience ultraviolet/titanium dioxide photocatalytic oxidation processes. Advanced oxidation processes (AOPs) with ultraviolet radiation and photocatalyst titanium dioxide are gaining immense importance in the domain of environmental engineering science today.[15] A comprehensive review of UV-TiO_2 photocatalytic oxidation process was conducted with an insight into the mechanism involved, catalyst TiO_2, irradiation sources, types of reactors, comparison between effective modes of TiO_2 application as immobilized on surfaces or as suspension, and photocatalytic hybrid membrane system.[15] Technological profundity and deep scientific vision are the cornerstones of research endeavor in AOPs today. Environmental engineering science is witnessing immense challenges today. A two-stage coagulation and sedimentation process coupled with microfiltration hollow fiber membrane process was found to achieve complete removal of TiO_2, and the recovered TiO_2 can be reused for a photocatalytic process after regeneration. This paper vastly targets the need of novel separation processes such as membrane science and AOPs in environmental protection and for the true emancipation of environmental engineering science.[15]

Loganathan et al.[16] delineated with deep and cogent insight in a review defluoridation of drinking water using adsorption processes. Excessive intake of fluoride, mainly through drinking water is a serious health hazard throughout the world whether it is developed or developing nations. Today there are several methods used for defluoridation of drinking water, of which, adsorption techniques are generally considered attractive and important because of their effectiveness, convenience, and ease of operation. In this paper, the authors deeply discussed various adsorbents used for defluoridation, their relative effectiveness, mechanisms and thermodynamics of adsorption, and the varied suggestions made on the choice of adsorbents for various circumstances and immense scientific vision. The authors in this paper deeply discussed factors influencing adsorption: (1) pH, (2) coexisting anions, (3) temperature, and (4) adsorption kinetics.[16] Adsorption thermodynamics, fluoride desorption, and adsorption regeneration are the other hallmarks of this paper.[16]

Elimelech et al.[17] discussed with lucid and cogent insight the future of seawater desalination and the vast domain of energy, technology, and the environment. In recent decades, numerous large-scale seawater desalination plants have been built and reenvisioned in water-stressed countries around the world. Seawater desalination is more energy intensive compared to conventional technologies for the treatment of fresh water and drinking water. In this paper, the author deeply discussed and reviewed the possible reductions in energy demand by the state of the art seawater desalination technologies, the visionary role of advanced materials, and innovative technologies in improving performance. Vast and visionary scientific ingenuity, the needs of scientific validation, and the futuristic vision of desalination technique will surely lead a long and effective way in the true emancipation of environmental sustainability today. Water scarcity is one of the serious and enigmatic challenges of human civilization today.[17] Today, over one-third of the world's population lives in water-stressed countries and by 2025, this figure is predicted to rise to nearly two-thirds.[17] The authors with deep scientific vision elucidates and enumerates on the economic vitality, public health, national security, and eco-system health behind global research and development initiatives in water science and technology. The mankind and human scientific progress today stands in the midst of scientific vision and vast scientific forbearance.[17] The authors in this treatise also stressed upon (1) current energy efficiency of desalination technique, (2) the contribution of novel materials in reducing energy consumption, (3) the imminent need of innovative systems and technologies in reducing energy demand, and (4) the status of desalination technique as a sustainable technological solution. The overarching goal of this treatise is to target environmental and energy sustainability, energy security, and the immediate need for innovations in water technology.[17]

Hong et al.[18] discussed and deliberated with vast scientific conscience chemical and physical aspects of natural organic matter (NOM) fouling in nanofiltration membranes. Membrane fouling is an enigma of membrane separation processes today. Scientific validation, technological profundity, and the futuristic vision of novel separation processes will all today lead a long and visionary way in the true emancipation of environmental protection. The role and contribution of chemical and physical interactions in NOM fouling of nanofiltration membranes are vastly investigated. Fouling is a vexing issue and a difficult research question in the scientific research pursuit in membrane science.[18] Diffusion characteristics in the field of membrane science are still unexplored and thus this area needs to be reenvisioned and reenvisaged with the course of scientific history, scientific challenge, and the visionary time frame.[18] Results of fouling experiments with three humic acids demonstrate that membrane fouling increases with increasing electrolyte (NaCl) concentration, decreasing solution pH, and addition of divalent cations. Nanofiltration membranes are usually made of polymeric films with a molecular cut-off between 300 and 1000 molecular weight. Success of nanofiltration technology depends on efficient control of membrane fouling. Technology and engineering science of membrane science are highly advanced today.[18] The chemical composition of feed water greatly influences the fouling rate of nanofiltration membranes by NOM. The role of solution ionic strength, pH, and divalent cations in NOM membrane fouling needs to be reenvisioned and reenvisaged as the science of membrane separation processes surpasses vast scientific boundaries.[18]

Human scientific vision and scientific far-sightedness in the field of environmental engineering science are the pillars of chemical process engineering today. The academic and scientific rigors in the field of chemical engineering science are veritably connected with the imminent need of environmental protection and industrial pollution control. The challenge and the vision of engineering science and technology are immense and far-reaching today. In this paper, the author reiterates the immediate need of environmental pollution control and environmental science as a vision toward furtherance of global science and engineering initiatives.

7.10 SCIENTIFIC INGENUITY AND THE VAST SCIENTIFIC ACUITY IN RESEARCH PURSUIT IN NANOMATERIALS

Nanomaterials are the next-generation smart materials. CNTs applications are the areas of scientific vision and scientific might. The vast scientific might and scientific fortitude in nanotechnology and environmental engineering applications need to be readdressed and reenvisaged as science and engineering trudges a difficult path of vision and scientific determination. Nanomaterials and ENPs are today opening up new windows of scientific innovation and scientific instinct in years to come. Nanotechnology is connected to chemical engineering by an unsevered umbilical cord. Research pursuit in the field of nanomaterials should be targeted to the health impacts of nanotechnology to human society. This is a visionary avenue of scientific endeavor. Today nanotechnology is connected to almost all branches of scientific endeavor. The author pointedly focuses in this paper on the scientific potential, scientific ingenuity, and the vast needs of nanotechnology to human society. Novel separation processes such as membrane science and nanofiltration are today challenging the vast scientific firmament. Human civilization's immense scientific prowess, the scientific ingenuity, and the immediate needs of human society will all lead a long and visionary way in the true emancipation of nanotechnology, chemical engineering, and environmental engineering science. This paper opens up and unfolds the scientific intricacies and the scientific ingenuity in research pursuit in the field of nanomaterials and ENPs. Science and technology are today huge colossus with a definite and purposeful vision of its own. The march of science thus needs to be reenvisioned and reenvisaged with the passage of scientific history and time.

7.11 ENVIRONMENTAL SUSTAINABILITY, WATER PURIFICATION, AND NANOTECHNOLOGY

Nanotechnology today is integrated with many diverse branches of science and engineering such as environmental engineering science and water purification. Environmental sustainability is the utmost need of human civilization today. Sustainable development should be the over-arching goal of all scientific research pursuit and human scientific progress today. Provision of pure drinking water and industrial wastewater treatment are the needs of human scientific endeavor in present-day human civilization. Scientific prudence and scientific ingenuity in the field of environmental engineering need to be vastly envisioned as mankind moves forward. Nanotechnology, nanofiltration, and membrane science are the boons of human civilization and human scientific progress today. Technology and engineering science needs to answer positively to the alarming global issue of groundwater heavy metal and arsenic contamination. In these areas of environmental protection, environmental sustainability should be the cornerstones of research and development initiative. The visionary words of Dr. Gro Harlem Brundtland, former Prime Minister of Norway on the science of "sustainability" need to be reenvisioned and reframed as mankind trudges a weary path toward vision and scientific forbearance.[19–24]

7.12 THE WORLD OF CHALLENGES IN THE FIELD OF APPLICATION OF NANOTECHNOLOGY TO HUMAN SOCIETY

Human scientific ingenuity and the vast scientific acuity in the field of nanotechnology applications are in the process of a newer scientific rejuvenation. There are lots of health-related concerns as regards applications of nanomaterials or CNTs in human scientific progress. Nanotechnology and nanomaterials are highly challenged branches of human scientific endeavor today. The scientific endeavor and the scientific ingenuity in nanotechnology applications needs to be reframed and reinvestigated with the march of science and engineering. The Royal Society, United Kingdom Report (2004) with deep and cogent insight detailed opportunities and uncertainties in nanoscience and nanotechnologies. The authors in this report elucidated on nanomaterials, nanometrology, electronics, optoelectronics, and information and communication technology. Bionanotechnology and nanomedicine are the salient features of this well-researched report. Nanomanufacturing and the industrial applications of nanotechnologies has a pivotal role in this report (The Royal Society and The Royal Academy of Engineering, UK, 2004).[1] The upshots of this study are:

- Define what is meant by nanoscience and nanotechnologies.
- Summarize the current state of scientific discerning in nanotechnologies.
- Identify the specific applications of new technologies.
- A greater emancipation of the science of nanomedicine.
- Identify the health and safety issues of nanotechnology and nanomaterials applications.
- Regulatory measures which need to be taken in the application of nanotechnology in human society.

The challenges and the vision of nanotechnology and nanomaterials application are vast and versatile today. Human scientific progress in the field of nanotechnology is in the similar manner surpassing vast scientific boundaries. This paper opens up new windows of innovation and scientific instinct in diverse areas of nanotechnology applications. Technology of nanoscience and nanoengineering thus needs to be more addressed and reenvisioned as human civilization surges forward toward a newer era.

7.13 MODERN SCIENCE, SCIENTIFIC PROFUNDITY, AND THE CHALLENGES AHEAD

Modern science and modern-day human civilization today stand in the midst of deep introspection and immense scientific vision. Frequent environmental disasters, global warming, and global climate change are challenging the vast scientific fabric globally. Today science and technology are huge colossus with a definite vision and a definite scientific understanding of its own. The vision and the challenge of modern science are immense as human civilization moves forward. Today the scientific world stands in the midst of difficulties, barriers, and immense scientific and technological comprehension. Nanotechnology today in the similar manner stands in the midst of scientific introspection and vast acuity. Applications of nanotechnology and nanomaterials such as CNTs to human society and engineering science are gearing toward an era of vast scientific regeneration. In this paper, the author vastly with wide vision elucidates the necessities and scientific imperatives in the application of CNTs toward the emancipation of science and engineering. Modern science is highly advanced today with vast scientific research forays in the field of nuclear science and space technology. Nations around the globe are gearing toward newer vision and a newer scientific profundity in the field of nanotechnology. Human scientific history as well as scientific research pursuit needs to be refurbished and reenvisioned as nanotechnology applications in human society moves forward. Nanotechnology is a highly advanced branch of engineering and science but research forays are still unexplored. Nanomaterials and CNTs applications need to be readdressed and reframed if the world's fundamental problems are to be mitigated. The challenges in human scientific history are immense and scientifically intricate. The author in this treatise deeply enumerates the scientific challenges, the vast scientific imagination, and the immense scientific needs of CNTs and nanomaterials with the sole vision of furtherance of science and engineering.

7.14 HUMAN SCIENTIFIC REGENERATION AND THE VISION BEHIND NANOTECHNOLOGY

Human civilization and scientific rigor today stands in the midst of vision and scientific forbearance. Scientific regeneration and the vast world of scientific validation need to be readdressed and reenvisioned with the passage of scientific history and time. Nanotechnology today is in the path of newer scientific regeneration. There are immense scientific barriers and deep scientific intricacies in the application of nanotechnology and nanomaterials to diverse areas of science and engineering. Chemical process engineering and environmental engineering science are two vibrant examples of scientific vision and scientific profundity of nanoscience and nanotechnology. Validation of science is of utmost necessity as chemical process engineering, environmental engineering, and material science that move toward a newer scientific paradigm. Today, nanotechnology and nanomaterials are shaping the global scientific destiny and are ushering in a new era in the field of chemical process engineering and environmental engineering science. The main upshot of this paper is to address the scientific needs, the scientific difficulties and barriers in the application of nanomaterials such as CNTs to human scientific progress and vision.

7.15 FUTURE RECOMMENDATIONS AND FUTURE FLOW OF THOUGHTS

The science of chemical process engineering needs to be reenvisioned and reorganized with the progress of scientific history and time. Chemical process engineering, material science, and nanotechnology are today leading a long and visionary way in the true realization of science and engineering to human society. Technology and engineering science has today few answers to the scientific travails and the scientific intricacies of nanomaterials and CNTs applications. The immense challenges and the vast vision of nanotechnology today stand unparalleled. Future recommendations in research endeavor in chemical process engineering and nanotechnology should focus toward more green engineering, the needs to protect ecological biodiversity, and the challenge to overcome global warming. Global water shortage and lack of pure drinking water should be the areas of grave scientific concern. Global research and development initiatives should be more technology-driven and targeted toward grassroot scientific research pursuit. Green chemistry and green engineering are the utmost needs of human civilization and human scientific endeavor today. Future flow of thoughts in the similar manner should evolve toward a newer scientific understanding and a newer scientific genre in the field of nanotechnology and material science.[19–24]

7.16 FUTURE DIRECTIONS OF SCIENTIFIC ENDEAVOR

Scientific research pursuit today stands in the midst of vision, scientific fortitude, and vast scientific advancement. Technology and engineering science have today few answers to the scientific intricacies and the scientific barriers of global water issues and global climate changes. Here comes the importance of nanomaterials, CNTs, and nanotechnology. Future scientific endeavor and future scientific genre should be directed toward more scientific emancipation in provision of basic needs of human society such as energy, water, food, and shelter. Industrial wastewater treatment and water purification technologies should be the cornerstone of all scientific endeavors in environmental engineering science. The world today stands in the midst of deep scientific introspection and vast technological profundity. The validation of science and engineering to the grave concerns of human society needs to be revamped and deeply streamlines. Arsenic and heavy metal groundwater contamination is an issue of immense concern as science and engineering surges forward.

7.17 CONCLUSION AND FUTURE SCIENTIFIC PERSPECTIVES

Technology and engineering science in today's modern scientific civilization needs to be reorganized and revamped as human scientific progress in chemical engineering, environmental engineering, nanotechnology, and material science assumes immense importance. Future scientific perspectives should be targeted toward green technology, green engineering, and green chemistry. Global environmental concerns, global water crisis, and global climate change are veritably challenging the scientific scenario. The needs of science and engineering are immense, thought-provoking and veritably inspiring. Technology of chemical process engineering and water process engineering needs to be readdressed and revamped as mankind faces environmental disasters of immense scientific proportions. Arsenic and heavy metal groundwater and drinking water contamination are challenging the vast global scientific fabric. The author in this paper rigorously points toward the scientific success of nanomaterials and CNTs applications to human scientific progress. This chapter opens up new avenues of futuristic vision and futuristic thoughts in the field of nanotechnology and nanomaterials with the sole aim of furtherance of science and technology in modern-day human civilization. Mankind's immense scientific girth, the civilization's vast scientific prowess, and man's need of basic provisions such as energy and water will go a long and visionary way in the true emancipation of energy and environmental sustainability. CNTs are the visionary nanomaterials of the future. Scientific cognizance, the vast scientific doctrine, and the world of technological validation are the torchbearers toward a newer visionary era in nanoscience and nanotechnology.

ACKNOWLEDGMENT

The author deeply acknowledges the contribution of his late father Shri Subimal Palit, an eminent textile engineer from India who taught the author rudiments of chemical engineering science.

KEYWORDS

- **carbon**
- **nanotubes**
- **vision**
- **chemical**
- **engineering**
- **nanotechnology**

REFERENCES

1. The Royal Society and The Royal Academy of Engineering Report. *Nanoscience and Nanotechnologies: Opportunities and Uncertainties*; United Kingdom, 2004.
2. Lloyd's Emerging Risks Team Report. *Nanotechnology, Recent Developments, Risks and Opportunities*; United Kingdom, 2007.
3. Chaudhry, Q.; Watkins, R.; Castle, L. Nanotechnologies in the Food Arena: New Opportunities, New Questions and New Concerns (Chapter 1). In *Nanotechnologies in Food*; (RSC Nanoscience and nanotechnology, No. 14) The Royal Society of Chemistry: United Kingdom: 2010.
4. The Royal Society Report, United Kingdom. *Knowledge, Networks and Nations, Global Scientific Collaboration in the 21st Century*; The Royal Society, UK, 2011.
5. United Nations Educational, Scientific and Cultural Organization (UNESCO), Paris, France Report. *The Ethics and Politics of nanotechnology*; France, 2006.
6. Royal Commission on Environmental Pollution Report, London, United Kingdom (Twenty-seventh Report). *Novel Materials in the Environment: The Case of Nanotechnology*; London, United Kingdom, 2008.
7. Borm, P. J. A.; Berube, D. A Tale of Opportunities, Uncertainties, and Risks. *Nano Today* **2008,** *3* (1–2), 56–59.
8. Ong, Y. T.; Ahmad, A. L.; Zein, S. H. S.; Tan, S. H. A Review on Carbon Nanotubes in an Environmental Engineering and Green Engineering Perspective. *Brazilian J. Chem. Eng.* **2010,** *27* (2), 227–242.

9. Rafique, M. M. A.; Iqbal. J. Production of Carbon Nanotubes by Different Routes—A Review. *J. Encapsul. Adsorpt. Sci.* **2011,** *1*, 29–34.
10. Kusworo, T. D.; Ismail, A. F.; Budiyono; Widiasa, I. N.; Johari, S.; Sunarso. The Uses of Carbon Nanotubes Mixed Matrix Membranes (MMM) for Biogas Purification. *Int. J. Water Res.* **2012,** *2*, (1), 5–10.
11. Shannon, M. A.; Bohn, P. W.; Elimelech, M.; Georgiadis. J. G.; Marinas. B. J.; Mayes. A. M. Science and Technology for Water Purification in the Coming Decades. *Nature* **2008,** *452*, 301–310 (Nature Publishing Group).
12. Houmann, M. L. Water Purification Device for a Developing Country Constructed from Local materials. Southern Scholars Senior Thesis, Senior Research Projects, Paper 15, Southern Adventist University, USA, 2012.
13. Shon, H. K.; Phuntso, S.; Choudhary, D. S.; Vigneswaran, S.; Cho, J. Nanofiltration for Water and Wastewater Treatment—A Mini Review. *Drink. Water Eng. Sci.* **2013,** *6*, 47–53.
14. Choudhary, D. S.; Vigneswaran, S.; Ngo, H-H.; Shim, W. G.; Moon, H. Biofilter in Water and Wastewater Treatment. *Korean J. Chem. Eng.* **2003,** *20* (6), 1054–1065.
15. Thiruvenkatachari, R.; Vigneswaran, S.; Moon, I. S. A Review on UV/TiO_2 Photocatalytic Oxidation Process. *Korean J. Chem. Eng.* **2008,** *25* (1), 64–72.
16. Loganathan, P.; Vigneswaran, S.; Kandasamy, J.; Naidu, R. Defluoridation of Drinking Water Using Adsorption. *J. Hazard. Mater.* **2013,** *248–249*, pp 1–19.
17. Elimelech, M.; Phillip, W. A. The Future of Seawater Desalination: Energy, Technology and the Environment. *Science* **2011,** *333*, 712–717.
18. Hong, S.; Elimelech, M. Chemical and Physical Aspects of Natural Organic Matter (NOM) Fouling of Nanofiltration Membranes. *J. Membr. Sci.* **1997,** 132, 159–181.
19. Palit, S. Advanced Environmental Engineering Separation Processes, Environmental Analysis and Application of Nanotechnology—A Far-reaching Review (Chapter 14). In *Advanced Environmental Analysis: Application of Nanomaterials*, Vol. 1; Chaudhery, M. H., Boris, K. Eds.; Royal Society of Chemistry: United Kingdom, 2017; pp 377–416.
20. Palit, S. Application of Nanotechnology, Nanofiltration and Drinking and Wastewater Treatment—A Vision for the Future (Chapter 17). *Water Purification*; Grumezescu, A. M. Ed.; Academic Press: USA, 2017; pp 587–620.
21. Hashim, M. A.; Mukhopadhyay. S.; Sahu, J. N.; Sengupta, B. Remediation Technologies for Heavy Metal Contaminated Groundwater. *J. Environ. Manage.* **2011,** *92*, 2355–2388.
22. Palit, S. Nanofiltration and Ultrafiltration—The Next Generation Environmental Engineering Tool and a Vision for the Future. *Int. J. Chem. Tech. Res.* **2016,** *9* (5), 848–856.
23. Palit, S. *Filtration: Frontiers of the Engineering and Science of Nanofiltration—A Far-reaching Review*; CRC Concise Encyclopedia of Nanotechnology; Ubaldo, O.-M., Kharissova. O. V., Kharisov. B. I., Eds.; Taylor and Francis: Boca Raton, FL, USA, 2016; pp 205–214.
24. Palit, S. Advanced Oxidation Processes, Nanofiltration, and Application of Bubble Column Reactor. In *Nanomaterials for Environmental Protection*; Kharisov. B. I., Kharissova. O. V., Rasika Dias, H. V. Eds.; Wiley: USA, 2015; pp 207–215.

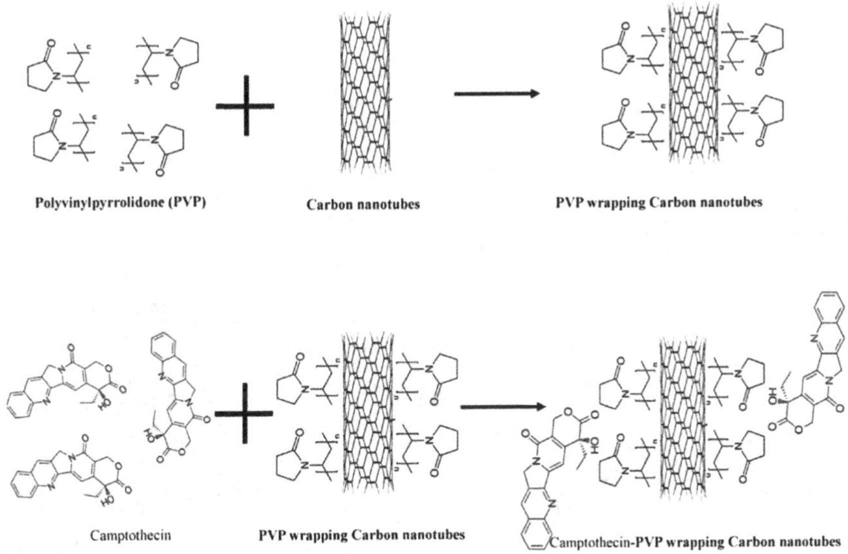

Polyvinylpyrrolidone (PVP) Carbon nanotubes PVP wrapping Carbon nanotubes

Camptothecin PVP wrapping Carbon nanotubes Camptothecin-PVP wrapping Carbon nanotubes

FIGURE 1.4 Enhancement of solubility and stability of camptothecin by camptothecin–PVP wrapping CNT complexes.

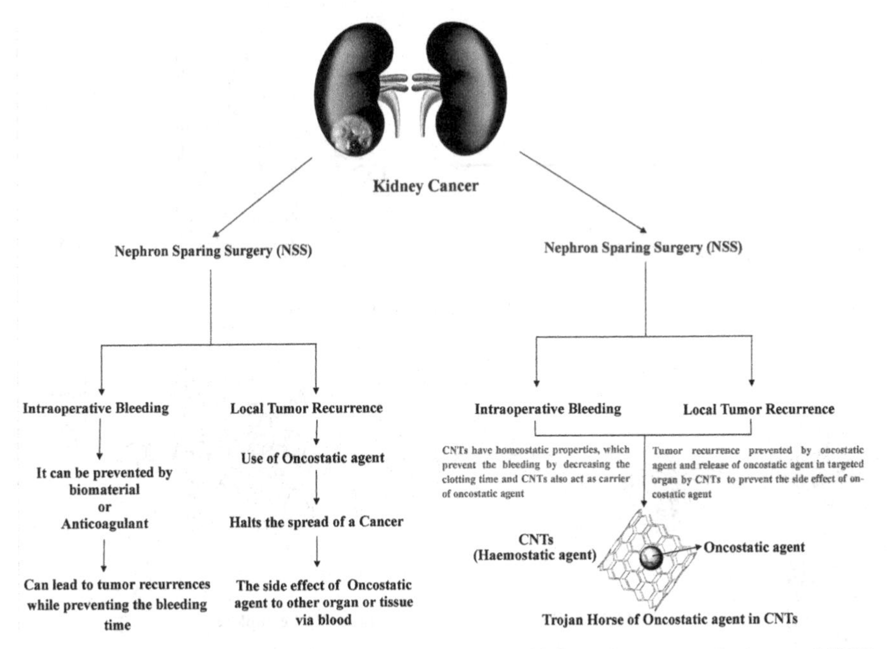

FIGURE 1.5 Carbon nanotubes act as Trojan horse which carries oncostatic drug and CNTs act as homeostatic carbon nanotubes.

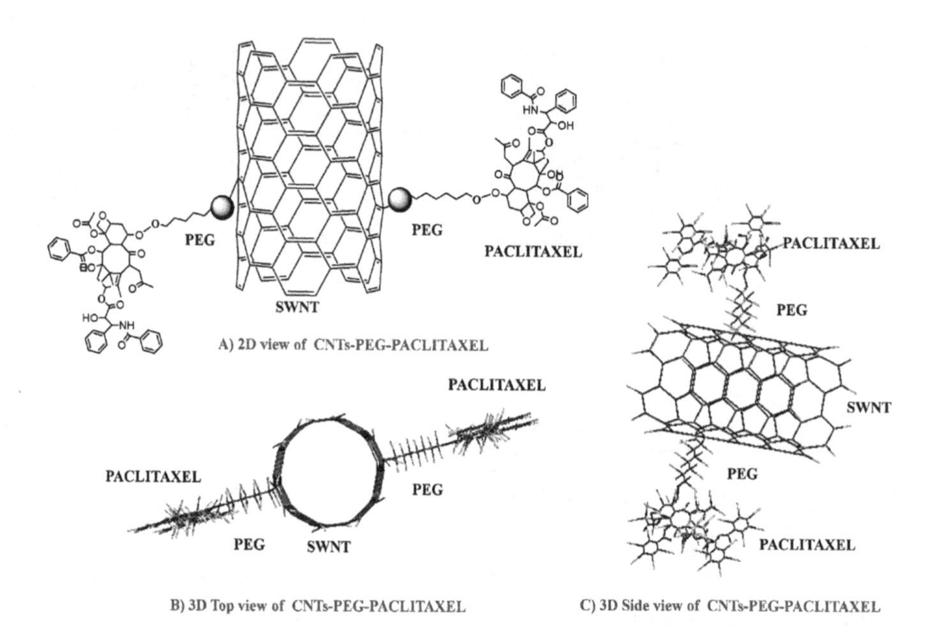

FIGURE 1.6 2D and 3D views of the CNT–PEG–paclitaxel as drug carrier.

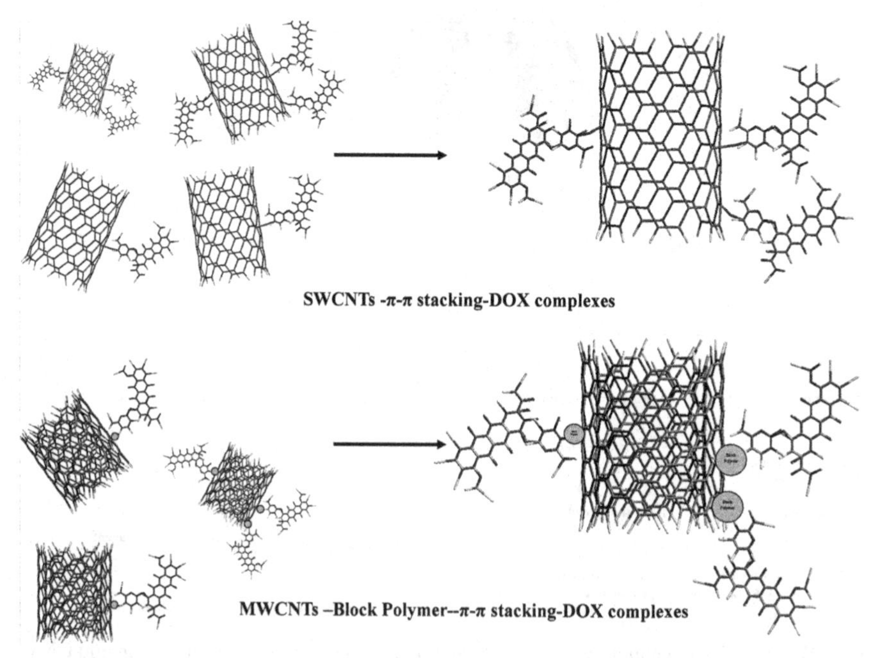

FIGURE 1.7 SWCNT and MWCNT as carriers for the doxorubicin through π–π stacking.

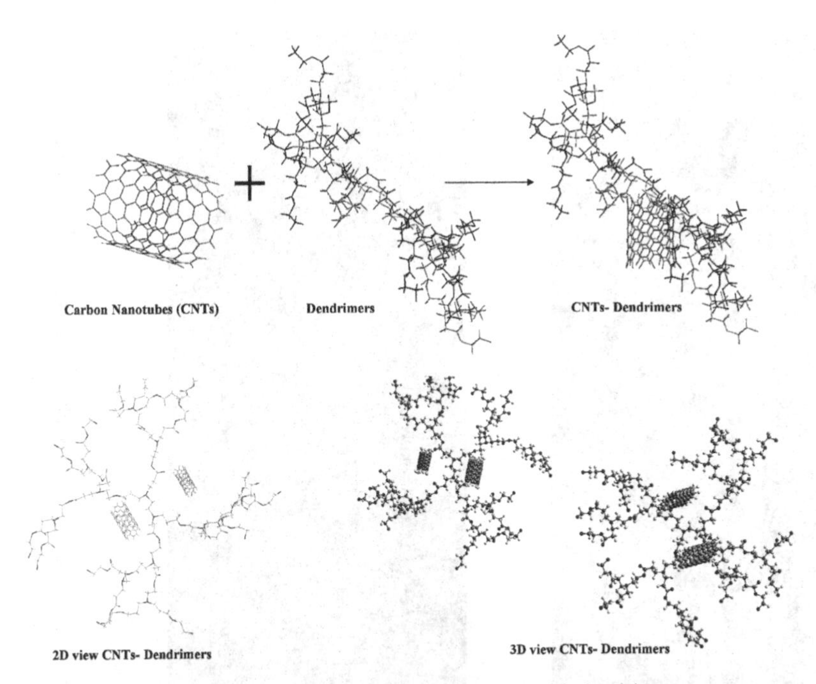

FIGURE 1.9 Using convergent method to link the CNTs with dendrimers.

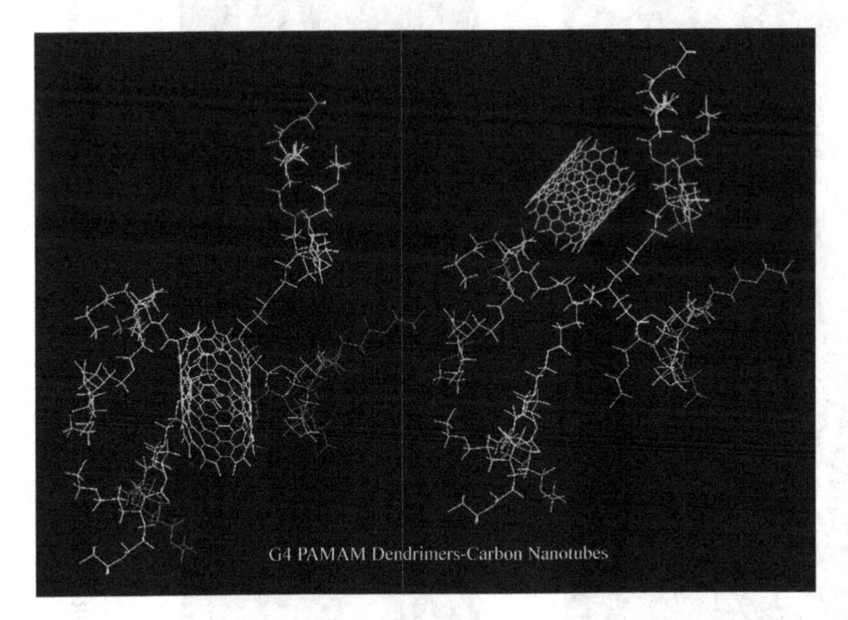

FIGURE 1.10 G4 PAMAM dendrimers are enrolled to carbon nanotubes to form a versatile carrier.

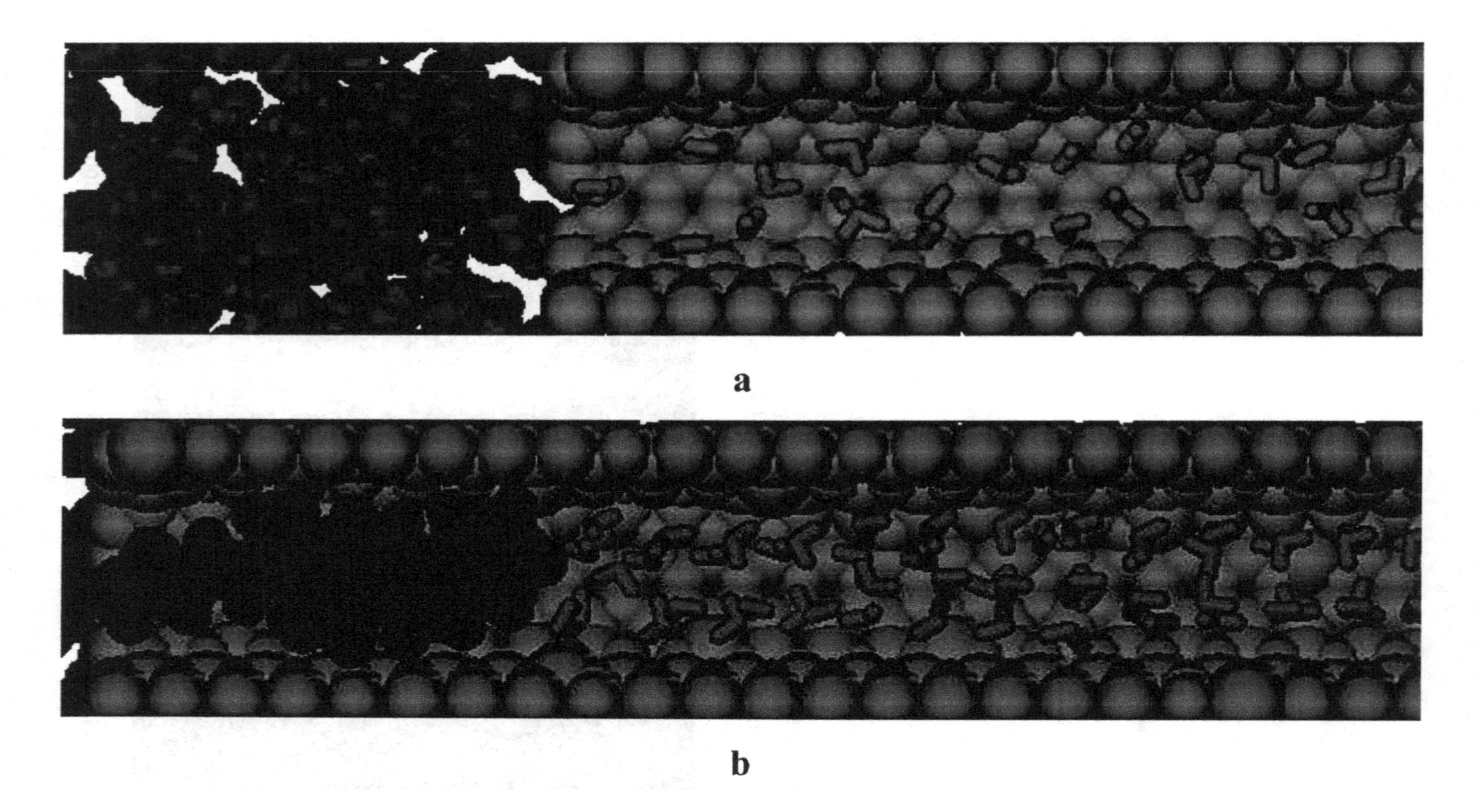

FIGURE 4.12 The structure of molecular water inside a carbon nanotube: (a) in a free state and (b) after the adsorption of cholesterol.

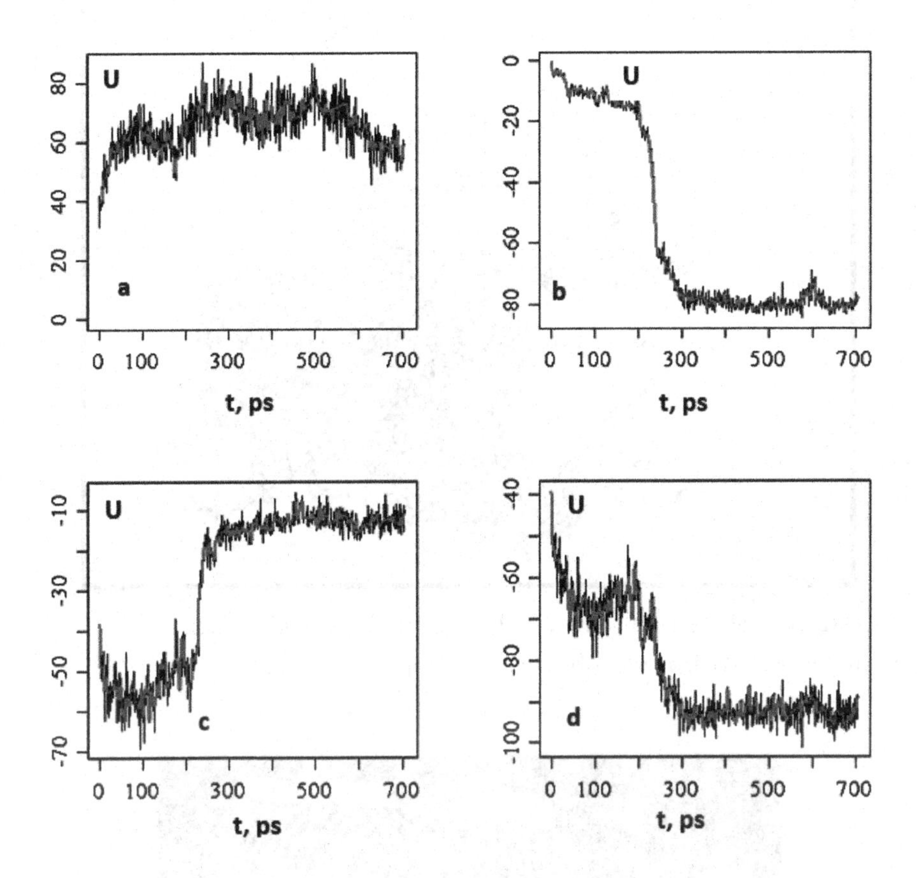

FIGURE 4.13 The interaction energies U (kcal/mole) of the cholesterol–nanotube–water constituents in the system: (a) cholesterol–cholesterol; (b) cholesterol–a nanotube; (c) cholesterol–water; and (d) cholesterol–the surrounding system.

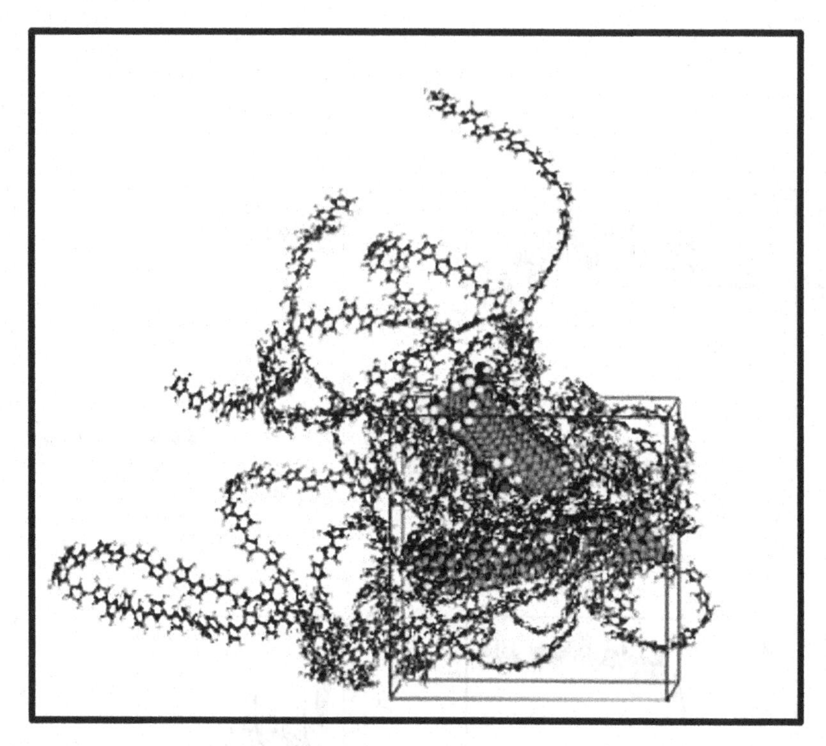

FIGURE 6.8 PPy/CNT-CA 90:10 (w/w) bulk model after refinement stage.

Source: Reproduced from permission from Ref. [56].

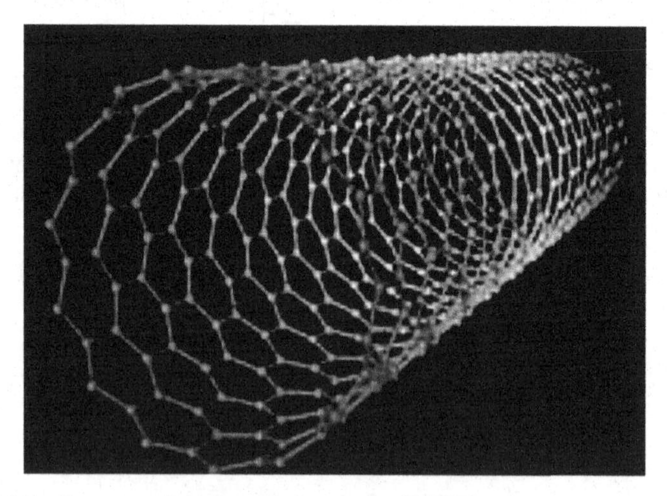

FIGURE 10.1 Structure of single-walled nanotube (SWNT).

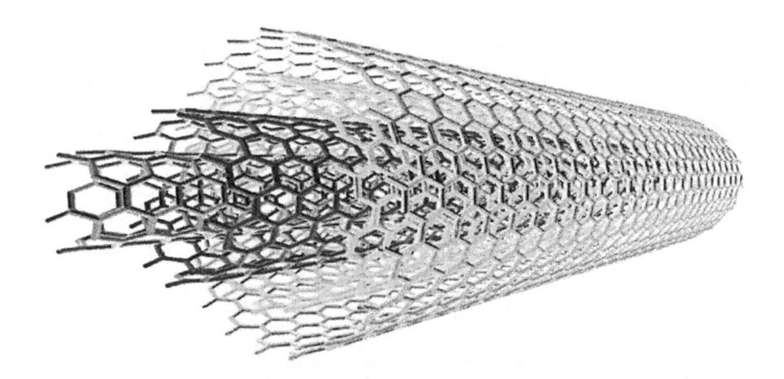

FIGURE 10.2 Structure of multiwalled nanotube.

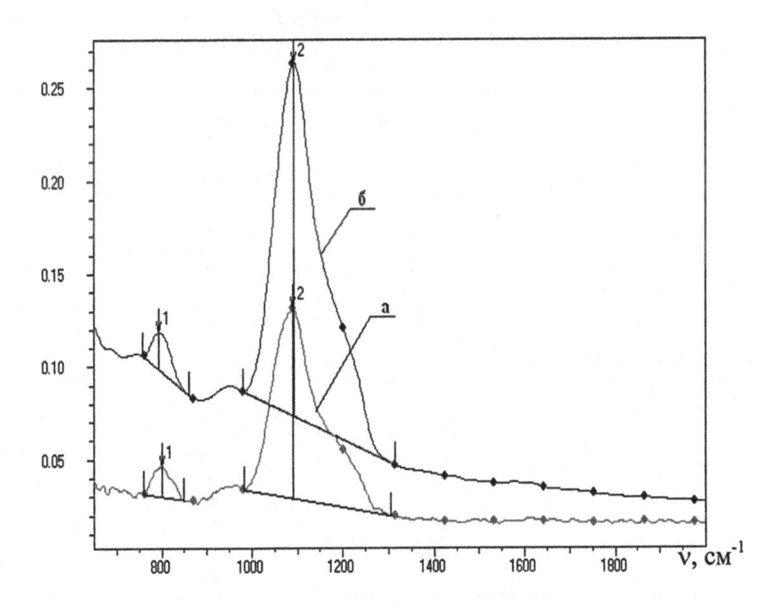

FIGURE 12.7 IR spectra of samples: (a) silica and (b) silica + Cu/C nanocomposites.

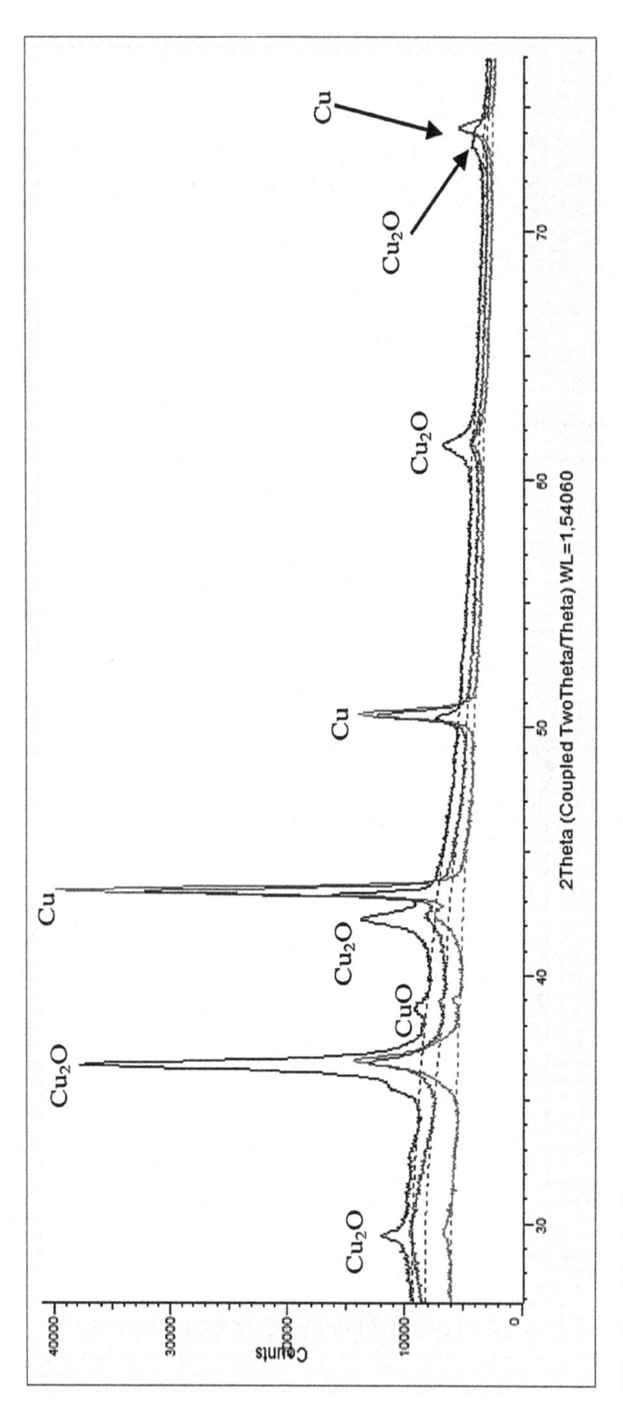

FIGURE 15.2 Diffractogram of copper/carbon nanocomposites.

CHAPTER 8

THE HYDROGEN BATTERIES BASED ON CARBON NANOTUBES

I. A. SHESTAKOV[1,2], S. S. VIDRINA[2], and
ALEXANDER V. VAKHRUSHEV[1,2*]

[1]*Department of Mechanics of Nanostructures, Institute of Mechanics, Udmurt Federal Research Center, Ural Division, Russian Academy of Sciences, Izhevsk, Russia*

[2]*Department of Nanotechnology and Microsystems, Kalashnikov Izhevsk State Technical University, Izhevsk, Russia*

[*]*Corresponding author. E-mail: vakhrushev-a@yandex.ru*

ABSTRACT

The chapter deals with the safe design of the hydrogen storage tank, which includes a set of cylindrical metal containers made of stainless steel filled with carbon nanotubes; a cooling jacket made of liquid nitrogen in the likeness of a Dewar vessel into which metal containers with carbon nanotubes (single-walled carbon nanotubes, SWNT) are dipped and filled with gaseous hydrogen. The analyses of the scheme of the experiment are presented and the results of the study of hydrogen storage at a liquid nitrogen temperature are shown. It is confirmed that hydrogen at the temperature of liquid nitrogen is retained by the SWNT in the accumulator tank, and at normal temperature, it is released. The data of microscopic (Phenom Pro microscope) and sorption studies of SWNT conglomerates of scaly structure are presented in this chapter.

8.1 INTRODUCTION AND PROBLEM STATEMENT

One of the potential causes of global warming is the hydrocarbon energy mode.[1] A pioneer and romantic of the hydrogen era, academician Veziroglu in the 20th century proposed one of the concepts of hydrogen energy, which found wide application among scientists, engineers, and businessmen of the 20th and 21st centuries.[2] The most environmental-friendly energy carrier is hydrogen. All over the world, there is a search for schemes for the use of hydrogen as a fuel for vehicles: road, rail, aviation, marine, and river, so this topic of research is relevant. It is important to create a safe hydrogen storage tank with a high hydrogen content.

The first successful experiments on storage of hydrogen in carbon nanotubes in the cryogenic temperature range (77 K) were performed by Nobel laureate Richard Errett Smalley in the last century, when it was possible to reach a capacity of 7.7% by mass at the temperature of cryogenic nitrogen. The concept of a hydrogen storage tank based on carbon nanotubes with liquid nitrogen for transport applications was first proposed in Refs. [3 and 4]. In the years 2007–2008, a project was initiated to develop a concept of a storage tank for storing hydrogen in carbon nanostructures at liquid nitrogen temperature and the concept of a nitrogen–hydrogen vehicle commissioned by the Japanese corporation "Nissan Motors."[5]

The use of hydrogen fuel in transport has different principles.[6] In our opinion, a promising storage tank for hydrogen is a high-pressure tank of 15–50 MPa at a liquid nitrogen temperature of 77 K, while a bulk nano-material-CNT (single-walled carbon nanotubes) is used in the hydrogen accumulator tank, which allows additionally accumulating hydrogen at low temperatures. The layout of the hydrogen storage tank for surface transport systems and aircraft of vertical takeoff for rapid climb, and speed in upper layers of the atmosphere are proposed in Ref. [7].

The tasks of the investigation were to study the potential of this hydrogen storage scheme, to develop, based on the studies carried out, a design of a storage tank with a hydrogen injection volume of 50 m^3 for surface transport systems and a vertical takeoff aircraft. For this, it was necessary not only to compile an adequate scheme of the experiment but also to investigate the fillers for hydrogen, which represented the CNT agglomerates. Also it is nessecery to perform a spectral analysis of agglomerates and metal particles contained in mixture of bulk material and conclude that it is advisable to use a particular bulk material. This experimental study and engineering design of hydrogen accumulators is a continuation of the theoretical computer simulations of the accumulation problems by nanostuctures detailed description of which is given in the works of Refs. [8–13].

8.2 EXPERIMENTS ON THE ACCUMULATION OF HYDROGEN

Using the developed accumulator tank and carbon nanotubes, experiments were carried out on hydrogen sorption by carbon nanotubes at a temperature of 77 K. The stand for the experiment included (Fig. 8.1): a standard cylinder with hydrogen volume of 0.040 m^3 at a pressure of 15 MPa; reducer, adapter from the reducer to the valve; the STG 400 valve, withstanding 50 MPa; cell section of the hydrogen storage tank with a volume of 75 cm^3, filled with carbon nanotubes; a Dewar vessel at 0.016 m^3 with liquid nitrogen.

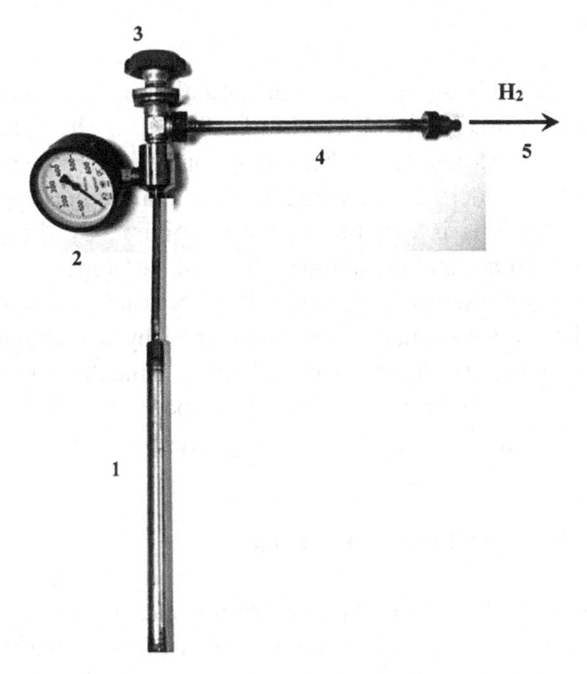

FIGURE 8.1 Scheme for the determination of sorbed hydrogen: 1—hydrogen storage tank, 2—manometer MPZ-UV 60 MPa, 3—gate STG-400, 4—adapter, and 5—to the flowmeter.

The sequence of the experiment is described below.

1. The capacity of the Dewar vessel is filled with liquid nitrogen; the cell of the hydrogen storage tank is inserted into it (held for 5 min). The tank temperature drops to 77 K.
2. A high-pressure valve opens and hydrogen is pumped into the tank-battery section at a pressure of 12–15 MPa. Hydrogen penetrates

into the carbon nanotubes, adsorbing in them, as well as into the space between them.

3. The valve opens and no absorbed hydrogen leaves, while the battery-hydrogen tank section remains in the Dewar vessel.
4. The valve is closed, and the battery-tank section is maintained under normal conditions for 60 min until the temperature is completely equalized.
5. Connect to a container of water to measure the volume of the released hydrogen.
6. Release of hydrogen and determination of the volume of displaced water.

The volume of released hydrogen at normal atmospheric pressure and temperature was 0.0035 m^3 and weight of 0.0003 kg. The mass of carbon nanotubes was 0.0015 kg. Thus, the absorption of hydrogen by mass was 2%. The relative low density of hydrogen accumulation in comparison with theoretical calculations is explained, in our opinion, by the presence of impurities in nanotubes and the nonideal form of nanotubes.

A series of experiments was carried out; the pressure of hydrogen supply to the cooled tank accumulator was established by a manometer on the reducer BRVD-250 and corresponded to 12 MPa, while the volume of sorbed hydrogen was 0.0034, 0.0035, and 0.0036 L, respectively. The average value was 0.0035 L; the deviation from the average was 3%.

8.3 INVESTIGATION OF THE FILLER

Let us consider the CNTs used in experiments on the accumulation of hydrogen. The images (Figs. 8.2–8.9), made with the SEM Phenom Pro desktop scanning electron microscope, prove that the agglomeration of CNTs has a "scaled-up" structure with a number of "depressions" and "closed spaces" (an increase of 310 and 38,000 times). In Figure 8.7, the nanotubes themselves are visible.

These "scales" and hollows repeatedly increase the surface. At present, the task is to study a hydrogen storage tank with a better bulk material. So, carbon nanotubes of Belgorod production have been studied. A series of spectral analysis of "coils" of carbon nanotubes and metallic inclusions is carried out (Figs. 8.8–8.11). A series of experiments on the spectral analysis of "coils" (Figs. 8.10 and 8.11) of carbon nanotubes and metallic inclusions were carried out. Spectral analysis (Figs. 8.12–8.15) showed that in this bulk material, aluminum and nickel are present as impurities.

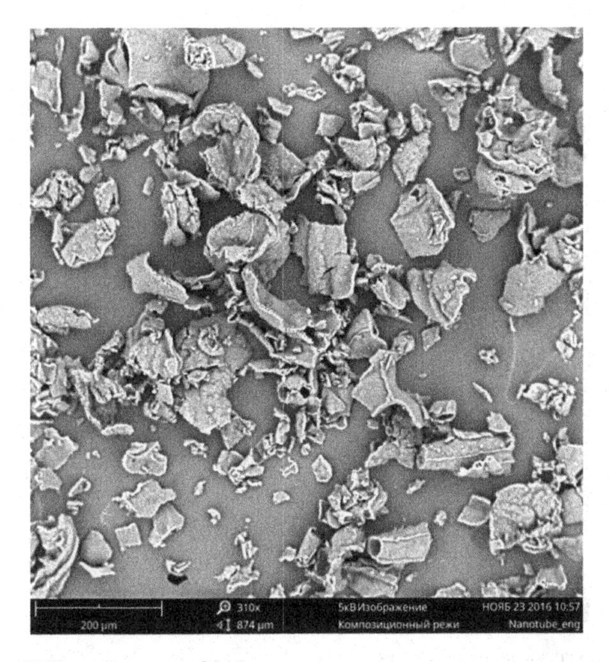

FIGURE 8.2 CNT, an increase of 310×.

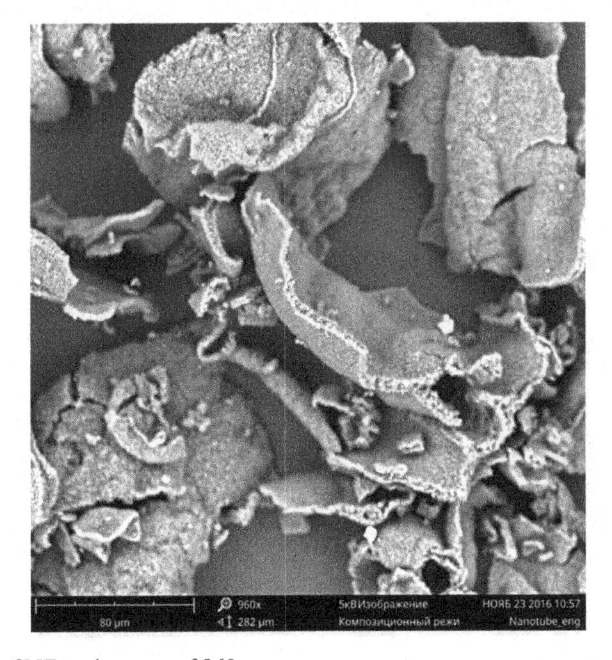

FIGURE 8.3 CNT, an increase of 960×.

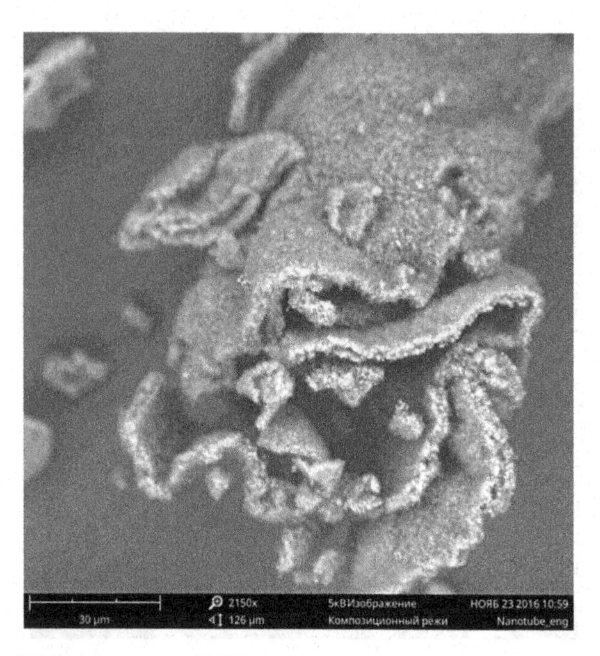

FIGURE 8.4 CNT, an increase of 2150×.

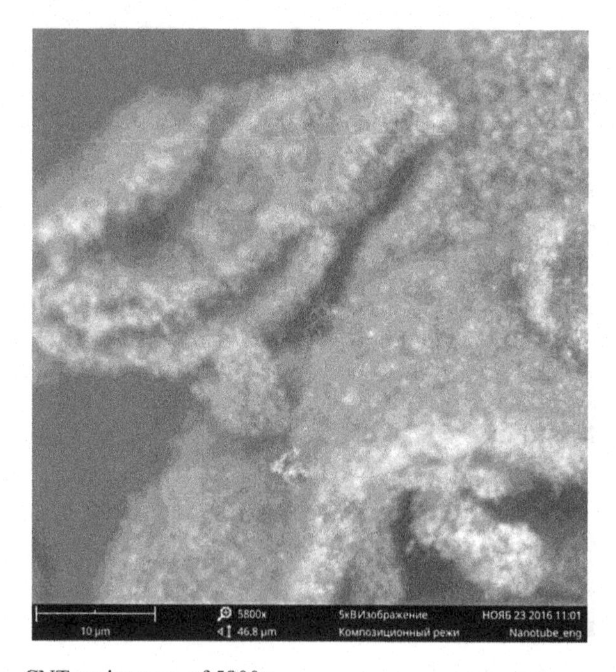

FIGURE 8.5 CNT, an increase of 5800×.

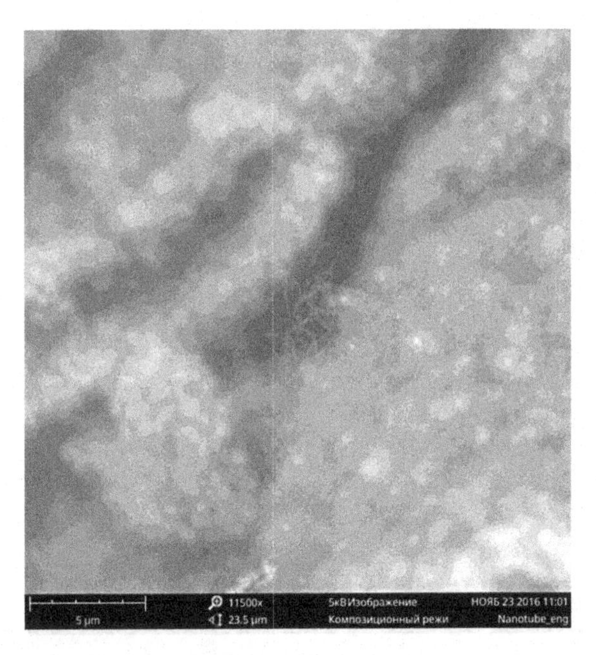

FIGURE 8.6 CNT, an increase of 11,500×.

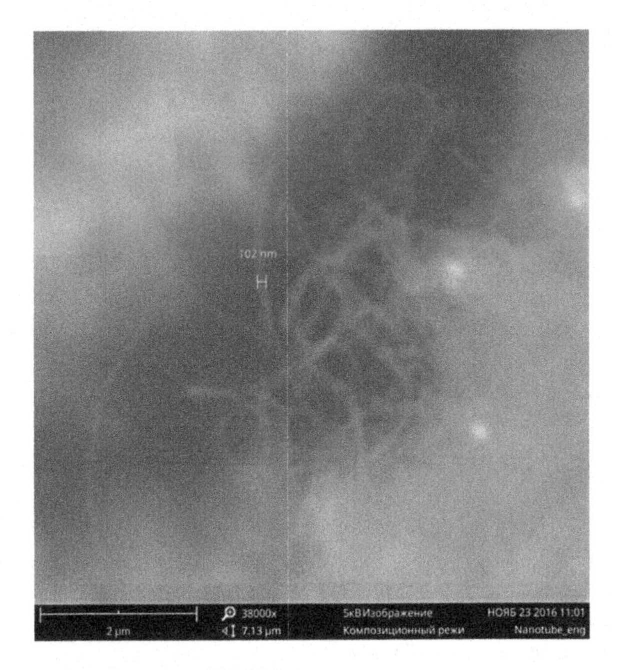

FIGURE 8.7 CNT, an increase of 38,000×.

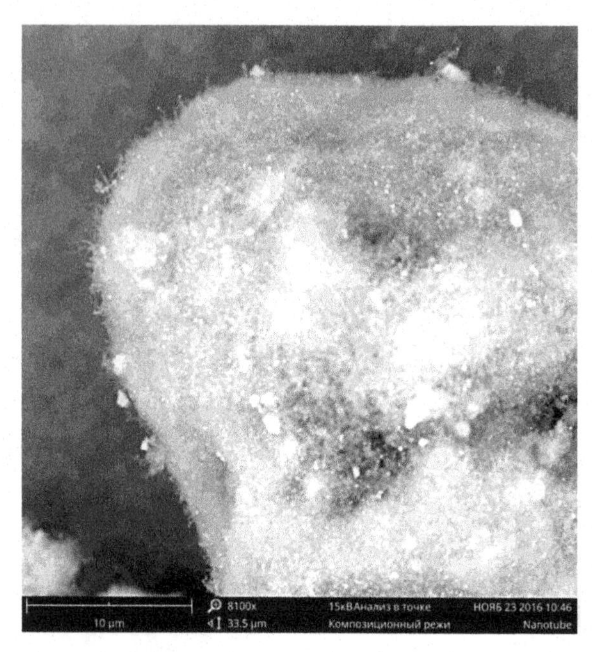

FIGURE 8.8 CNT, an increase of 8100×.

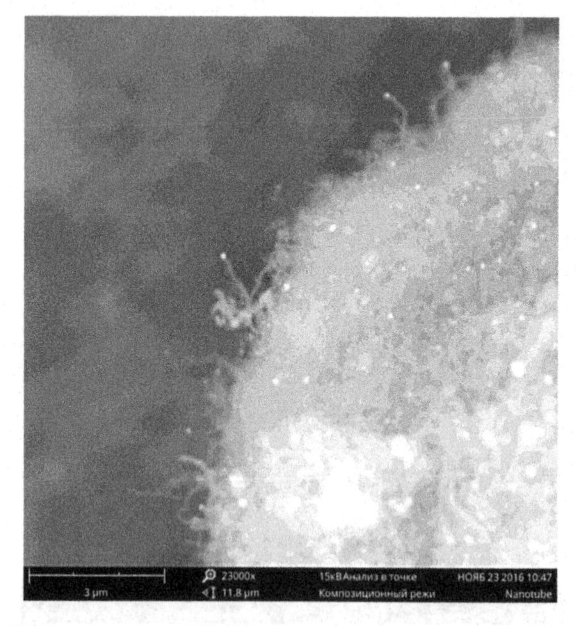

FIGURE 8.9 CNT, an increase of 23,000×.

FIGURE 8.10 CNT (point **1**).

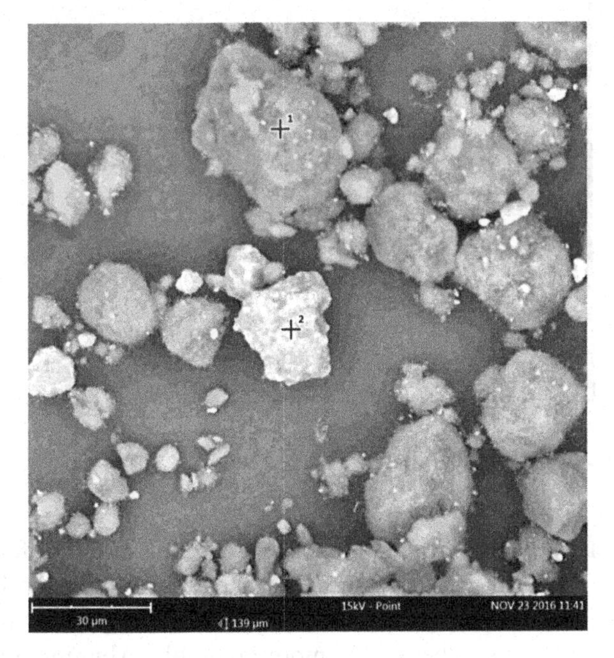

FIGURE 8.11 CNT (point 2).

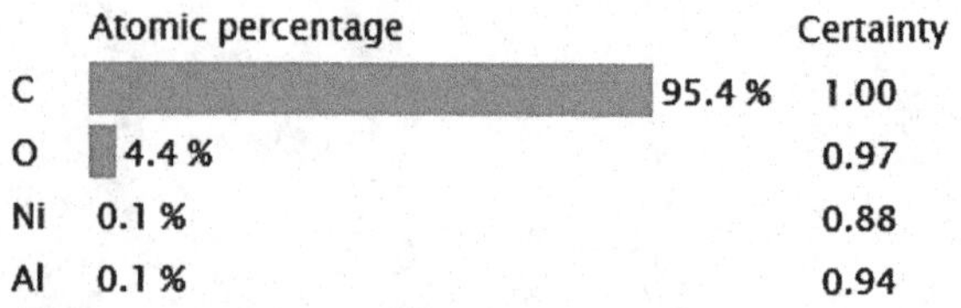

Atomic percentage
Certainty
C
95.4 %
1.00
O
4.4 %
0.97
Ni
0.1 %
0.88
Al
0.1 %
0.94

FIGURE 8.12 Spectral analysis CNT (point 1).

Atomic percentage
Certainty
C
67.6 %
0.98
Al
17.4 %
0.99
Ni
8.6 %
0.98
O
6.4 %
0.97

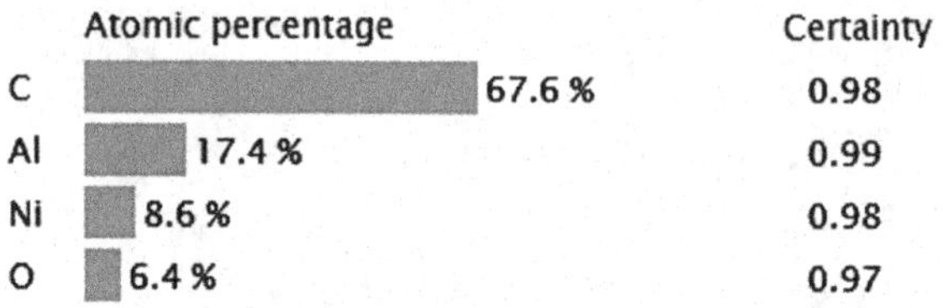

FIGURE 8.13 Spectral analysis CNT (point 2).

C
O
Ni
Al
Ni
Ni
0
1
2
3
4
5
6
7
8
132,403 counts in 30 seconds

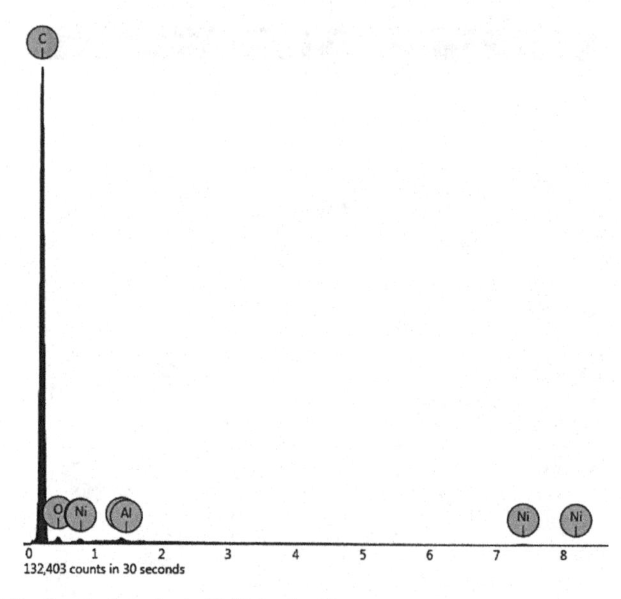

FIGURE 8.14 Spectral analysis CNT (point 1).

Despite that the coils of nanotubes have a better structure for the sorption of hydrogen than the scaly structure, they increase the weight of the bulk material. Therefore, the coils of nanotubes should be used when these nanostructures are chemically more favorable. Aluminum and nickel agglomerates (in a hydrogen tank you can use both the first type and the second one). Carbon bulk material with "scaly" structure has a low density (0.2 g/cm^3), and, also, has many and closed spaces. It is very loose, so it is better to use it in tank accumulators for aircraft.

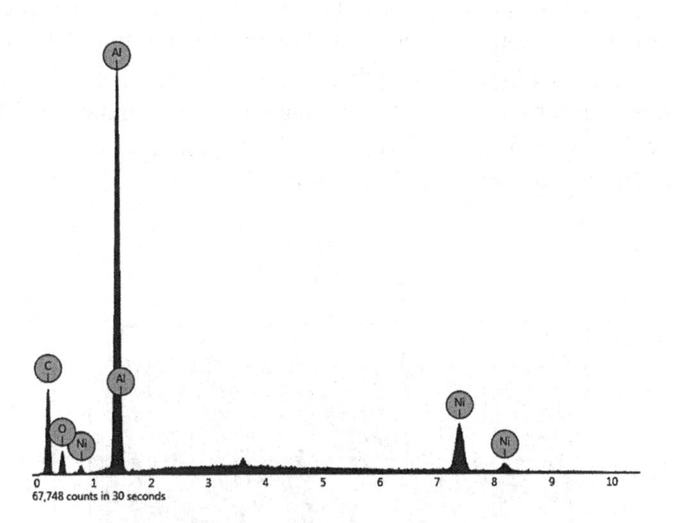

FIGURE 8.15 Spectral analysis CNT (point 2).

Carbon bulk material with a "coil" structure has a large active surface. In this case, the agglomerates of aluminum and nickel have a high thermal conductivity, so the mixture is rapidly cooled in large volumes, and the hydrogen storage tank enters the operating mode much faster. High density of bulk material (510 kg/m^3) is compensated by a high active surface. Due to the large mass of such bulk material, it is better to use it in land transport hydrogen accumulator tank.

8.4 HYDROGEN ACCUMULATORS

The accumulator should provide storage of a large mass of hydrogen in comparison with classical cylinders. At the same time, the battery should guarantee two times more refueling cycles, which would make it cost effective for land transport systems—cars and buses. This result was achieved due to a new set of essential design features in accordance with the invention.[7,17] Based on the studies carried out, a model of a hydrogen storage tank with a mass of 90 kg is developed for a fuel tank for ground equipment (Figs. 8.16 and 8.17).

This tank is an alternative to nine standard cylinders, the total mass of which is 460 kg. Volume of hydrogen will be shown after full vaporization of the liquid nitrogen and heating of CNT to ambient temperature.[10] This tank battery has the following specifications at temperature 77 K and pressure 15 MPa: length is 797 mm; width is 712 mm; height is 600 mm; number of cylinders are 91; the volume of one cylinder is 0.0009 m^3; total volume of cylinders is 0.0819 L; the volume of injected hydrogen under normal conditions is 50 m^3; volume of liquid nitrogen is 0.030 m^3; weight of one cylinder is 0.7 kg; total weight of cylinders is 63 kg; the weight of the accumulator tank is 90 kg; and weight of filling of storage battery is 120 kg.

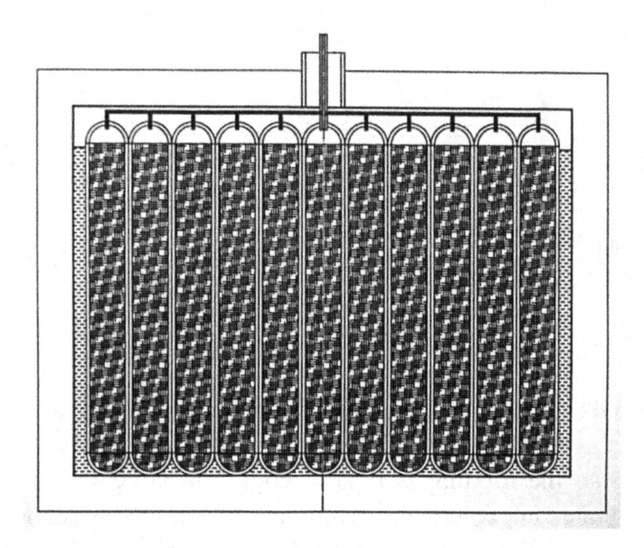

FIGURE 8.16 Hydrogen accumulator tank for the car (side view).

The proposed hydrogen accumulator, in comparison with the known technical solutions[14–16,18–21] of a similar purpose, provides the storage of hydrogen of a much larger mass, which makes it possible, when using a battery for motor vehicles, to have about a hundred times the number of refueling cycles of the engine.

Figure 8.18 shows a diagram of the vertical takeoff and landing aircraft.[22] Based on this configuration, a model of an aircraft for vertical takeoff and landing in a scale of 1:3 was developed and created (Fig. 8.19). The main technical characteristics of the model are: mass—14 kg, main rotor thrust—250 N, screw diameter—635 mm, engine weight—2.5 kg, working volume of the engine—85 cc. cm, volume of fuel to be filled—1 L, used fuel gasoline in a mixture with oil 1:50, height—1250 mm, and fuselage diameter—807 mm.

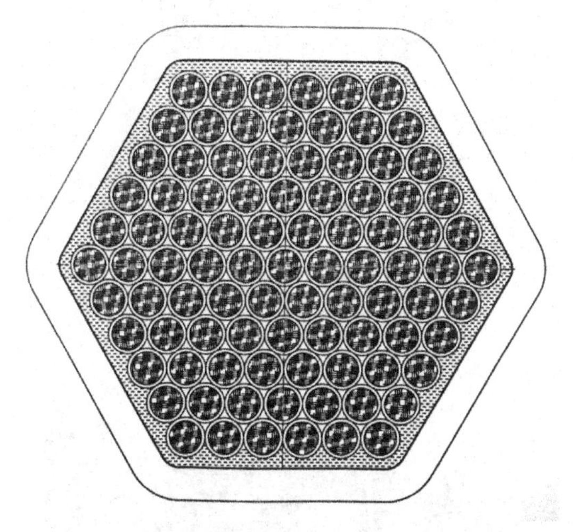

FIGURE 8.17 Hydrogen accumulator tank for the car (top view).

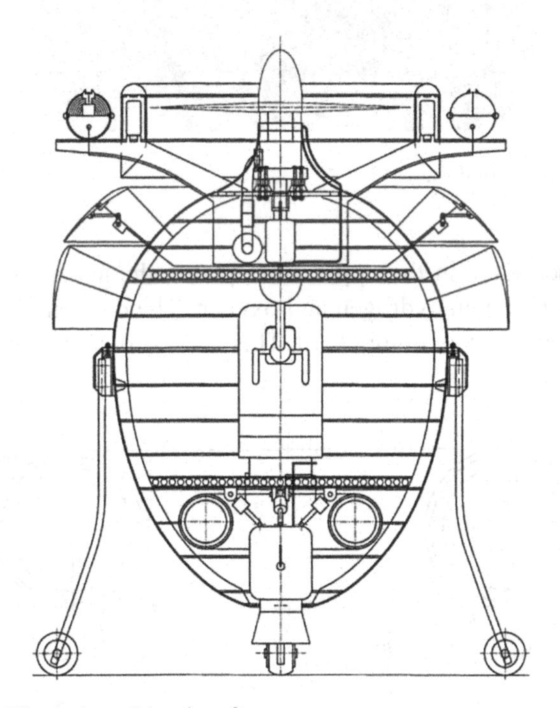

FIGURE 8.18 The design of the aircraft.

FIGURE 8.19 Model of the aircraft.

As an upper stage in the upper atmosphere, the apparatus contains an additional jet engine on hydrogen and oxygen. The construction of the tank for hydrogen is shown in Figure 8.20. This tank is filled with nanotubes.

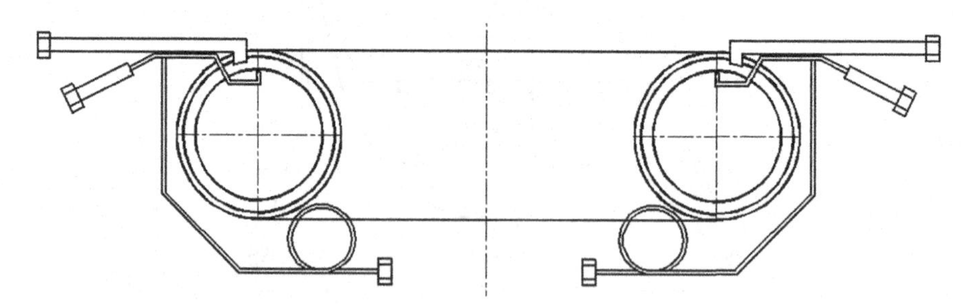

FIGURE 8.20 Hydrogen accumulator tank (side view).

8.5 CONCLUSIONS

The use of a hydrogen storage tank on transport will significantly improve safety and expand the scope of cryogenic technology, thanks to its simplicity of construction and reliability. A safe design of the hydrogen storage tank has been developed, which, compared to the known technical solutions, can have two orders of magnitude more refueling cycles and nine times less mass compared to standard steel cylinders. The developed technique of experimental research of processes of accumulation of hydrogen by these systems with nanotubes together with computer modeling will allow developing engineering designing and optimization of such systems.

ACKNOWLEDGMENT

The investigation was carried out in the frameworks of the state task to Kalashnikov ISTU (project no. 1239) and financial support of Russian Science Foundation (project no. 15-19-10002). In the frameworks of the state task, Kalashnikov ISTU corrected and extended the mathematical model, carried out computational experiments, and obtained main research results.

KEYWORDS

- **hydrogen storage tank**
- **carbon nanotubes**
- **hydrogen transport**
- **Dewar vessel**
- **hydrogen accumulators**

REFERENCES

1. Manaev, O. I. The Dynamics and Structure of Energy [E-resource]. *Energy Secur. Energy Sav. Sci. Tech. J.* **2008,** *2.* http://www.endf.ru/23_1.php (accessed Sept 21, 2017) (in Russian).
2. Bokris, J. O'M.; Veziroglu, T. N.; Smit, D. *Solar-hydrogen Energy. The Power That Can Save the World.* MEI Publ.: Moscow, 2002; p 164.

3. Gusev, A. L.; Spitsyn, B. V.; Kazaryan, M. A. Principles of Hydrogen Storage and Rational Design of Novel Materials for Hydrogen Accumulation. *Int. Sci. J. Altern. Energy Ecol. (ISJAEE)* **2007,** *4*, 202–203.
4. Mishhenko, A. I.; Belogub, A. V.; Savickij, V. D.; Talda, G. B.; Shatrov, E. V.; Kuznetsov, V. M.; Ramenskiy, A. Yu. *The Use of Hydrogen for Motor Vehicles. Atomic-hydrogen Power Engineering and Technology;* Collection of Articles, Issue 8. Energoatomizdat Publ.: Moscow, 1988 (in Russian).
5. Gusev, A. L.; Spitsyn, B. V.; Kazaryan, M. A. Principles of Hydrogen Storage and Rational Design of Novel Materials for Hydrogen Accumulation. *Alternative Energy Ecol.* **2007,** *4*, 202–203 [Grant: ISTC №. 3658p. ISTC Partner Project (STC "TATA"—The Company "Nissan Motor")].
6. Ramenskiy, A. Yu.; Shelishh, P. B.; Nefedkin, S. I.; Rychakov, A. A.; Starostin, M. V. In *Application of Hydrogen in Road Transport: Prospects in the Russian Market.* Proceedings of the International Symposium on Hydrogen Energy. Moscow, November 1–2, 2005, MEI, pp 169–174 (in Russian).
7. Yanovskiy, Yu. G.; Shestakov, I. A.; Vahrushev, A. V.; Lipanov, A. M. Battery Hydrogen. Patent 2,498,151 RF, MPK F17C11/00, November 10, 2013 (in Russian).
8. Suyetin, M. V.; Vakhrushev, A. V. Nanocapsule for Safe and Effective Methane Storage. *Nanoscale Res. Lett.* **2009,** *4* (11), 1267–1270.
9. Volkova, E. I.; Suyetin, M. V.; Vakhrushev, A. V. Temperature Sensitive Nanocapsule of Complex Structural Form for Methane Storage. *Nanoscale Res. Lett.* **2010,** *5*(1), 205–210.
10. Suyetin, M. V.; Vakhrouchev, A. V. Guided Carbon Nanocapsules for Hydrogen Storage. *J. Phys. Chem. C.* **2011,** *115*(13), 5485–5491.
11. Suyetin, M. V.; Vakhrouchev, A. V. Nanocapsule with Pump for Methane Storage. *Phys. Chem. Chem. Phys.* **2011,** *13*(20), 9863–9870.
12. Volkova, E. I.; Vakhrushev, A. V.; Suyetin, M. V. Improved Design of Metal-organic Frameworks for Effi-cient Hydrogen Storage at Ambient Temperature: A Multiscale Theoretical Investigation. *Int. J. Hydrog. Energy* **2014,** *39*, 8347–8350.
13. Volkova, E. I.; Vakhrushev, A. V.; Suyetin, M. V. Triptycene-modified Linkers of MOFs for Methane Sorption Enhancement: A Molecular Simulation Study. *Chem. Phys.* **2015,** *459*, 14–18.
14. Ying, J. Y. Nanoporous Systems and Templates the Unique Self-Assembly and Synthesis of Nanostructures. *Sci. Spectra* **1999,** *18*, 56–63.
15. Chebak, A. F. Storage Tank for Hydrogen. Patent 2,267,694 RF, MPK F17C11/00, October 27, 2009 (in Russian).
16. Fedorov, A. S.; Kuzubov, A. A. The Method and Apparatus for Storing Gas Inside a Nanopore of a Solid Carrier. Patent 2,319,893 RF, MPK F17C11/00, March 20, 2008 (in Russian).
17. Vahrushev, A. V.; Lipanov, A. M.; Suetin, M. V. A Container for Storing Various Liquid and Gaseous Substances. Patent 2,347,135 RF, MPK F17C11/00, February 20, 2009 (in Russian).
18. Gorjachev, I. V.; Voshhinin, S. A. Battery Hydrogen. Pat. 2,346,202 RF, MPK F17C11/00; February 10, 2009 (in Russian).
19. Hultquist, S. J.; Tom, G. M.; Kirlin, P. S.; McManus, J. V. Sorbent-based Gas Storage and Delivery System for Dispensing of High-purity Gas, and Apparatus and Process for Manufacturing Semiconductor Devices, Products and Precursor Structures Utilizing Same. Patent 6,132,492 MPK F17C11/00 USA, October 17, 2000.

20. Klos, H.; Schütz, W. Device for Storing Compressed Gas. Patent 6,432,176 MPK F17C11/00 USA, August 13, 2000.
21. Yaghi, O. M.; Eddaoudi, M.; Li, H.; Kim, J.; Rosi, N. Isoreticular Metal-organic Frameworks, Process for Forming the Same, and Systematic Design of Pore Size and Functionality There in, with Application for Gas Storage. Patent 6,930,193 MPK F17C11/00 USA, August 16, 2005.
22. Shestakov, I. A.; Vakhrushev, A. V.; Lipanov, A. M. Investigation of the Additional Lifting Force of the Aircraft of Vertical Take-off and Landing. *Aviat. Ind.* **2016,** *3*, 1–4 (in Russian).

CHAPTER 9

CARBON NANOTUBES AND MATERIAL SCIENCE: A CRITICAL OVERVIEW AND A VISION FOR THE FUTURE

SUKANCHAN PALIT*

Department of Chemical Engineering, University of Petroleum and Energy Studies, Energy Acres, Post Office Bidholi via Premnagar, Dehradun 248007, Uttarakhand, India

E-mail: sukanchan68@gmail.com; sukanchan92@gmail.com

ABSTRACT

The world of science and engineering today stands in the midst of deep scientific vision and vast scientific comprehension. Carbon nanotubes (CNTs) and material science, in the similar manner, stand in the crucial juncture of scientific forbearance and deep scientific ingenuity. Nanomaterials and engineered nanomaterials are the scientific innovations of today and are surpassing vast and versatile scientific frontiers. CNTs are the smart material of today's modern human civilization. In this chapter, the author deeply comprehends the utmost need, the vast scientific vision, and the scientific profundity in the application of CNTs in diverse areas of science and engineering. Nanotechnology and material science are the two opposite sides of the visionary coin in today's modern civilization. This chapter unravels the immense importance of nanomaterials and engineered nanomaterials such as CNTs in the furtherance of science and engineering today. Scientific research pursuit today stands in the crucial juncture of deep scientific vision and vast scientific ingenuity. The author, with vast scientific truth, opens a new chapter in the field of research pursuit in CNTs and delineates in deep details the future flow of scientific thoughts in the field of its applications.

9.1 INTRODUCTION

The world of science and engineering of nanotechnology is moving ahead at a rapid pace. Human scientific vision, the scientific verve, and ingenuity are the forerunners toward a newer era in the field of nanoscience and nanotechnology today. Nanomaterials and engineered nanomaterials are the smart materials and the green technology of today's scientific regeneration. Nanotechnology is today connected with every branch of science and engineering. Carbon nanotubes (CNTs) are the wonders of science and technology and are surpassing vast and versatile scientific frontiers. Technological ingenuity and verve, scientific validation, and the imminent needs of the human society will all lead a long and visionary way in the true emancipation and the true realization of nanoscience and nanotechnology in modern science. Today, modern science stands in the crucial juncture of vision and scientific fortitude. The world of science is highly challenged with vast scientific forays in the field of nuclear science and space technology. Nanoscience, nanotechnology, and CNTs are the scientific imperatives and deep scientific vision of science and engineering today. Technological validation and scientific motivation are the veritable forerunners of a new dawn in the field of engineering science and technology today. This treatise veritably opens up a new window of scientific innovation and scientific instinct in the area of CNTs and nanomaterials in years to come. Nanotechnology and application of CNTs are today linked by an unsevered umbilical cord. The success of human scientific endeavor, the vast technological verve, and the utmost needs of human society will all lead a long and visionary way in the true emancipation of nanotechnology, nanomaterials, and engineered nanomaterials.

CNTs, long, thin cylinders of carbon, were discovered in 1991 by Sumio Iijima.[1] These are large macromolecules that are unique for their size, shape, and remarkable physical properties. Human scientific regeneration after 1991 stands in the midst of deep vision and forbearance. CNTs and their applications, thus in the similar vein, need to be reenvisioned and vastly reenvisaged as science and engineering treads, a visionary path toward scientific verve and immense might. CNTs can be thought of as a sheet of graphite (a hexagonal lattice of carbon) rolled into a cylinder.[1]

These interesting structures have ushered in much excitement in recent years and a large amount of research pursuit has been dedicated to their understanding. Human scientific challenge, the vast scientific profundity, and fortitude will all lead a long and visionary way in the true understanding and true realization of nanoscience and nanotechnology today. The vision of CNTs applications are immense, far-reaching, and in the process of true realization of nanotechnology and human society.[1] Nanotubes have a broad range of electronic, thermal, and structural properties that change depending on the different kinds of nanotube (defined by their diameter, length, and chirality or twist).[1] To make things more scientifically robust, besides having a single cylindrical wall (single-walled carbon nanotubes; SWNTs), nanotubes have multiple walls (multiwalled carbon nanotubes; MWNTs)—cylinders inside the other cylinders.[1] Scientific policy vision, nanotechnology's vast scientific discernment, and the imminent needs of scientific validation in nanoscience and nanotechnology will all lead a long and effective way in the true emancipation of nanomaterials application today.

9.2 THE VISION OF THE STUDY

Human scientific ingenuity and deep scientific cognizance in the field nanotechnology and nanomaterials are changing the face of mankind and human civilization today. Technology and engineering science are highly advanced today as civilization trudges a weary path toward a newer era and a newer knowledge dimension. The vision of the study is to bring before the readers the importance of CNTs and their diverse applications. CNTs and nanomaterials have applications in every area of science and engineering. Engineered nanomaterials are the other avenues of scientific endeavor and deep scientific profundity. The world of science and engineering today stands in the midst of deep crisis and scientific introspection with the growing concerns for global climate change, global warming, and depletion of fossil fuel resources. Chemical process engineering, environmental engineering science, and petroleum engineering are the target areas of scientific revamping and scientific enigma today. Technology and engineering science need to be reenvisioned and reorganized as petroleum engineering and environmental engineering trudge a visionary and erudite path toward a newer scientific and knowledge dimension. The necessity of science and this research pursuit is immense and inspiring. The author delves deep into the scientific barriers and difficulties of CNTs applications and nanotechnology applications in the furtherance of science and engineering in modern civilization. The challenge of mitigating global climate change is immense and thought provoking. The author also deeply delineates the scientific success, the scientific ardor, and the innovations in the field of CNTs, nanomaterials, and engineered materials. Material science and the area of ecomaterials are the next-generation innovative materials today. Human scientific regeneration and vast scientific divination are the necessities of innovation and scientific farsightedness today. Today, material science and chemical process engineering are linked by an unsevered umbilical cord. In this well-researched treatise, the author deeply pronounces on the scientific success, the vast scientific profundity, and the scientific necessities of CNT applications in engineering science today. The author deeply relates the success of science and immense triumph of engineering science in the field of both nanomaterials and material science in decades to come.

9.3 SCIENTIFIC VISION AND THE SCIENTIFIC DOCTRINE OF CNT SCIENCE

Scientific vision and deep scientific profundity are the necessities of human innovations and deep scientific introspection today. Inspiration in scientific research pursuit is of highest order today as human civilization moves toward a newer eon in the field of space technology and nuclear engineering. Applications of CNTs and nanotechnology emancipation are of highest order today as mankind witnesses immense challenges in the pursuit of knowledge. Chemical engineering science and environmental engineering are today linked with the science of nanotechnology and nanomaterials with immense vision and scientific might. CNT science applications are today in the path of newer scientific regeneration and a newer scientific vision. Global environmental engineering scenario is immensely grave and shocking to the scientific domain today. The vision of scientists and engineers has immensely changed with the ushering of a new century in global human civilization today. Science and technology of nanoscience and nanotechnology need to be veritably envisioned with the passage of scientific history, scientific vision, and visionary timeframe. The engineering and technology of CNTs are today surpassing vast and versatile scientific frontiers and are in the process of newer scientific regeneration. Energy security and energy sustainability are the pillars of human scientific endeavor in engineering science in present day human civilization.[1] Sustainable development, whether it is energy, electricity, environment, or water, is the utmost need of human civilization today. This treatise unfolds the human scientific research pursuit in the field of CNTs, nanomaterials, and nanotechnology with the sole aim of furtherance of science and engineering.[1]

9.3.1 SYNTHESIS: ARC AND LASER VAPORIZATION AND HEAT TREATMENT METHODS

The worldwide excitement surrounding CNTs was originally sparked by Iijima's production of highly perfect multiwalled tubes using arc vaporization in 1991.[1] Scientific vision and scientific fortitude witnessed immense changes after that discovery. The first synthesis of single-walled nanotubes in 1993 also involved arc evaporation, this time with metal impregnated electrodes. This technique stands visionary and pathbreaking as science and engineering of CNT application and synthesis trudge forward. The original method used by Iijima to prepare nanotubes differed slightly from the Kratschmer–Huffman technique for C_{60} production, in that the graphite electrodes were held at a short distance apart during arcing, rather that kept in contact.[1] The world then witnessed a new beginning in the field of CNT synthesis.[1] A significant advancement came in July, 1992 when Thomas Ebbesen and Pulickel Ajayan, working in the same laboratory as Iijima, discovered with profound scientific conscience that increasing the pressure of helium in the arc-evaporation chamber dramatically improved the yield of nanotubes.[1] A variety of different arc-evaporation reactors have been employed for nanotube production but a stainless steel vacuum chamber with a viewing port is probably the most common type. A number of factors have been shown to be important in producing a good yield of high-quality nanotubes. Perhaps the most important is the pressure of helium in the evaporation chamber. A striking increase in the number of tubes is highly evident as the pressure is increased.[1] At pressures above 500 Torr, there is no obvious change in sample quality but there is a substantial fall in total yield.[1] Scientific regeneration, scientific vision, and the vast scientific discernment are the pillars of research endeavor in CNTs today. Another important factor in the arc-discharge method is the current, as investigated in several studies. Too high a current will result in a hard, sintered material with free nanotubes.[1] Therefore, the current should be kept as low as possible, consistent with maintaining a stable plasma. There have been a number of variations on the classic "arc-evaporation" method since the beginning of research pursuit in nanotubes in 1990s. Several scientific groups have deeply experimented with using alternatives to helium for arc evaporation.[1] Technological innovations and vast scientific validation are thus ushered in a new era of science and engineering of nanotubes. Harris[1] deeply investigated the domain of growth mechanisms of multiwalled nanotubes in the arc, vapor phase growth, liquid phase growth, solid phase growth, and production of multiwalled nanotubes by high-temperature heat treatments. Thus, the immense importance of arc and laser vaporation methods.

9.3.2 SYNTHESIS: CATALYTIC CHEMICAL VAPOR DEPOSITION AND RELATED METHODS

The synthesis of CNTs by catalysis has a number of innovative advantages over the arc and laser methods discussed earlier. In particular, catalysis (or chemical vapor deposition, CVD, as the process is often called) is much more relevant to scale-up than arc or laser evaporation and many promising processes for the large-scale catalytic synthesis of both SWNTs and MWNTs have been widely developed.[1] Thus, it is possible, using catalytic methods to grow arrays of aligned nanotubes and substrates. Such arrays are showing immense and potential promise as field emission displays. Human scientific research pursuit is thus in the midst of immense scientific acuity and deep articulation. It may also be possible to construct nanoelectronic circuits by using catalysis and chemical reaction kinetics. It is vastly believed that the main disadvantage of catalytic methods is that the nanotubes produced in this way are structurally inferior to those made by high-temperature treatment techniques.[1] Scientific research forays in the catalytic methods of nanotubes synthesis began in 1970s and concentrated mainly on post-1991 research pursuit. In the early 1970s, significant work was envisioned in the synthesis of nanotubes by Terry Baker's group in Harwell Laboratories, United Kingdom and by Tom Baird, John Fryer and co-workers at University of Glasgow, United Kingdom. Morinobu Endo carried out his doctoral studies with Agnes Oberlin in France and began working on the catalytic growth of CNTs in the early 1970s.[1] Thus ushered in a newer knowledge dimension in the future of nanotubes synthesis. The travails and vision of nanoscience and nanoengineering thus laid path toward a newer visionary era in science.[1]

9.3.3 PURIFICATION AND PROCESSING

Techniques described in previous sections do not produce pure CNTs. When multiwalled nanotubes are produced by arc evaporation, for example, they are always accompanied by nanoparticles and disordered carbon. When produced catalytically, nanotubes are contaminated with catalyst particles and support materials. It is often necessary to purify these impurities in CNTs synthesis. Harris[1] described and elucidated in deep details the need of purification in processing. The author[1] has discussed in deep details purification of MWNT. When produced catalytically, MWNT samples inevitably contain residual catalyst nanoparticles. Here comes the vast necessity of purification and processing innovations. The most successful methods involve high-temperature annealing. Rodney Andrews and his colleagues at University of Kentucky, USA propounded the purification of catalytically produced MWNTs by annealing at "graphitization" temperatures (1600–3000°C).[1] Acid treatment, often combined with heat treatments, has also been used to purify CVD-produced MWNTs. Acid treatment and oxidation, physical techniques, assessment of purity, and processing of MWNTs are the today's innovations in chemical purification of CNTs.[1]

9.3.4 THE VAST SCIENTIFIC DOCTRINE OF THE CHEMISTRY AND BIOLOGY OF NANOTUBES

The vast insights into CNTs occurred in the early 1990s. The basic aspect of nanotube chemistry was that they are reactive at the tips. Harris[1] discussed with deep and cogent insight covalent functionalization, functionalization of sidewalls, noncovalent functionalization, and biological functionalization. The other important areas covered are toxicities of CNTs. The scientific triumph and the benefits of the advances in biological functionalization can be today felt across the whole field of nanotube science. Functionalization is often used in the preparation of CNT composites in order to improve the bonding between the nanotubes and the matrix material. Chemistry and biology of CNTs are in the path of newer scientific rejuvenation today. This entire chapter depicts poignantly the contribution of chemistry and biology in the synthesis and functionalization of nanotubes.

9.4 RECENT SCIENTIFIC ENDEAVOR IN THE FIELD OF CNTs SCIENCE

Scientific ingenuity and vast scientific vision are the imminent needs of scientific research pursuit today. CNTs application is changing the face of human civilization today. Research forays in the field of CNTs are challenging the vast scientific firmament of nanoscience and nanotechnology today. Scientific validation and technological innovation are the pillars of research and development initiatives in CNTs, nanotechnology, material science, and diverse areas of science and engineering.

Zhou et al.[2] discussed with immense lucidity the material science of CNT and its fabrication, integration, and properties of macroscopic structures. This is a watershed text in the field of CNTs and nanotechnology. In this paper, the author summarizes some of the recent studies on the material properties of the CNTs. The main focus is on SWNTs.[2] The authors have described and discussed experiments on synthesis of SWNTs with controlled molecular structures and assembly of functional macroscopic structures. Science of nanotechnology and nanoengineering are today in the path of newer scientific regeneration. The scientific travails, the deep scientific profundity, and the vast scientific potential of nanoengineering are the forerunners toward a newer visionary era in global research and development initiatives. The authors in this treatise have also discussed in deep details electronic field emission properties of macroscopic CNT cathodes.[2] The last decade has been an exciting time for the vast and versatile field of carbon materials. The discoveries of fullerenes and CNTs have attracted worldwide attention and imagination of scientists and engineers. Technology and engineering science in this decade is slowly under a scientific metamorphosis and vast scientific regeneration. Tremendous of research work have been devoted to the studies and the academic rigor of these new carbon allotropes. In this paper, the authors have discussed with lucid and cogent insight fabrication and processing of single-walled nanotubes, purification, opening and closing the SWNTs tips, integration, electrophoretic deposition, electron field emission properties, and the vast and holistic field of nanomaterials technology.[2] From the materials point of view, the nature of CNTs differ significantly from that of the fullerenes.[2] C_{60}, as a molecule, can be made and fabricated with precise and uniform composition and a molecular structure that is essentially defect-free. Functional C_{60} solids such as single crystals and epitaxial films can be readily fabricated. Technology and engineering science of carbon nanomaterials thus is ushering in a newer eon. The case of

CNTs in fabrication is more complicated. A CNT is not a well-defined molecule. It has different structures, weight, dimension, and as a result, different properties. Fabrication of homogeneous materials requires control of not only the building blocks but also the higher level architecture by which the "molecules" can assemble. This treatise gives a vast scientific understanding and a deep scientific discerning of the domain of fabrication of CNTs. Vast scientific vision, immense scientific conscience, and the scientific doctrine of nanotechnology are the pivots of this paper. Nanotechnology is linked with every branch of science and engineering and is in the path of newer scientific revamping. CNTs have immense variation in the molecular structure.[2] The vast variation in molecular structure provides an additional material design parameter. The polydispersity also leads to nonuniform and unpredictable properties. In this treatise, the authors vastly proclaim some recent works and research and development forays carried out in the authors' group on the materials science aspects of CNTs. This well-researched treatise opened up new challenges and new scientific ingenuity in the field of CNTs and their vast and varied applications.[2]

Popov[3] discussed and described with deep scientific conscience and lucidity CNTs and their properties and applications. The scientific prowess of research and development initiatives in CNTs and nanotechnology are challenging the vast scientific firmament today. CNTs are unique tubular structures of nanometer diameter and large length/diameter ratios.[3] The nanotubes may consist of up to tens and hundreds of concentric shells of carbons with adjacent shells separation of approximately 0.34 nm. Vast scientific vision and scientific comprehension are the hallmarks of this paper.[3] The carbon network of the shells is closely related to the honeycomb arrangement of the carbon atoms in the graphite sheet. CNTs have amazing mechanical and electronic properties.[3] They have a quasi one-dimensional structure and the graphite-like arrangement of the carbon atoms in the shell. Thus, the CNTs have high Young's modulus and tensile strength which makes them preferable for composite materials with vastly improved mechanical properties. Scientific vision, deep scientific profundity, and the vast domain of scientific validation are the veritable pillars of this well-researched treatise. The areas of research pursuit elucidated in this paper are synthesis of CNTs, arc discharge, laser ablation, catalytic growth, and growth mechanisms of multiwalled and single-walled nanotubes. The other covered areas are optical properties and electrical transport in perfect nanotubes. An instinctive discussion on vibrational properties of CNTs is given in this paper. The vast challenges of science and engineering, the immense need of scientific validation, and the scientific profundity in the field of nanoscience and nanoengineering will surely lead a long and visionary way in the true emancipation of science and technology today.[3] Nanotechnology is the immediate need of human society and human scientific progress. This treatise with immense scientific conscience and scientific adjudication proclaims this immense need of mankind today.[3]

Harris[1] discussed with deep and cogent insight synthesis, properties, and applications of CNT science. In the introduction, the author dealt with buckminsterfullerene, fullerene-related CNTs, SWNT and double-walled CNTs, and catalytically produced CNTs. Human scientific regeneration and scientific vision are the veritable pillars of this well-researched scientific treatise.[1] The author in this book deeply delved into the synthesis; arc and laser vaporization; heat treatment methods; catalytic CVD; purification and processing; physical properties such as electronic, mechanical, optical, and thermal; and chemistry and biology of CNTs. The other areas covered are CNT composites, heterogeneous nanotubes, probes, and sensors. Nanotechnology is the need of human scientific rigor today. Carbon in its miscellaneous form has been used in art and technology since prehistoric times. It is the forerunner to the present scientific endeavor in nanoengineering. The history of carbon science has witnessed illustrious personalities and immense scientific emancipation. By the early 1980s, however, carbon science was widely considered to be an extremely matured discipline. The situation is different now due to the synthesis in 1985 by Harry Kroto of the University of Surrey, UK and Richard Smalley of Rice University, USA and their colleagues.[1] The world was then ushered in a new era of scientific forbearance and scientific fortitude. Richard Buckminster Fuller of USA was one of the earliest discoverers and a scientific proponent of CNTs and fullerenes. The Kroto–Smalley experiments were miniscule.[1] So a more robust experiment was done by Wolfgang Kratschmer of the Max Planck Institute in Heidelberg, Germany and Donald Huffman of the University of Arizona, USA. They used a simple carbon arc to vaporize graphite, again in an atmosphere of helium, and collected the soot which settled on the walls of the vessel. Human scientific regeneration and vision witnessed an immense challenge as nanotechnology entered in a newer eon. Sumio Iijima, an electron microscopist then working at the NEC Laboratories in Japan extended the prevailing research work to a much larger research and development initiative. Thus, the world was in the threshold of a newer visionary eon. In this book, the author deeply comprehends the immense necessity and the vast scientific potential behind nanoengineering and nanotechnology.

Khan et al.[4] discussed with deep scientific conscience CNTs and their possible applications. CNTs are the closed tabular structures consisting of nested cylindrical graphitic layers capped by fullerene-like ends with a hollow internal cavity, which was first discovered by Iijima in 1991.[4] It consists of either one cylindrical graphene sheet, that is, SWNT or of several nested cylinders with an interlayer spacing of 0.34–0.36 nm, that is, MWNT. The lengths of SWNTs and MWNTs are usually over 1 μm and diameters range from approximately 1 nm (SWNTs) to approximately 50 nm (for MWNTs).[4] CNTs show excellent electronic and mechanical properties. This well-researched article gives a wider view of the current state of research on CNTs. Technological motivation, scientific validation, and vast scientific doctrine are the forerunners of scientific research endeavor today. In this review, the authors have focused on different synthesis routes for CNTs growth, used during the last 12 years, and possible future applications of CNTs especially in fuel cell applications.[4] Carbon is an element with unique properties and is the lightest among all the elements of Group IV of the periodic table. The science of the CNTs is visionary as well as pathbreaking. It differs in many ways from other elements such as Si, Ge, Sn, and Pb of the same group.[4] The vast vision of science of nanotechnology, the challenges and scientific travails of nanoscience applications, and the vast needs of human society will all lead a long and visionary way in the true emancipation of global energy and environmental sustainability. Carbon in the condensed phase has a hexagonal ground state graphite with sp^2 bonding and is a highly anisotropic, nearly two-dimensional semimetal. The most instrumental and vibrant recent discovery was the identification of C_{60} as a molecule having the shape of a regular truncated icosahedron by the eminent scientist Dr. Kroto of United Kingdom, which ushered in an enigmatic research and development initiative in nanotechnology. In the mid-1980s, the same group of erudite scientists discovered the chemical compound of fullerenes.[4] Technological and scientific rigor, the serendipity of scientific endeavor, and the vast necessities of human civilization will surely lead a long and effective way in the true realization of the world of nanotechnology. This paper presents a wide overview of the science of CNTs, its vast and versatile applications, and the challenges and the vision in futuristic vision in CNTs application.[4] The authors discussed with deep scientific conscience structure of CNTs, MWNTs and SWNTs, synthesis of CNTs, and the current status of research in the field of CNTs and other nanomaterials. The success of science and technology and the immense potential of nanomaterials and engineered nanomaterials are the veritable forerunners toward a newer emancipation and a newer scientific vigor in the field of nanotechnology.[4]

Khare et al.[5] discussed with immense scientific farsightedness in a well-researched review CNT-based nanocomposites. Carbon nanofibers and nanotubes are immensely promising to revolutionize several fields in material science and are the major constituent of the domain of nanotechnology.[5] Today the vision and the challenges of the science are immense and pathbreaking. Human scientific and academic rigor in the field of chemical process engineering and nanotechnology are in the path of newer scientific regeneration. In this paper, the author deeply comprehends on the further market development and its dependence on material availability at reasonable prices. Nanotubes have diverse applications in a wide range of technical areas such as aerospace, energy, automobile, medicine, and the vast chemical industry. Nanocomposites have applications as gas adsorbents, templates, actuators, composite reinforcements, catalyst supports, probes, chemical sensors, nanopipes, nanoreactors, etc. The vast and versatile vision of nanocomposites applications, the futuristic vision of nanoengineering, and the scientific potential of nanoscience applications are the veritable forerunners toward a newer scientific regeneration in the field of nanotechnology. Recent scientific research and development initiatives in CNTs nanocomposites are extensively reviewed.[5] The world of applied chemistry, the interfacial bonding properties, mechanical performance, and electrical properties of polymer and ceramic are reviewed in instinctive details.[5]

Hone et al.[6] discussed with deep and cogent insight thermal properties of CNTs and nanotube-based materials. The thermal properties of CNTs are directly related to their unique structure and small size. Because of these properties, nanotubes may prove to be an ideal material for the study of low-dimensional phonon physics and for thermal management, both on micro- as well as macroscale.[6] Scientific validation, deep technological motivation, and scientific stewardship are the torchbearers toward a newer era in the field of nanoscience today. Scientific vision and deep scientific ingenuity are the pillars of today's research pursuit in nanotechnology. In this paper, the authors also discussed the thermal properties of nanotubes by measuring the specific heat and thermal conductivity of bulk SWNT samples.[6] In addition, the authors also synthesized nanotube-based composite materials and measured their thermal conductivity. The measured specific heat of single-walled nanotubes differs from that of both 2D graphene and 3D graphite, especially at low temperatures, where 1D quantization of the phonon band-structure is deeply observed.[6] Scientific and academic rigor in the field of nanotechnology are at its helm as human civilization and human scientific progress trudges forward. The authors have touched upon the importance of specific heat and thermal conductivity in the scientific progress in CNTs.[6]

Shanov et al.[7] reviewed synthesis and characterization of CNT materials. Today, the world of science and engineering stands in the midst of deep scientific and technological vision and vast scientific prudence. This paper provides a survey of synthesis, purification, and characterization of CNTs. Recent scientific research pursuit on growth and study of CNT arrays are presented and discussed in minute details. This watershed text opens up new vistas and newer avenues in research and development initiatives in CNTs. Human scientific stewardship, scientific ingenuity, and the challenges of nanoengineering are changing the vast scientific landscape of nanotechnology. The number of publications and patents on CNT synthesis is rapidly increasing. Still there are many visions and many goals in the pursuit of science in CNTs today. The world of science deeply remains challenged and devastated with the growing concerns for energy and environmental sustainability. Here come the imminent and immediate concerns for CNT research and global nanotechnology research and development initiatives. One of the major challenges in the field of nanotechnology is the production of large-scale and low-cost SWNT and MWNT.[7] Another field of immense interest comprises the pursuit of controlled CNT growth in terms of selective deposition, orientation, and preselected metallic or semiconducting properties. Our vast understanding of CNTs is slowly emerging and needs to be envisioned and envisaged with the passage of scientific history and the visionary timeframe. CNTs growth mechanism has been rapidly evolving but more consideration is still envisioned to investigate the variety of the observed growth features and the vast visionary world of experimental results. There are three principal tools to produce high-quality SWCNT, laser ablation, electric arc discharge, and CVD.[7] CVD has become a visionary technique in the field of material science and nanotechnology. This well-researched treatise profoundly depicts the scientific succor, the scientific foresight, and the scientific serendipity in nanotubes applications in human society.[7,11–13]

9.5 RECENT SCIENTIFIC RESEARCH PURSUIT IN THE FIELD OF MATERIAL SCIENCE

Material science and nanotechnology are two opposite sides of the visionary coin today. Nanoengineering is a revolutionary branch of scientific research pursuit today. Scientific and technological validation is the veritable pillar of nanotechnology and material science today. Material science and nanomaterials are today linked by an umbilical cord. The world of science and engineering stands challenged as material science enters a newer eon in this scientific century. Smart materials, engineered nanomaterials, and nanomaterials are the scientific vision and a deeper scientific revolution of tomorrow.

Wood[8] discussed with deep and cogent insight the top 10 advances of material science. The author has vastly assembled the top 10 advances in material science over the last 50 years. Scientific prudence, deep scientific foresight, and scientific wisdom stand as pillars of human scientific progress today.[8] Today, the domain of material science is in the path of newer scientific rejuvenation and also in the path of a newer beginning. The author in this paper deeply discusses with profound impact and scientific conscience the defining discoveries, moments of inspiration, and the shifts in understanding that have veritably shaped the dynamic field of material science. Some of the visionary global research initiatives are the field of semiconductors, scanning probe microscopes, giant magnetoresistive effect, semiconductor lasers and LEDs, carbon fiber reinforced plastics, materials for lithium batteries, CNTs, soft lithography, and metamaterials.[8] Scientific vision and vast scientific farsightedness are the pivots of this well-researched treatise. The results of the research pursuit in the field of nanotechnology are thought provoking and awe inspiring. Human factor engineering and the vast world of nanoengineering are the other avenues of this well-researched treatise. The authors have also discussed the international scientific roadmap for semiconductors. Human civilization's immense scientific prowess, man's vast scientific girth and determination, and the futuristic vision of nanotechnology and material science are the forerunners toward a newer eon in the field of science and engineering.[8]

National Institute of Materials Science, Japan Report[9] discussed with immense lucidity the vision of materials research in 2020. Generally speaking, it takes a long time before a new material or substances can be put into practical use, from the basic study and fundamental R&D until final development. The domain of material science today stands in the crucial juncture of vision and immense scientific acuity. Stages redefine the material processing.[9] The report discussed with deep and cogent insight the vision for entire science and technology; the vision for materials research in Japan and the world; and research fields to be prioritized by 2020, the world of nanomeasurement technology, nanocreation technology, nanosimulation technology, nanobiotechnology, and post-nanotechnology initiative.[9] Material science is a marvel of science today. The needs of material science, the vast technological profundity, and the needs of engineering science to human society are the forerunners toward a newer world of nanoscience and nanoengineering today.[9]

Mueller et al.[10] discussed in deep details recent progress and emerging applications in the field of machine learning in materials science. Scientific vision is of utmost necessity in the path toward true emancipation of material science and nanotechnology. Data-to-knowledge ideas are beginning to show immense potential within material science.[10] The innovative concept of data-driven methods forms the core of the United States' Materials Genome Initiative. Scientific paradigm in the field of data-driven technologies is witnessing immense overhauling today. To significantly accelerate the vast pace of scientific innovation, efficient and effective methods to (1) generate, (2) manage, and (3) utilize relevant information are immensely necessary. The last of these visionary tasks can be accomplished in a systematic way through an approach known as "machine learning," a branch of artificial intelligence. The authors with deep scientific lucidity explain and investigate the world of machine learning and artificial intelligence.[10]

Human scientific endeavor and human scientific genre are moving from one visionary paradigm toward another. Material science and nanotechnology are today in the path of newer scientific rejuvenation. Engineering science and technology have few answers toward the scientific intricacies and the scientific travails of nanotechnology. This entire treatise opens up newer scientific understanding and wisdom in the field of nanotechnology.

9.6 CNTs AND CHEMICAL ENGINEERING SCIENCE

Chemical engineering science and CNTs are the pathbreaking avenues of engineering science today. Mankind's immense scientific prowess, the vast world of scientific acuity, and the farsightedness of technology will all lead a long and visionary way in the true emancipation and the true realization of nanotechnology and chemical engineering science in modern day human civilization. CNTs are the wonders of science today. Chemical engineering science today stands in the midst of deep and vast scientific introspection. Environmental engineering science is an area which needs to be reenvisioned and reorganized as human civilization marches forward toward a newer era of scientific might and vision.[11–13]

9.7 CNTs AND THE VAST WORLD OF ENVIRONMENTAL PROTECTION

CNTs and nanotechnology have applications in almost every branch of human scientific endeavor today. The world today is drifting toward a monstrous crisis of climate change and depletion of fossil fuel resources. The domain of environmental protection and environmental engineering science today is in the path of newer scientific rejuvenation as science and engineering surges forward toward a newer era. Nanotechnology is a wonder of science and is a huge colossus with a definite and purposeful vision of its own. Material science and nanotechnology are two opposite sides of the visionary scientific coin. This treatise opens up new chapters of scientific innovation and scientific instinct in the domain of both material science and nanotechnology. Global warming, global climate change, and the vast world of water purification encompasses environmental protection today. The applications of CNTs in environmental protection are vast, versatile, and pathbreaking. Frequent environmental catastrophes and depletion of fossil fuel resources are causes of immense scientific concerns in modern human civilization. The author reiterates the importance of technological validation and scientific motivation in the field of nanomaterials application in almost every area of science and engineering.

9.8 ENERGY SUSTAINABILITY AND NANOTECHNOLOGY

Human scientific progress and the vast academic rigor in the field of nanotechnology are advancing at a rapid pace. Scientific fortitude and scientific wisdom are the necessities of scientific innovation and invention today. Today, the world of science stands in the midst of deep scientific travails, scientific barriers, and unending scientific calamities. Global climate change, global warming, and frequent environmental catastrophes are changing the face of science and engineering. Water purification, drinking water treatment, and industrial wastewater treatment are the need of the hour as human civilization and human scientific progress move forward. Man's immense scientific girth, mankind's scientific grit and determination, and the travails of science will all lead a long and visionary way in the true realization of energy and environmental sustainability. Energy sustainability, environmental sustainability, and nanotechnology are the veritable needs of human civilization and human scientific progress today. The world stands in the crucial juncture of scientific introspection and deep scientific comprehension. Global environmental crisis and environmental engineering disasters have challenged the vast scientific landscape and changing the course of human civilization and human scientific progress today. Thus the need of the hour is sustainable development. Heavy metal and arsenic groundwater poisoning are the causes of immense concern throughout the scientific fraternity today. Technological innovations, scientific rebuilding, and deep scientific discernment are the necessary tools of research pursuit today. In this paper, the author repeatedly proclaims and pronounces the ultimate need of human civilization, that is, true realization of sustainability. Nanoscience and nanotechnology are the revolutionary branches of scientific pursuit today. Man's immense scientific prowess and ingenuity, the world of scientific validation, and the necessities of human society are the forerunners toward a newer visionary era of sustainable development.[11–13]

9.9 WATER PURIFICATION, ENVIRONMENTAL SUSTAINABILITY, AND THE ROADS TO FUTURE

Mankind today stands in the crucial juxtaposition of introspection, deep comprehension, and vision. Scientific challenges are immense in this century of science. Drinking water treatment and industrial wastewater treatment are the veritable challenges of human scientific progress today. Human civilization stands devastated as global water crisis and global climate change challenge the veritable scientific firmament. Thus, the need and the vision of environmental sustainability. The roads to the distant future in science and engineering are wide and bright. Science of sustainability today is a huge colossus with a definite and purposeful vision of its own. Global water crisis is in a state of immense disaster. Arsenic and heavy metal groundwater contamination are leading the civilization to a dismal state. South Asia, particularly India and Bangladesh, are in the true grip of immense crisis and an unending environmental catastrophe. Thus, the need of research forays in water science and a global research and development initiative in groundwater remediation. In this paper, the author deeply discusses the need of nanotechnology, material science, and water science in the furtherance of science and engineering globally. Today, technology has few answers to the growing concerns for environment and water. The vision and the challenges of water research are pathbreaking and painstaking as science trudges forward. Global vision in the field of nanotechnology needs to be revamped and envisioned as human civilization witnesses the challenges of human progress. This research work opens a newer window in the field of nanotechnology and water science with the sole aim of true emancipation of nanomaterials.[11–13]

9.10 FUTURE SCIENTIFIC RECOMMENDATIONS AND FUTURE FLOW OF THOUGHTS

Science and engineering are the necessities of human civilization and human scientific progress today. Future recommendations and future flow of scientific thoughts should be targeted toward more emancipation and realization in the field of nanomaterials, engineered nanomaterials, and CNTs. Science and technology today are highly retrogressive as regards application of nanotechnology and the ever-growing concerns of the future of petroleum engineering science, environmental engineering science, and energy and environmental sustainability. Sustainable development, whether it is economic, social, environmental, or energy, is the utmost need of the hour in modern science today. Water process engineering, the vast world of water purification, drinking water treatment, and industrial wastewater treatment are today immensely challenging the vast scientific firmament. Thus the immediate and the future need of environmental sustainability and the challenging world of environmental protection. Nanotechnology, CNTs, and nanomaterials have equal contribution in both fields of environmental engineering science and petroleum engineering. Technological profundity, scientific verve, and deep determination are the necessities of innovation and vast scientific prognosis today. Today nanotechnology is immensely replete with scientific and academic rigor. This chapter targets the imminent needs of scientific vision and scientific farsightedness in the field of nanomaterials and engineered nanomaterials. Frequent global environmental disasters, the issues of global climate change and global warming, and the depletion of fossil fuel resources are changing the face of mankind today. Technology and engineering science are highly challenged today and need to be revitalized as regards application of nanotechnology to human society. Future recommendations, scientific conscience, and flow of thoughts thus should be toward greater emancipation of nanoscience and nanotechnology in present day human civilization. This chapter will surely and veritably open new windows of challenge and verve in the areas of nanomaterials and engineered nanomaterials in decades to come. The author deeply pronounces and justifies the need of nanotechnology in the furtherance of human civilization.

9.11 CONCLUSION AND SCIENTIFIC PERSPECTIVES

Scientific research pursuit in the modern civilization is highly challenged today as the civilization witnesses immense scientific travails and vast barriers. Scientific perspectives in this present century are being vastly envisioned than in the previous century. Technology and engineering science of nanotechnology, nanomaterials, and engineered nanomaterials today stands in the visionary platform of scientific might, scientific verve, and deep scientific rejuvenation. Global environmental and energy policies in this century are frequently revised and reframed as science and engineering surges forward. Today, the frontiers of science are vastly reenvisaged as mankind moves from one scientific dilemma toward another. The future of science is vast and versatile today as far as the energy and environmental sustainability are concerned. In this entire chapter, the author rigorously points toward the need of CNT and nanotechnology applications in human scientific progress. Green chemistry and green engineering are the other sides of the visionary coin in the scientific research pursuit toward nanoscience and nanotechnology today. Scientific profundity, scientific ingenuity, and deep scientific determination and girth are the forerunners toward a newer eon in the field of nanomaterials, nanotechnology, chemical process engineering, and environmental engineering science. Science and technology are two huge colossi with a definite and purposeful vision and goal of their own. The author in this treatise repeatedly pronounces the importance, the vast scientific imperative, and the scientific necessity of the domain of CNTs and engineered nanomaterials in building and revamping the concept of green engineering. Technology and engineering science of nanoscience are in the similar vein not immensely advanced and need to be reorganized and revamped as human civilization and mankind's vast scientific ingenuity surges forward. This chapter unfolds the answers toward the scientific travails and scientific barriers in CNT and nanomaterials applications in human scientific progress. The world of challenges and the scientific grit and determination will surely open up new doors of innovation and scientific instinct in the areas of nanomaterials and engineered nanomaterials in decades to come.

ACKNOWLEDGMENT

The author with deep respect acknowledges the contribution of his father, late Mr. Subimal Palit, an eminent textile engineer from India who taught the author the rudiments of chemical engineering.

KEYWORDS

- **carbon**
- **material**
- **science**
- **vision**
- **nanotechnology**

REFERENCES

1. Harris, P. J. F. *Carbon Nanotube Science: Synthesis, Properties and Application*; Cambridge University Press: United Kingdom, 2009.
2. Zhou, O.; Shimoda, H.; Gao, B.; Oh, S.; Fleming, S.; Yue. G. Materials Science of Carbon Nanotubes: Fabrication, Integration and Properties of Macroscopic Structures of Carbon Nanotubes. *Account. Chem. Res.* **2002,** *35* (12), 1045–1053.
3. Popov, V. N. Carbon Nanotubes: Properties and Applications. *Rep. Rev. J.* **2003,** *R 43*, 61–102.
4. Khan, Z. H.; Husain, M. Carbon Nanotube and Its Possible Applications. *Indian J. Eng. Mater. Sci.* **2005,** *12*, 529–551.
5. Khare, R.; Bose, S. Carbon Nanotube Based Composites: A Review. *J. Miner. Mater. Charact. Eng.* **2005,** *4* (1), 31–46.
6. Hone, J.; Llaguno. M. C.; Biercuk, M. J.; Johnson, A. T.; Batlogg, B.; Benes, J.; Fischer, J. E. Thermal Properties of Carbon Nanotubes and Nanotube Based Materials. *Appl. Phys. A* **2002,** *74*, 339–343.
7. Shanov, V.; Yun, Y-H.; Schulz, M. J. Synthesis and Characterization of Carbon Nanotube Materials (Review). *J. Univ. Chem. Technol. Metall.* **2006,** *41* (4), 377–390.
8. Wood, J. The Top Ten Advances in Material Science. *Mater. Today* **2008,** *11* (1–2), 40–45.
9. National Institute of Materials Science Report. *A vision of Materials Science in the Year 2020* (Excerpts from the 2020 NIMS Policy Paper), Japan, 2007.
10. Mueller, T.; Kusne, A. G.; Ramprasad, R. Machine Learning in Material Science: Recent Progress and Emerging Applications. In *Reviews in Computational Chemistry*, 1st ed.; Parrill, A. L., Lipkowitz, K. B., Eds.; John Wiley and Sons: USA, 2016; Vol. 29.

11. Palit, S. Nanofiltration and Ultrafiltration—The Next Generation Environmental Engineering Tool and A Vision for the Future. *Int. J. Chem. Tech. Res.* **2016,** *9* (5), 848–856.
12. Palit, S. Filtration: Frontiers of the Engineering and Science of Nanofiltration—A Far-reaching Review; CRC Concise Encyclopedia of Nanotechnology; Ubaldo O.-M., Kharissova, O. V., Kharisov. B. I., Eds.; Taylor and Francis: Boca Raton, FL, USA, 2016; pp 205–214.
13. Palit, S. Advanced Oxidation Processes, Nanofiltration, and Application of Bubble Column Reactor. In *Nanomaterials for Environmental Protection*; Kharisov, B. I., Kharissova, O. V., Rasika Dias, H. V., Eds.; Wiley: USA, 2015; pp 207–215.

IMPORTANT WEBSITES FOR REFERENCE

https://en.wikipedia.org/wiki/Carbon_nanotube
http://www.understandingnano.com/what-are-carbon-nanotubes.html
https://www.azonano.com/article.aspx?ArticleID=1381
https://www.sciencedaily.com/terms/carbon_nanotube.htm
https://www.ncbi.nlm.nih.gov/pmc/articles/PMC4141964/
http://science.sciencemag.org/content/339/6119/535.full
https://www.nanowerk.com/nanotechnology/introduction/introduction_to_nanotechnology_22.php
https://science.howstuffworks.com/nanotechnology2.htm
https://www.intechopen.com/books/carbon-nanotubes-polymer-nanocomposites/functionalization-of-carbon-nanotubes
https://www.newscientist.com/article/2093356-carbon-nanotubes-too-weak-to-get-a-space-elevator-off-the-ground/
https://www.sciencedirect.com/science/article/pii/S1878535210001747
http://nano-c.com/technology-platform/what-is-a-nanotube/

CHAPTER 10

CARBON NANOTUBES: PROPERTIES AND APPLICATIONS

SONIA KHANNA*

Department of Chemistry, School of Basic Science and Research, Sharda University, Greater Noida, India

**E-mail: sonia.khanna@sharda.ac.in*

ABSTRACT

Carbon nanotubes are one of the most sought topics in the field of nanotechnology. Carbon nanotubes are the cylindrical carbon molecules with novel properties of extraordinary strength, flexibility, sensing, conductance, etc. They have the potential to be used in electronics due to their unequalled electronic properties, in medical and pharmacological fields due to their unique surface area and stiffness. They have been observed in various forms, each having its characteristic behavior. Different methods of their preparations such as arc discharge, laser ablation, and chemical vapor deposition to produce nanotubes in sizeable amounts are also discussed. They have an added advantage of being a potential device for drug delivery. The properties and applications of nanotubes are also discussed in this chapter.

10.1 INTRODUCTION

Carbon nanotubes (CNTs) are the basis of nanotechnology. Carbon with an atomic number of 6 plays a pivotal role in nanotechnology. They were discovered by Iijima in 1991[1] accidentally while studying the surface of graphite electrode used in electric arc discharge (AD). This accidental observation laid a foundation of exciting field of nanotechnology and started a new direction in the carbon research. They have been researched and explored in various fields because of their nanometric dimensions.

A CNT is a hexagonal array of carbon atoms rolled up into a long, thin, hollow cylinder[2] and are known for their size, shape, and remarkable physical properties. Tremendous research has been carried out to modify the behavior of nanotubes in terms of their application for different applications. Processes, such as doping and functionalizing nanotubes result in altogether change in behavior of properties of nanotubes. The insoluble nature of nanomaterials has also been addressed by modifying the surface of nanotubes by modifying the nanotubes for their application in biological field. All these studies are discussed in this chapter.

10.2 STRUCTURE OF NANOTUBES

A CNT is a tube-shaped material, made of carbon, and has diameter on the nanometer scale. Two types of CNTs are observed—single-walled nanotubes (SWNT) and multiwalled nanotubes (MWNT), each having different characteristics and properties.

SWNT: SWNT can be visualized as rolled version of graphene sheet made up of sp^2 hybridized sheet of carbon atoms. The SWNT generally contains 10 atoms around the circumference and is one-atom thick and have one-dimensional structure. They have diameter close to 1 nm and tube length may vary millions of times longer. They can be formed in three different forms—armchair, zigzag, and chiral on the basis of rolling of graphene sheet in to a seamless cylinder (Fig. 10.1). Each form has a special effect on electrical property of nanotube.

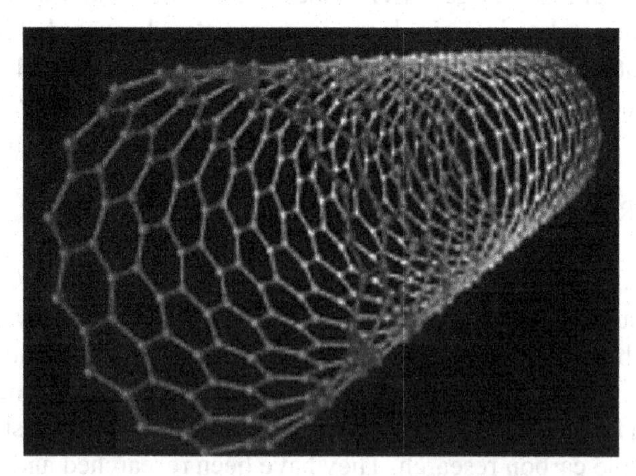

FIGURE 10.1 **(See color insert.)** Structure of single-walled nanotube (SWNT).

MWNT: The discovery of C60 motivated the researchers to search for other carbon compounds containing curved graphenes. This led to the discovery of MWNTs made of concentric cylinders of rolled-up graphene sheets, capped with semifullerenes. They are large molecules with many SWNT stacked inside the other. The length of a tube is in the range of a few μm, the diameter was 10–20 nm (Fig. 10.2). As the size increases, these structures exhibit properties between fullerenes and graphite.

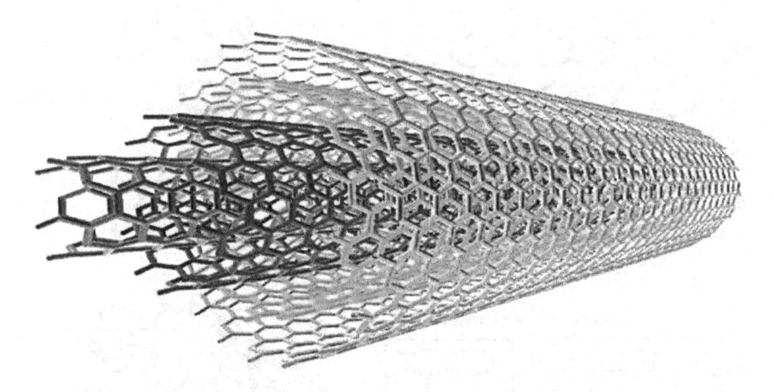

FIGURE 10.2 (See color insert.) Structure of multiwalled nanotube.

Two models can be used to describe the structure of MWNTs (Fig. 10.3) depending on the manner of rolling of nanotubes during its formation. In Russian doll model, sheets of graphite are arranged in concentric cylinders within a larger SWNT. In the parchment model, single sheet of graphite is rolled in around itself, resembling a scroll of parchment or rolled newspaper. The interlayer distance in MWNT is close to distance between graphene layers, approximately 3.3 Å. They have greater tensile strength than SWNT.

10.3 SYNTHESIS OF NANOTUBES

In light of increasing demand of high-quality and mass production of nanotubes, various research studies are going on. A variety of techniques have been developed to synthesize CNTs. There are different preparation methods for CNTs—AD, laser ablation, chemical vapor deposition (CVD) as well as some of the more recent methods working with high pressure of the carbon monoxide or some unique catalytic mixture. Carbon nanostructures such as fullerenes, graphene, and nanotubes are of great interest for the current research as well as for future industrial applications. The reason for this is that the band gap of single-walled CNTs can vary from 0 to about 2 eV and hence their electrical conductivity can be the one of a metal or the semiconductor. These methods are discussed below.

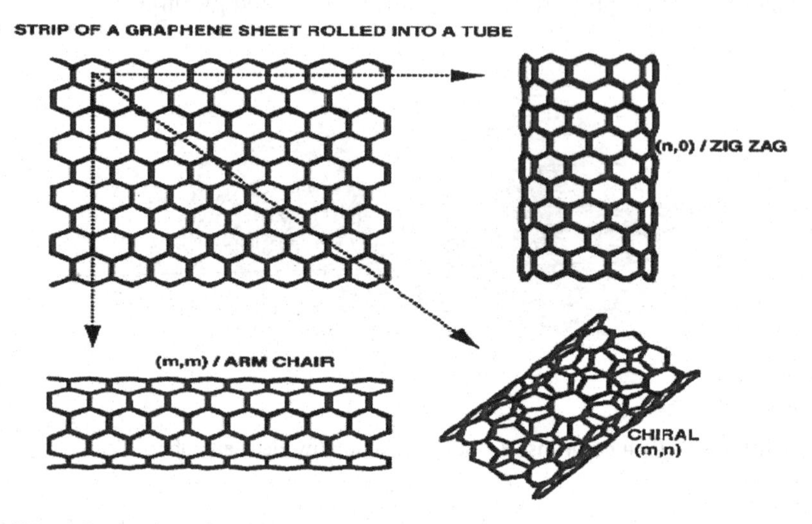

FIGURE 10.3 Models of multiwalled nanotube arrangement.

10.3.1 ARC DISCHARGE

It was one of the first methods to synthesize SWNT in large amounts. In 1990, Kratschmer and Huffman developed an AD technique and reported the synthesis of new form of solid, consisting of somewhat disordered hexagonal packing of soccer ball-shaped C60 molecule. It produces the best quality nanotubes.[3] During AD method, a current of about 50 A is passed between two graphite rods placed in an enclosure filled with some inert gas (like helium or argon) at low pressure (between 50 and 700 mbar). The carbon rods act as electrodes which are kept at different potentials. The anode is moved close to the cathode until an arc appears and the electrodes are kept at the distance of 1 mm for the whole duration of the process that takes about 1 min. After the depressurization and cooling of the chamber, the nanotubes together with the by-products can be collected. Most nanotubes deposit on the cathode. SWNT are produced when Co and Ni or some other metal is added to the anode (Fig. 10.4). Different dimensions of nanotubes can be generated depending on mixture of helium and argon. These mixtures have different diffusion coefficients and thermal conductivities which affect the speed of interaction between carbon and catalyst. The single layer of nanotubes nucleates and grows on metal particles at different rates highlighting the effect of temperature densities of carbon and catalyst.[4] Bethune et al. synthesized SWNTs by covaporizing carbon and cobalt in an arc generator. The tubes had about 1.2 nm diameter. The tubes formed a web-like deposit woven through the fullerene-containing soot, giving it a rubbery texture.[5]

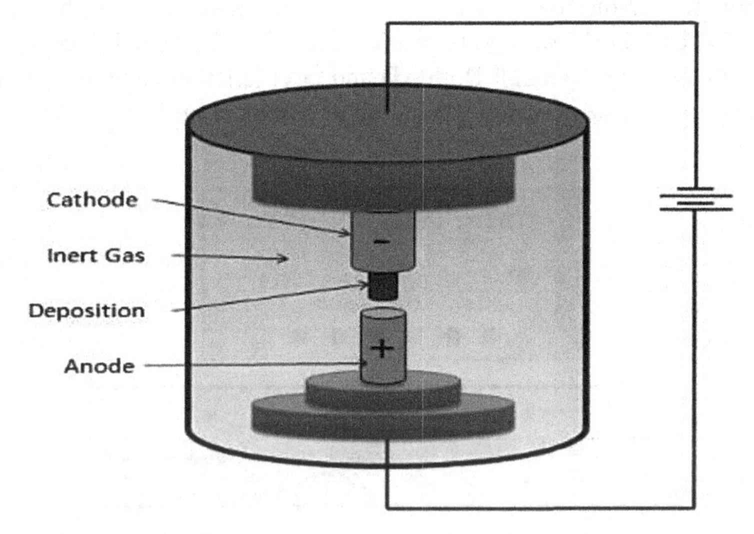

FIGURE 10.4 Arc discharge method.

One can tune the process to produce primarily multiwalled CNTs using two standard graphite electrodes to a process that produces single-walled CNTs using metal catalysts such as Ni, Fe, Mo, or Co doped into the electrode. These metals catalyze the breakdown of gaseous molecule into carbon and then tube starts to grow with a metal particle at the tip.[6,7] AD is the most straightforward approach to synthesize CNTs, but its application as a large-scale production technique suffers due to the moderate yield of CNTs. Ebbensen reported large scale of MWNT by variant of standard AD technique.[8,9]

10.3.2 LASER ABLATION

In 1995, Smalley and his group used for the first time laser ablation to grow high-quality nanotubes. Intense laser pulses ablate a carbon target which is placed in a tube furnace heated to 1200°C. During the process, some inert gas such as helium or argon flows through the chamber to carry the grown nanotubes to the copper collector. After the cooling of the chamber, the nanotubes and the by-products, such as fullerenes and amorphous carbon overcoating on the sidewalls of nanotubes can be collected (Fig. 10.5). The use of pure carbon leads for both methods to the synthesis of MWNTs and the addition of a catalyst such as iron, yttrium, sulfur, nickel, and molybdenum leads to the formation of the SWNTs. The SWNTs form bundle together by van der Waals forces. There are reports of production of SWNT using continuous wave of CO_2 laser system.[10] Both AD and laser ablation techniques require high amount of energy to reorganization of carbon atoms.

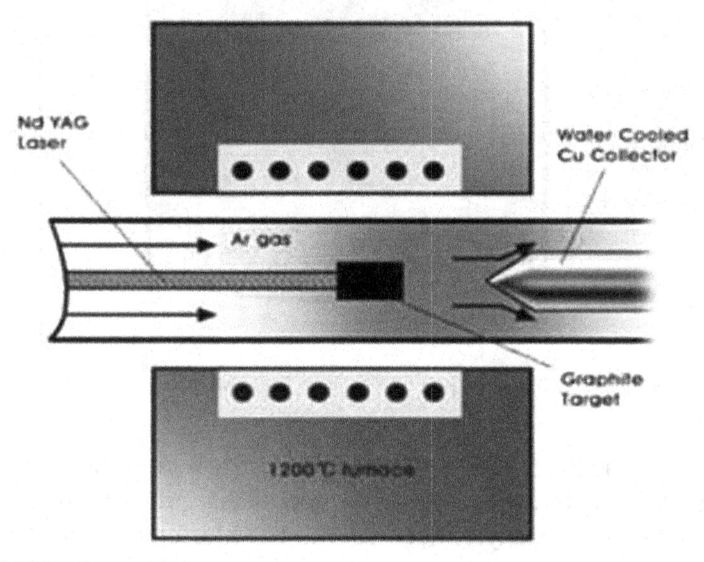

FIGURE 10.5 Laser ablation setup.

Laser ablation processes can also be tuned to yield either single-walled CNTs or multiwalled CNTs based on the variation of parameters in the growth process such as the laser wavelength, laser power, laser pulse duration, furnace temperature, and graphite target composition. In particular, a low metal: graphite ratio in the target and a high furnace temperature (typically ≈ 1200°C) is generally associated with good-quality crystalline single-walled CNTs, whereas greater amounts of metal yields multiwalled CNTs, and lower furnace temperatures compromise crystallinity of the CNT tube walls. Thess et al. (1996) presented an efficient route for the synthesis of arrays of SWNT. Their process involved condensation of a laser-vaporized cobalt mixture at 1473 K. The yield was more than nickel–carbon 70%. X-ray diffraction and electron microscopy showed that these SWNT were almost uniform in diameter and they self-organized into "ropes" which consisted of 100–500 nanotubes in a two-dimensional triangular lattice.

Growth of nanotubes by AD and laser ablation method involves "root growth" rather than "tip growth" with the tubes growing away from the metal particles, with carbon being continuously supplied to the base. Loiseau et al.[11] proposed a diffusion segregation model occurring at the catalyst particle surface to explain the mechanism of growth of a nanotube. The initiation of growth of nanotube occurs by segregation of cluster of Co and carbon atoms. The starting point was said to be a small nanotube portion capped by a fullerene (Fig. 10.6).

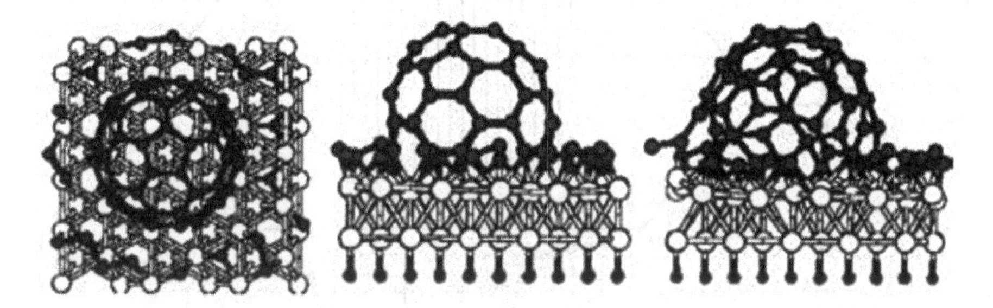

FIGURE 10.6 Simulation of root growth mechanism.

The growth of nanotubes can be visualized as formation of a round or pear-shaped precursor on the metal catalyst surface. This is followed by diffusion of carbon on sides of precursor leaving the top free of the resulting nanotube. It results in the formation of rodlike nanotube. The nanotube grows upward by staying attached to the surface of precursor and finally detaches itself from the metal surface. SWNT or MWNT are grown depending on the size of catalyst particles. Both the above methods are nowadays not used for synthesis due to various problems such as requirement of very high temperature, constant evaporation of carbon source and formation of highly disordered and tangled nanotubes requiring high purification. In light of all these shortcomings, other methods were adopted.

In 1999, Richard Smalley and coworkers developed a high-pressure carbon monoxide method for the synthesis of CNTs. A continuous-flow gas phase carbon monoxide was used as feed stock and iron pentacarbonyl [$Fe(CO)_5$] was used as a catalyst. It resulted in formation of very thin SWNT with minimum defects and high selectivity.[12]

10.3.3 CHEMICAL VAPOR DEPOSITION

During CVD, a carbon substrate covered with metal catalysts, like nickel, cobalt, iron, or a combination, is heated to approximately 700°C and promotes the growth of CNTs (Fig. 10.7). Commonly used gaseous carbon substrate source include methane, carbon monoxide, and acetylene.

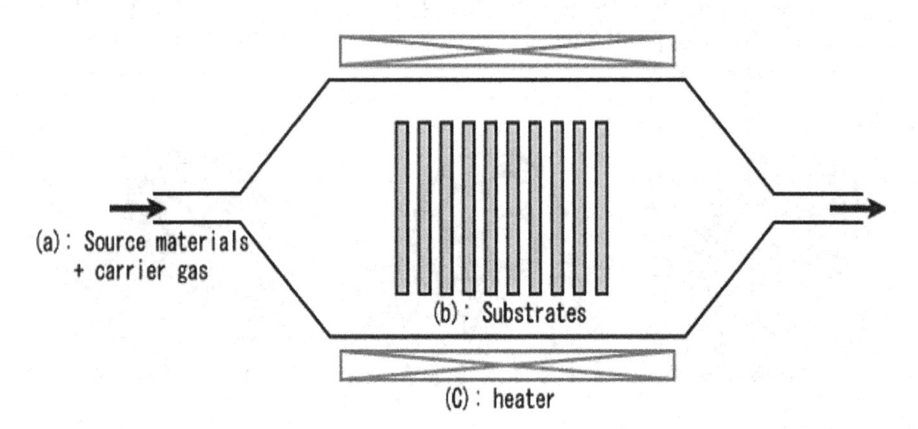

FIGURE 10.7 Chemical vapor disposition.

The CVD method extends this idea by embedding these metallic particles (iron, in the case of the seminal paper) in properly aligned holes in a substrate causing the carbon to convert into active atomic carbon. It is followed by diffusion of this active carbon toward the substrate which is heated and coated with catalyst. The growth starts after two gases are passed through the chamber, a carrier gas such as nitrogen, hydrogen, or argon, and some hydrocarbon gas such as acetylene (C_2H_2) or methane (CH_4). Correct alignment of carbon substrate and catalyst can result in desired growth of nanotubes. Appropriate metal catalyst can grow SWNT as well as MWNT.

The synthesis production yield, which indicates the amount of CNTs in the converted carbon, reaches 90%. CVD is commonly used for the industrial purposes because the method is already well investigated and offers acceptable results on the industrial scale. This method has advantage of low cost and mild operation. It is believed that upon thermal decomposition of organometallic compounds, it is the metal particles that first begin to take shape on the substrate surface. As more metal particles aggregate, the size of the catalytic center increases. Once the catalytic center reaches the optimal size for CNT nucleation, the dissolved carbon precipitates in the form of CNTs. The effect of hydrogen in the CVD process is to raise the yield of CNTs. CVD has the advantage of large-scale production of CNTs. CVD method allows the pretreatment of the substrate surface which can lead to impressive, controlled MWNT and SWNT architectures.[13]

MWNT self-assemble into aligned structures because of the van der Waals interaction between the single tubes. The substrates are catalytically patterned by the photolithography which enables selective deposition of the catalyst and consequently selective and controlled growth of CNTs. During CVD growth, the outer walls of the tubes interact with their neighbors via van der Waals forces, forming a rigid bundle which grows perpendicularly to the substrate.[14,15] This technique has been explored and developed for industrial scale of nanotubes. Su et. al. reported bulk yield of SWNT from methane using Al_2O_3.[16] Similar synthesis has also been reported with Co–Mo catalyst using CO as carrier gas. Maruyama et al.[17] reported high-purity SWNT without any amorphous carbon-coating product. In 2004, Iijima introduced another modification, the supergrowth CVD or water-assisted CVD. Here, the activity and lifetime of the catalyst are enhanced by addition of water into the CVD reactor. With this method, dense well-aligned nanobunches, perpendicular to the substrate with heights up to 2.5 mm can be produced.[18] With the low-temperature CVD, that was firstly reported in 2006, tungsten filament increases the decomposition of the precursor gases. SWNT were observed at temperatures as low as 350°C.[19]

10.4 SIZE AND TEXTURAL STUDIES OF CARBON NANOMATERIALS

The size of the nanoparticles is a very important parameter as there is an optimal size for each application. For example, for in vivo experiments, it must be taken into account that to cross the blood–brain barrier, the nanoparticles have to be in a range of 15–50 nm, whereas to pass through the endothelium, they must be smaller than 150 nm. Thus, particles between 30 and 150 nm are retained in the heart, stomach, and kidney, whereas particles between 150 and 300 nm usually stay in liver and spleen. Another example is when magnetic particles are used as carriers for the purification of biomolecules. In this case, sizes above 40 nm are necessary in order to have a good migration toward the magnet.

10.5 PURIFICATION OF CNTs

A large amount of impurities remain in nanotubes after their synthesis and need to be purified before their application. These impurities may vary from metal particles, amorphous carbon, fullerenes, etc. The fullerene is purified by chromatographic separation and fractional crystallization on the basis of solubility or volatility in organic solvents. In case of CNTs, the carbonaceous impurities are insoluble and nonvolatile making its purification a complex procedure. The metal catalysts can be removed most easily by dissolving them in inorganic acids; nevertheless, there is still a certain amount of metal catalysts encapsulated in multishell carbon particles, which are very difficult to remove. Amorphous carbon and multishelled carbon particles can be removed by oxidation, but the CNTs are also susceptible to oxidation. The key for developing an efficient purification method is to optimize the reaction conditions to allow selective oxidation of the carbonaceous impurities while protecting CNTs. The major impurities in laser, CVD, and HiPco-produced CNTs are metal catalysts and amorphous carbon. Generally, reflux in an inorganic acid or thermal treatment can achieve a partially efficient purification.[20,21] SWNTs produced in HiPco process can also be purified by a low-temperature oxidation (200°C) followed by concentrated HCl sonication.[22]

10.6 FUNCTIONALIZATION OF CNTs

Many applications require covalent modification to meet specific requirements, that is, in case of biosensors, the biomolecules require electron mediators to promote electron transfer. Similarly, electrochemical metal ion sensors require specific functional groups which show potential affinity toward particular metal ion. CNTs also do not disperse in organic matrices due their inert nature and form bundles with each other. The poor solubility of CNTs in organic solvents restricts them to be used as drug delivery agents into living systems in drug therapy. Hence, many modification approaches such as physical, chemical, or combined have been exploited for their homogeneous dispersion in common solvents to improve their solubility. The surface of nanotubes can be modified by various ways to enhance their dispersion in organic media. The solubilization of CNTs by functionalization generated novel applications of nanotubes which were otherwise not possible. Many applications require covalent modification to meet specific requirements, that is, in case of biosensors, the biomolecules require electron mediators to promote electron transfer.[23,24] Loung et al. reported the production of dangling bonds in nanotubes upon sonication of nanotubes in organic solvents.[25] This led to development of versatile chemical methods to modify and functionalize CNTs. This surface functionalization of nanotubes made it possible to solubilize and disperse CNTs.

Similarly, electrochemical metal ion sensors require specific functional groups which show potential affinity toward particular metal ion. The modification protocol was generally achieved by attaching specific molecule or entity which imparts chemical specificity to the substrate material. These chemical modifications can be easily achieved in many ways as discussed below.

10.6.1 COVALENT INTERACTIONS

Covalent modification involves attachment of a functional group onto the CNT. The functional groups can be attached onto the side wall or ends of the CNT. The end caps of the CNTs have the highest reactivity due to its higher pyrimidization angle and the walls of the CNTs have lower pyrimidization angles which has lower reactivity (Fig. 10.8). Although covalent modifications are very stable, the bonding process disrupts the sp^2 hybridization of the carbon atoms because a σ bond is formed. The disruption of the extended sp^2 hybridization typically decreases the conductance of the CNTs.[26–28]

10.6.2 OXIDATION

Oxidation of CNTs also functionalizes the surface by breaking the carbon–carbon bonded network of the nanolayers under acidic conditions. It allows the introduction of oxygen units in the form of carboxyl, phenolic, and lactone groups. In liquid phase reactions, CNTs are treated with oxidizing solutions of nitric acid or a combination of nitric and sulfuric acid to the same effect. However, over oxidation may occur causing the CNT to break up into fragments, which are known as carbonaceous fragments. Xing et al.[29] reported sonication-assisted oxidation, with sulfuric and nitric acid, of CNTs and produced carbonyl and carboxyl groups. Pan et al.[30] reported the MWCNT nanohybrids prepared initially by the oxidative pretreatment of CNTs with 3:1 H_2SO_4/HNO_3 mixture, then they were activated using $SOCl_2$, and finally, acyl chloride was coupled with ethylene diamine. The resulted MWCNTs were again modified with mercaptoacetic acid-coated QDs.

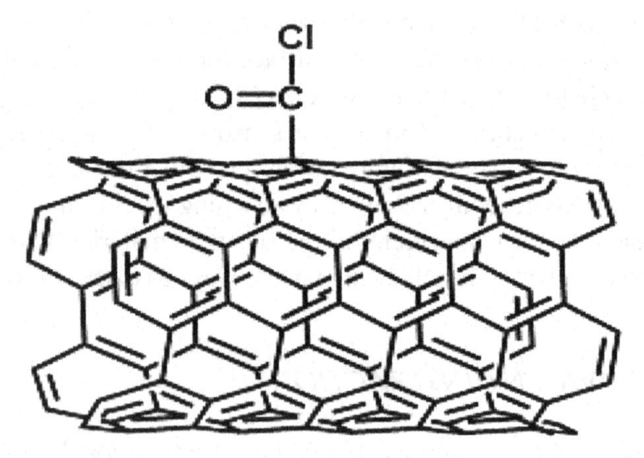

FIGURE 10.8 Functionalization of nanotube by attachment of functional group.

10.6.3 ESTERIFICATION/AMIDATION

The functionalization of nanotubes is carried out by using carboxylic groups, which acts as a precursor for most esterification and amidation reactions. The carboxylic group is converted into an acyl chloride with the use of thionyl or oxalyl chloride which is then reacted with the desired amide, amine, or alcohol. CNTs modified with acyl chloride react readily with highly branched molecules such as poly(amidoamine), which acts as a template for silver ion and later being reduced by formaldehyde. Amino-modified CNTs can be prepared by reacting ethylene diamine with an acyl chloride-functionalized CNTs. In a similar way, thiol-stabilized ZnS-capped CdSe QDs were protected with 2-aminoethanethiol and linked to the acid-terminated CNTs in presence of a coupling agent such as 1-ethyl-3-(3-dimethylaminopropyl) carbodiimide hydrochloride (Fig. 10.9).[31]

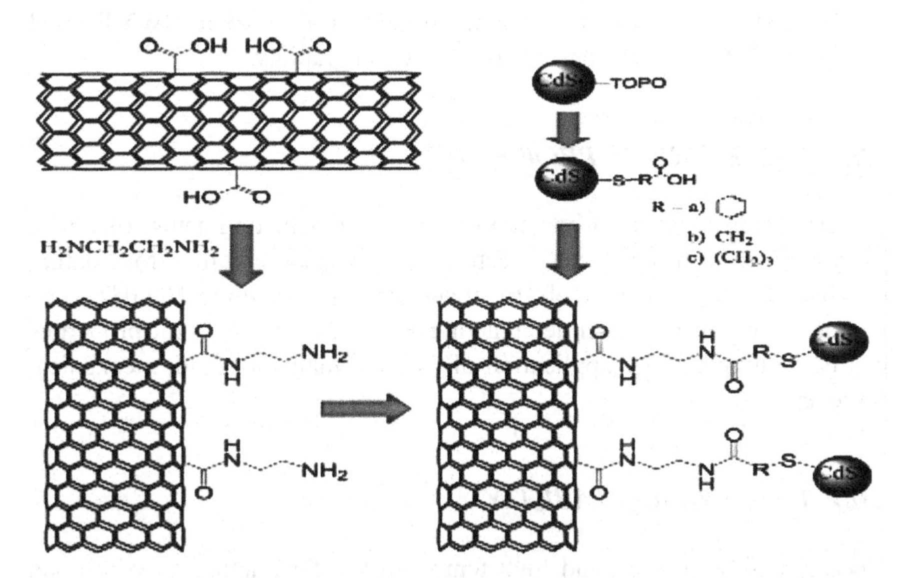

FIGURE 10.9 Functionalization of CNT via amide bonds.

10.7 PROPERTIES

10.7.1 STRENGTH

CNTs have a higher tensile strength than steel and Kevlar. The Young's modulus of the nanotubes can be as high as 1000 GPa (which is approximately five times higher than steel) while their tensile strength can be up to 63 GPa (around 50 times higher than steel). This strength originates from sp^2 bonds between the individual carbon atoms. CNTs are not only strong, they are also elastic. Upon application of force, nanotube can bend and returns to its original shape when the force is removed. A nanotube's elasticity does have a limit, and under very strong forces, it is possible to permanently deform the shape of a nanotube. A nanotube's strength can be weakened by defects in the structure of the nanotube. Defects occur from atomic vacancies or a rearrangement of the carbon bonds. Defects in the structure can cause a small segment of the nanotube to become weaker, which in turn causes the tensile strength of the entire nanotube to weaken. The tensile strength of a nanotube depends on the strength of the weakest segment in the tube similar to the way the strength of a chain. These properties, coupled with their low density, give nanotubes huge potential in a range of structural applications. Applications on the nanometer and micrometer scale, such as SWNT-based transistors and chemical sensors are progressing rapidly.

10.7.2 ELECTRICAL PROPERTIES

The sp^2 bonds between carbon atoms results in conducting nature of CNTs. They can also withstand strong electric currents because of the strong nature of bonds. SWNT can route electrical signals at speeds up to 10 GHz when used as interconnects on semiconducting devices. Their electronic properties can be manipulated by application of external magnetic field, mechanical force, etc.

10.7.3 THERMAL STABILITY

CNTs are able to withstand high temperatures, thus acting as very good thermal conductors. The temperature stability of CNTs is estimated to be up to 28,000°C and about 750°C in air. The CNTs are shown to transmit over 15 times the amount of watt per minute as compared to copper wires.[34] The measured thermal conductivity of an individual MWNT (~3000 W/m K) is larger than that of a natural diamond and the basal plane of graphite (both 2000 W/m K). The density of SWNTs can be as small as 0.6 g/cm^3, and the low density of MWNTs (range from 1 to 2 g/cm^3, 4–10 times lower than steel) makes them useful for potential application in light-weight high-strength materials. The pristine CNTs contain carbon atoms with extended π-network structure, which makes them chemically stable.

10.8 APPLICATIONS OF CNTs

Carbon-based technology has been emerged to develop highly sensitive, inexpensive, and low-power devices proposing an alternative to silicon-based conventional technology that involves rigorous fabrication steps using top–down approach. Also, it has already been speculated that the scaling of silicon-based devices will soon reach their limit; therefore, there is an urgent need to explore novel materials. Carbon materials such as CNT and graphene with inherent nanoscale characteristics have offered a great potential for presenting next-generation material for diverse applications such as field emission, electronics, sensors, and energy.

There are many potential applications of CNTs owing to their remarkable properties. They have potential to be used in electronics, textile industry as water-proof and tear-proof fabric, and sensor, based on the property of thermal conductivity and many more. They possess extraordinary heat and electrical conductivity behavior, making them a suitable candidate for numerous applications. Some of the important applications of CNTs are discussed below.

10.8.1 SENSORS

The smaller and uniform dimensions of CNTs have resulted in some interesting applications. They have been investigated for their application as sensors due to its extremely small size, high conductivity, high mechanical strength, and flexibility. A MWNT has conducting tips and has been used as probe device.[32] Since MWNT tips are conducting, they can be used in scanning instruments such as SEM and AFM. The nanotubes tips do not suffer crash on contact with substrate and have an advantage of slenderness and possible to sense those features which are otherwise impossible. These nanotubes can be selectively modified chemically through functional groups attachment.[33] Functionalized nanotubes were used as AFM tips to measure binding forces between protein–ligand pairs. They have also been reported to image biomolecules such as DNA and amyloid-b protofibrils which are otherwise not possible to be imaged by other methods.[34]

Recent research has shown that nanotubes can be used as advanced miniature chemical sensors.[35] Gas sensing is of great importance to environmental monitoring, control of chemical processes, as well as in many medical and agricultural applications. Traditional gas sensors such as silicon devices, semiconducting metal oxides, organic and composite materials are widely used, though they are limited by their operating parameters such as temperature, resistivity, and sensitivity. Individual semiconducting SWNT-based chemical sensors are reliable for the detection of small amounts of toxic gases such as NH_3 and NO_2. The electrical resistivities of SWNT were found to change sensitively on exposure to molecules of NO_2, NH_3, and O_2. By monitoring the change in the conductance of nanotubes, functionalized SWNTs experienced a much greater change of resistance upon exposure to NH_3 than did pristine tubes, giving them greater sensitivity as sensors.[36,37,40]

The detection efficiency of the chemical sensor is significantly important to accurately monitor the concentration of various toxic gases in the atmosphere. Ammonia (NH_3) and nitrogen dioxide (NO_2) adversely affect human health and pollute the environment, and sources of these gases are primarily agriculture, natural waste products, industrial products, as well as manufacturing of chemicals. Moreover, presence of high concentrations of hydrogen (H_2) and methane (CH_4) becomes explosive. Jang et al. reported a NH_3 gas sensor with laterally aligned MWNTs, with N-type heavily doped Si wafer used as a substrate.[38]

10.8.2 NANOELECTRONICS

One of the potential applications of nanotubes is in the field of electronics, due to its highly conducting nature. A great deal of research has been focused on the application of nanotubes in place of silicon-based devices to alter materials and technologies. A number of nanoelectronic devices using CNTs have been reported. Recently, CNTs-based nanoelectronic devices such as field-effect transistors and single-electron transistors have attracted considerable interest. Out of the two types, SWNT are the most conducting type. Twisting and bending of nanotube makes it highly conducting. With high conductivity and small size, nanotubes may be an alternate option to copper which is generally used but has limitation of ineffectivity at size less than 40 nm. The CNTs can also be a substitute for cooper for interconnecting transistors in chips owing to its higher conductivity and small dimensions. They can also be used alongside existing IC manufacturing processes.[39] Some of the applications of nanotubes in electronics are discussed below.

10.8.2.1 DIODES

Nanotube diodes have been reported by Dai et al.[14] by doping a p–n junction as in normal silicon diode. One-half of the semiconducting SWNT were doped with K, leaving another half undoped. Doped portion of nanotube acted as n-type semiconductor due to electron donation by K atoms, while undoped section acted as p type. This diode behaved as conventional diode.

Lee et al.[41] carried out the doping by applying different charges at different parts of nanotubes instead of doping different elements into tube, thereby initiating the n–p character.

10.8.2.2 TRANSISTORS

Nanotubes have shown their application as transistors, which is an important part of digital switches. Transistors form a basis of modern integrated circuits acting as digital switches. Different configurations of CNTs have allowed SWNT to act as transistors but under low-temperature conditions. In such a switch, a molecule can be positioned inside a CNT to affect the electronic current flowing across it. The result is a molecular scare gate in which the position of the molecule controls the flow of the electrical current. In this model, the gate is about 1 nm in size, or three orders of magnitude smaller than a silicon chip having SWNT supported on silicon wafer and shows the semiconducting behavior.[41]

10.8.3 BIOMEDICAL APPLICATIONS

CNTs have emerged as a potential carrier for drug delivery system for treatment of diseases such as cancer due to their physicochemical properties. The availability of chemical modification and biofunctionalization methods have made it possible to generate CNTs biofunctionalized with bioactive materials such as proteins, carbohydrates, or nucleic acids.[42,43] CNTs have been used as carrier for drug delivery. Small drug molecules can be covalently conjugated to CNT for in vitro delivery. Cycloaddition-functionalized CNTs have been reported for the delivery of an anticancer drug or antifungal drug into cells.[44] Platinum (IV) complex, a anticancer drug, is administered to cancer cells via functionalized CNTs.[45] Similar drug delivery is also reported for MWNTs[46] and nanographene oxide,[47,48] opening new opportunities for drug delivery. Many studies on use of CNT as pressure, thermal, gas, chemical, and biological sensors have been reported. Balavoine et al. reported the crystallization of streptavidin, a biotin-binding protein on the hydrophobic surface of MWNT.[49] SWNT has also been reported to act as molecular sensors via field-effect doping by gaseous species NH_3 and NO_2.[50] In addition to applications for drug delivery and treatment, the optical properties of CNTs make them useful for optical probes.

10.9 OTHER APPLICATIONS

Dirt-/water-resistance fabric: Nanotechnology has been used to develop fabric which is dirt and water resistant and show the behavior of self-cleaning (Fig. 10.10) This behavior is inspired from petals of lotus flower which has hydrophobic molar layer made up of nanometric size hairs which allows sliding of water droplets on it, removing any dust particle on petal surface. Fabrics are coated with layers of silver nanoparticles to prevent growth of dirt on the surface. Such type of cloth can be fabricated to make it dirt and water resistant. It would not be required to clean such type of cloth and it will be free from damage.

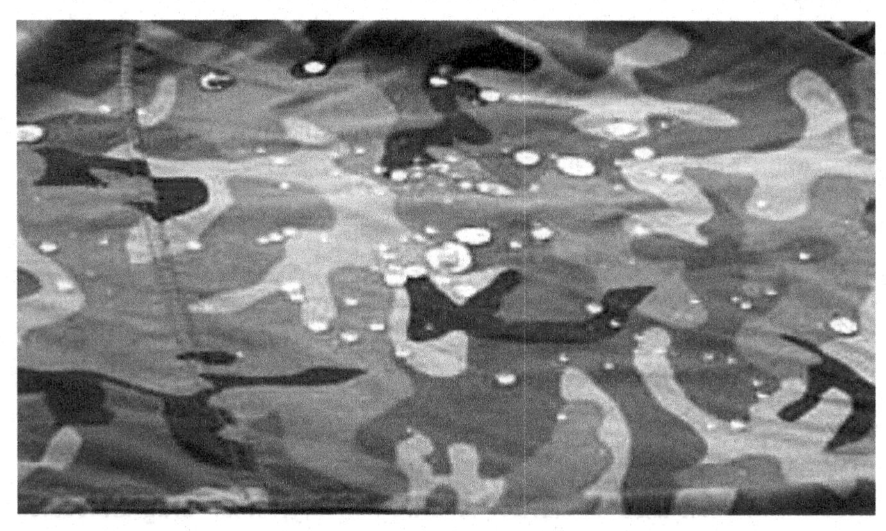

FIGURE 10.10 Self-cleaning water- and dirt-resistant fabric.

Other miscellaneous applications: In recent research, CNTs have been used to prepare in polymer-reinforced nanocomposites, showing excellent mechanical resistance and strength. CNTs have been mentioned as an ideal candidate for composite materials. The nanocomposites offer better strength, elasticity, toughness, durability, and conductivity. One of the hurdles in nanocomposites is the achievement of homogeneous dispersion of nanotubes through the matrix. This problem is overcome by mixing a suspension of matrix material with CNTs to create a uniform composition and by chemical or physical functionalization of nanosurface. Conducting polymers such as polypyrrole has been used to make composite films with MWNTs by electrochemistry.[51] CNTs application in sports materials, such as tennis rackets and bicycle frames, has shown better strength and performance. Nanotube-enabled fabric has also been used in textile for the production of bullet-proof jacket. Nanotubes incorporation in various skincare products has also given excellent skin results.

Energy storage devices: CNTs have also been investigated as a device for energy storage and production. Nanotubes have attracted attention due to their small dimensions, smooth surface topology, and faster electron transfer kinetics on nanotubes. They have shown superior performance as compared to other electrodes in terms of reaction rates and reversibility.[52] CNTs possess hollow cavities; creating excellent nanocontainers for hydrogen storage. Hydrogen molecules can be physically adsorbed on the exterior surface of CNTs or interstitial spaces between them. The hydrogen storage capacity of CNTs depends on several factors, such as tube diameter, length, and purity.

10.10 CONCLUSION

CNTs have been extensively studied making way for basic understanding and potential for various applications. This review has discussed several applications of CNTs. The remarkable physical properties of CNTs have created a host of application possibilities, based on electronic and mechanical behavior of nanotubes. CNTs have been proved to be safer option in biomedical applications such as drug delivery, implants, and genetic engineering. The properties and application behavior are still under search and still lot is yet to be explored. This review presented the details of the methods for synthesis and technique applied on the nanotubes for their applications.

KEYWORDS

- **carbon nanotubes**
- **single-walled nanotubes**
- **multiwalled nanotubes**
- **functionalization**
- **nanoelectronics**

REFERENCES

1. Iijima, S. Helical Microtubules of Graphitic Carbon. *Nature* **1991,** *354*, 56–58.
2. Masciangioli, T.; Zhang, W. X. Environmental Technologies at the Nanoscale. *Environ. Sci. Technol.* **2003,** *37*, 102A–108A.
3. Krätschmer, W.; Lamb, L. D.; Fostiropoulos, K.; Huffman, D. R. Solid C60: A New Form of Carbon. *Nature* **1990,** *347*, 354–358.
4. Bianco, A.; Kosteralos, K.; Prato, M.; Partidos, C. D. Carbon Nanotubes: On Road to Deliver. *Curr. Drug Deliv.* **2005,** *2*, 253–259.
5. Bethune, D. S.; Kiang, C. H.; Vries, M. S.; Gorman, G.; Savoy, R.; Vazquez, J.; Beyers, R. Cobalt-catalysed Growth of Carbon Nanotubes with Single-atomic-layer Walls. *Nature* 1993, *363*, 605.
6. Brenner, D. Emperical Potential for Hydrocarbons for Use in Simulating the Chemical Vapour Deposition of Diamond Films. *Phys. Rev. B* **1990,** *42* (15), 9458–9471.
7. Calvert, P. Strength in Unity. *Nature* **1992,** *357*, 365–366.
8. Ebbesen, T. W.; Ajayan, P. M. Large Scale Synthesis of Carbon Nanotubes. *Nature* **1992,** *358*, 220–222.
9. Journet, C.; Maser, W. K.; Bernier, P.; Loiseau, A.; Chapelle, M. L.; Lefrant, S.; Deniard, P.; Lee, R.; Fischer, J. E. Large-scale Production of Single-walled Carbon Nanotubes by the Electric-arc Technique. *Nature* **1997,** *388*, 756–758.
10. Paul, S.; Samdarshi, S. K. A Green Precursor for Carbon Nanotube Synthesis. *N. Carbon Mater.* **2011,** *26* (2), 85–88 (Xinxing Tan Cailao).
11. Gavillet, J.; Loiseau, A.; Ducastelle, F.; Thair, S.; Bernier, P.; Stephan, O.; Thibault, J.; Jean-Christophe, C. Microscopic Mechanisms for the Catalyst Assisted Growth of Single-wall Carbon Nanotubes. *Carbon* **2002,** *40*, 1649.
12. Smalley, R. E.; Yakobson, B. The Future of Fullerenes. *Solid State Com.* **1998,** *107* (11), 597–606.
13. Li, W. Z.; Xie, S. S.; Qian, L. X.; Chang, B. H.; Zou, B. S.; Zhou, W. Y.; Zhao, R. A.; Wang, G. Large Scale Synthesis of Aligned Carbon Nanotubes *Science* **1996,** *274*, 1701–1703.
14. Dai, H. Controlling Nanoworld. *Phys. World* **2000,** *13*, 43.
15. Li, J.; Papadopoulos, C.; Xu, J. M. Highly-ordered Carbon Nanotube Arrays for Electronics Applications *Appl. Phys. Lett.* **1999,** *75*, 367.
16. Su, M.; Zheng, B.; Liu, J. A Scalable CVD Method for the Synthesis of Single Walled Carbon Nanotubes with High Catalyst Productivity. *Chem. Phys. Lett.* **2000,** *322*, 321–326.
17. Maruyama, S.; Kojima, R.; Miyauchi, Y.; Chaishi, S.; Kehno, M. Low Temperature Synthesis of High Purity Single Walled Carbon Nanotubes from Alcohol. *Chem. Phys. Lett.* **2002,** *360*, 229–234.
18. Kuzmych, O.; Allen, B. L.; Star, A. Carbon Nanotube Sensors for Exhaled Breath Components. *Nanotechnology* **2007,** *18*, 375502/1–3775502/7.
19. Hata, K.; Futaba, D. N.; Mizuno, K.; Namai, T.; Yumura, M.; Iijima, S. Water-assisted Highly Efficient Synthesis of Impurity-free Single-walled Carbon Nanotubes. *Science* 2004, *306* (5700), 1362–1364.
20. Dillon, A. C.; Gennett, T.; Jones, K. M.; Alleman, J. L.; Parilla, P. A.; Heben, M. J. A Simple and Complete Purification of Single-walled Carbon Nanotube Materials. *Adv. Mater.* 1999, *11*, 1354–1358.

21. Borowiak-Panel, E.; Pichler, T.; Knupfer, M.; Graff, A.; Jost, O.; Pompe, W.; Kalenczuk, R. J.; Fink, J. Reduced Diameter Distribution of Single-walled Carbon Nanotubes by Selective Oxidation. *Chem. Phys. Lett.* **2002,** *363*, 567–572.
22. Zhou, W.; Ooi, Y. H.; Russo, R.; Papanek, P.; Luzzi, D. E.; Fischer, J. E.; Bronikowski, M. J.; Willis, P. A.; Smalley, R. E. Structural Characterization and Diameter-dependent Oxidative Stability of Single Wall Carbon Nanotubes Synthesized by the Catalytic Decomposition of CO. *Chem. Phys. Lett.* **2001,** *350*, 6–14.
23. Sampath, S.; Lev, O. J. Electrochemical Oxidation Of NADH On Sol–Gel Derived, Surface Renewable, Non-Modified And Mediator Modified Composite-Carbon Electrodes. *Electroanal. Chem.* **1998,** *446*, 57–65.
24. Rico, E. M.; Vidal, J. M.; Pelegrina, A. B.; Micor, J. R. L. Voltammetric Determination of Lead (II) Ions at Carbon Paste Electrode Modified with Banana Tissue. *J. App. Sci.* **2007,** *9*, 1286–1292.
25. Luong, J. H. T.; Hrapovic, S.; Wang, D.; Bensebaa, F.; Simard, B. Solubilization of Multiwalled Carbon Nanotubes by 3-aminopropyltriethoxysilane Towards the Fabrication of Electrochemical Biosensors with Promoted Electron Transfer *Electranalysis* **2004,** *6*, 132–139.
26. Hou, X. M.; Wang, L. X.; Zhou, F.; Li, L. Q.; Li. Z. High-density Assembly of Gold Nanoparticles to Multiwalled Carbon Nanotubes Using Ionic Liquid as Interlinker. *Mater. Lett.* **2009,** *63*, 697–699.
27. Guo, S. J.; Dong, S. J.; Wang, E. K. Constructing Carbon Nanotube/Pt Nanoparticle Hybrids Using an Imidazolium-salt-based Ionic Liquid as a Linker. *Adv. Mater.* **2010,** *22*, 1269–1272.
28. Wu, B.; Hu, D.; Kuang, Y.; Liu, B.; Zhang, X.; Chen, J. Functionalization of Carbon Nanotubes by an Ionic-liquid Polymer: Dispersion of Pt and PtRu Nanoparticles on Carbon Nanotubes and Their Electrocatalytic Oxidation of Methanol. *Angew. Chem. Int. Ed.* **2009,** *48*, 4751–4754.
29. Yangchuan, X.; Liang, L.; Charles, C. C.; Robert, V. H. Sonochemical Oxidation of Multiwalled Carbon Nanotubes. *Langmuir* **2005,** *21* (9), 4185–4190.
30. Pan, B.; Cui, D.; He, R.; Gao, F.; Zhang, Y. Covalent Attachment of Quantum Dots on Carbon Nanotubes. *Chem. Phys. Lett.* **2006,** *417*, 419–424.
31. Banerjee, S.; Wong, S. S. Synthesis and Characterization of Carbon Nanotube–Nanocrystal Heterostructures. *Nano Lett*. 2002, *2*, 195–200.
32. Dai, H. J.; Hafner, J. H.; Rinzler, A. G.; Colbert, D. T.; Smalley, R. E. Nanotubes as Nanoprobes in scanning Probe Microscopy. *Nature* **1996,** *384*, 147.
33. Chen, J.; Hamon, M.; Hu, H.; Chen, Y.; Rao, A.; Eklund, P. C.; Haddon, R. C. Solution Properties of Single-walled Carbon Nanotubes. *Science* **1998,** *282*, 95–98.
34. Wong, S. S.; Harpner, J. D.; Lansbury, P. T.; Lieber, C. M. Carbon Nanotube Tips: High-resolution Probes for Imaging Biological Systems. *J. Am. Chem. Soc.* **1998,** *120*, 603–604.
35. Kong, J.; Franklin, N. R.; Zhou, C.; Chaplin, M. C.; Peng, S.; Cho, K.; Dai, H. Nanotube Molecular Wires as Chemical Sensors. *Science* 2000, *287*, 622–625.
36. Bekyarova, E.; Davis, M.; Burch, T.; Itkis, M. E.; Zhao, B.; Sunshine, S.; Haddon, R. C. Chemically Functionalized Single-walled Carbon Nanotubes as Ammonia Sensors. *J. Phys. Chem. B* **2004,** *108*, 19717–19720.
37. Collins, P. G.; Bradley, K.; Ishigami, M.; Zettl, A. Extreme Oxygen Sensitivity of Electronic Properties of Carbon Nanotubes. *Science* **2000,** *287*, 1801–1804.

38. Jang, Y.-T.; Moon, S.-I.; Ahn, J.-H. Simple Approach in Fabricating Chemical Grown Multiwalled Carbon Nanotubes. *Sens. Actuators* **2004,** *99*, 118–122.
39. Close, G. F., Yasuda, S., Paul, B., Fujita, S., Wong, H.-S.P. A 1 GHz Integrated Circuit with Carbon Nanotube Interconnects and Silicon Transistors. *Nano Lett.* **2008,** *8* (2), 706–709.
40. Langer, L.; Bayot, V.; Grivei, E.; Issi, J. P.; Heremans, J.-P.; Olk, C. H.; Stockman, *L.*; Van Haesendonck, C.; Bruynseraede, Y. Quantum Transport in a Multiwalled Carbon Nanotube. *Phys. Rev. Lett.* **1996,** *76*, 479.
41. Lee, J. U.; Gipp, P. P.; Heller, C. M. Carbon Nanotube p–n Junction Diodes. *Appl. Phys. Lett.* **2004,** *85*, 145.
42. Zhou, W.; Rutherglen, C.; Burke, M. P. Wafer Scale Synthesis of Dense Aligned Array of Single Walled Carbon Nanotubes. *Nano Res.* **2008,** *1*, 158–165.
43. Kim, S. N.; Rusling, J. F.; Papadimitrakopoulos, F. Carbon Nanotubes for Electronic and Electrochemical Detection of Biomolecules. *Adv. Mater.* **2007,** *19*, 3214–3228.
44. Wu, W.; Wieckowski, S.; Pastorin, G.; Bennicasa, M.; Klumpp, C.; Briand, J. P.; Gennaro, R.; Prato, M.; Bianco, A. Targeted Delivery of Amphotericin B to Cells Using Functionalized Carbon Nanotubes. *Angrew. Chem. Int. Ed. Engl.* **2005,** *44*, 6358–6362.
45. Feazell, R. P.; Nakayama-Ratxhford, N.; Dai, H.; Lippard, S. J. Soluble Single Walled Nanotubes as Longboat Delivery Systems for Platinum(IV) Anticancer Drug Design. *J. Am. Chem. Soc.* **2007,** *129* (9), 8438.
46. Ali-Boucetta, H.; Al-jamal, K. T.; McCarthy, D.; Prato, M.; Bianco, A.; Kostarelos, K. Multiwalled Carbon Nanotube-doxorubicin Supramolecular Complex for Cancer Therapeutics. *Chem. Commun.* **2008,** *28*, 459–461.
47. Sun, X.; Liu, Z.; Welsher, K.; Robinson, J. T.; Goodwin, A.; Zaric, S.; Dai. H. Nano-graphene Oxide for Cellular Imaging and Drug Delivery. *Nano Res*. **2008,** *1*, 203–212.
48. Liu, Z.; Robinson, J. T.; Sun, X. M.; Dai, H. J. PEGylated Nanaograhene Oxide for Delivery of Water Insoluble Cancer Drugs. *J. Am. Chem. Soc.* **2008,** *130*, 10876–10877.
49. Balavoine, F.; Schulz, P.; Richard, C.; Mallouh, V.; Ebbensen, T. W.; Mioskowski, C. Helical Crystallization of Proteins on Carbon Nanotubes: A First Step Towards the Development of New Biosensors. *Angrew. Chem. Int. Ed.* **1999,** *38*, 1912–1915.
50. Kong, J.; Franklin, N. R.; Zhou, C. W.; Chapline, M. G.; Peng, S.; Cho, K.; Dai, H. J. Nanotube Molecular Wires as Chemical Sensors. *Science* **2000,** *287*, 622–625.
51. Hughes, M.; Chen, G. Z.; Shaffer, M. S. P.; Fray, D. J.; Windle, A. H. Electrochemical Capacitance of a Nanoporous Composite of Carbon Nanotubes and Polypyrrole *Chem. Mater*. **2002,** *14*, 1610–1613.
52. Britto, P. J.; Santhanam, K. V. S.; Ajayan, P. M. Carbon Nanotube Electrode For Oxidation Of Dopamine. *Biolelectrochem. Bioenerg.* **1996,** *41*, 121–125.

CHAPTER 11

CHANGE IN THE ELECTRONIC STRUCTURE AND MAGNETIC PROPERTIES OF MODIFIED COPPER/ CARBON NANOCOMPOSITES

VLADIMIR I. KODOLOV[1,2*], V. V. TRINEEVA[1,3], N. S. TEREBOVA[1,2], I. N. SHABANOVA[1,2], T. M. MAKHNEVA[1,3], R. V. MUSTAKIMOV[1,4], and A. A. KOPYLOVA[1,2]

[1]*Basic Research—High Educational Center of Chemical Physics and Mesoscopy, Udmurt Scientific Centre, Ural Division, RAS, Izhevsk, Russia*

[2]*Kalashnikov Izhevsk State Technical University, Izhevsk, Russia*

[3]*Institute of Mechanics, Udmurt Scientific Centre, Ural Division, RAS, Izhevsk, Russia*

[4]*Scientific Innovation Centre, Izhevsk Electromechanical Plant—"KUPOL," Izhevsk, Russia*

[*]*Corresponding author. E-mail: Vkodol.av@mail.ru*

ABSTRACT

The conditions are established for the synthesis of copper-doped carbon nanocomposites modified by the compounds of silicon, phosphorus, and metal oxides. The hypothesis for the mechanism of the change of the nanocomposite electronic structure and the copper atomic magnetic moment is offered according to which the increase of the metal atomic magnetic moment occurs due to a redox process and depends on the number of the electrons participating in the process.

The maximum atomic magnetic moment of copper is 4.2 μ_B at the transition from P^{+5} to P^0. The growth of the phosphorus-containing layer on a nanogranule leads to the decrease of the copper atomic magnetic moment because of the decrease in the wave propagation velocity of a quantized electron.

11.1 INTRODUCTION

Earlier in Refs. [1–3], the appearance and growth of the atomic magnetic moment of copper in copper/carbon (Cu/C) nanocomposites (NCs) as well as in Cu/C NC modified by substances containing p elements such as silicon, phosphorus, and sulfur have been noticed. When the above elements were introduced into the NC carbon shell, the increase of the atomic magnetic moment of copper was observed.[3]

In the present paper, the results on the change in the electronic structure are analyzed in comparison with the newly obtained data on the modification of Cu/C NC by nickel oxide.

The carried-out investigations have shown that Cu/C NC can be thought of as a nanogranule consisting of a "nucleus" (copper-containing clusters interacting with carbon fibers) and a "shell" (carbon fibers consisting of the fragments of polyene and carbyne with unpaired electrons) and, therefore, it is quite possible that on the "shell" surface a redox process can take place. During the process, the elements from the modifier are reduced and the "shell" electrons participating in the process are "replenished" owing to the electrons of a metal (copper) cluster. The analysis of the electronic structure of Cu/C NCs modified by sp elements is of great scientific interest.

In Ref. [4], an experiment is presented where magnetic copper was obtained in a specimen consisting of a 2.5-nm thick copper film between graphene layers of 14 nm in thickness. The obtained nanomaterial is similar to the Cu/C NCs previously developed by us.[1] In Ref. [4], it is stated that the magnetic properties of copper appear due to the "sucking out" of copper electrons by the graphene layers.

In fact, when electron acceptors are present, 10 paired d electrons of copper can unpair and move to the higher orbitals. In this case, the maximum atomic magnetic moment of copper can be larger than that of gadolinium and attain the value of 5 μ_B.

Cu/C NCs are obtained from copper oxide and polyvinyl alcohol with the use of a mechanochemical process owing to redox reactions; this leads to the formation of the reduced copper clusters and the carbon fiber layers and the growth of the copper atomic magnetic moment to 1.3 μ_B, which can be explained by the unpairing of two electrons in the d orbitals of copper. When Cu/C NC is modified with sp elements, the substances are used, the interaction of which with the nanogranule involves five, four, and two electrons. Such substances are ammonium polyphosphate (APP), silica (S), and nickel oxide (NO).

11.2 EXPERIMENTAL

The modification of Cu/C NCs by the above compounds was carried out using mechanochemical synthesis as described in Ref. [1]; however, at the final stage of the process, the temperature regime was changed.[5]

The redox process takes place at the phase interface between the substance containing an oxidized element and the electron-saturated shell of the Cu/C NC. The ratio of the reactants (Cu/C NC and element-containing substances) varied from 2 to 0.5. When the Cu/C NC was modified with nickel oxide, the ratio was 1:1. To activate the process, the reaction mixture was wetted. After the mechanochemical process (the trituration in a triturating machine at the energy consumption of about 260–270 kJ/mol), the prepared nanoproduct was dried and held in a closed crucible at about 150°C and after that in vacuum at 100–150°C for not more than 3 min.

The results of the mechanochemical process were evaluated with the use of X-ray diffraction (XRD), X-ray photoelectron spectroscopy (XPS), and electron paramagnetic resonance (EPR). XRD was conducted on a diffractometer DRON-3 using a copper cathode. The EPR study was carried out with the use of EPR spectrometer in Prof. M. Ya. Mel'nikov Laboratory in MSU.

The XPS investigations were conducted on an X-ray electron magnetic spectrometer with the resolution 10^{-4} and luminosity—0.085% at the excitation by AlK_α line 1486.5 eV, in vacuum of 10^{-8}–10^{-10} Torr. The technological advantage of X-ray electron magnetic spectrometers is the constructive separation of the magnetic energy analyzer from the spectrometer vacuum chamber.

Based on van Vleck theory, a model was developed describing the correlation of the parameters of the multiplet splitting of Cu3s spectra with the change in the number of uncompensated d electrons and atomic magnetic moment of metals in nanostructures.

The phase composition of the modified phosphorus-containing metal/C NC was investigated with the use of an analytical transmission electron microscope (TEM) FEI Tecnai G2F20 with an energy-dispersive spectrometer EDS or energy-dispersive spectrometer (EDAX). The elemental and phase compositions were determined by the EDS spectra of the excitation of the secondary X-ray radiation.

11.3 RESULTS AND DISCUSSION

The results of the XRD study of the mechanochemical interaction of Cu/C NC with phosphorus-, silicon-, and nickel-containing compounds indicate the reduction of the elements accompanied by the formation of phases containing elements with zero level of oxidation (Figs. 11.1–11.3).

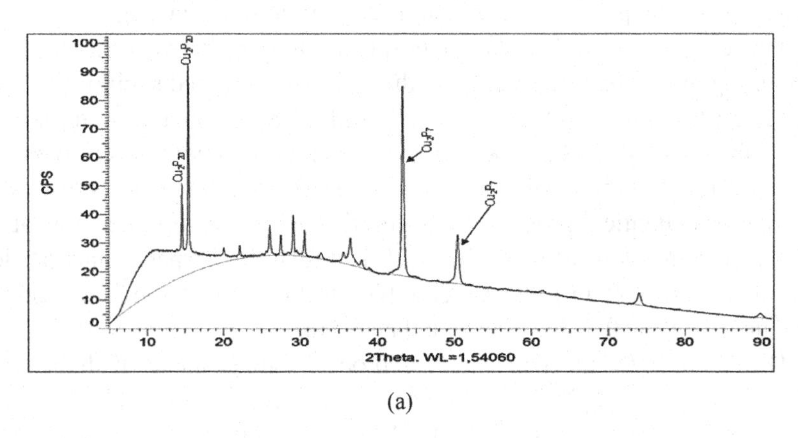

(a)

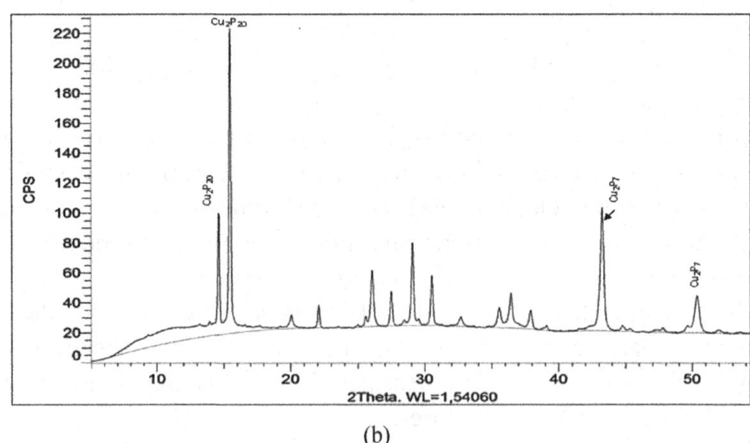

(b)

FIGURE 11.1 X-ray diffraction patterns of phosphorus-containing Cu/C NCs obtained at the ratio of NC:APP equal to 2 (a) and 1 (b).

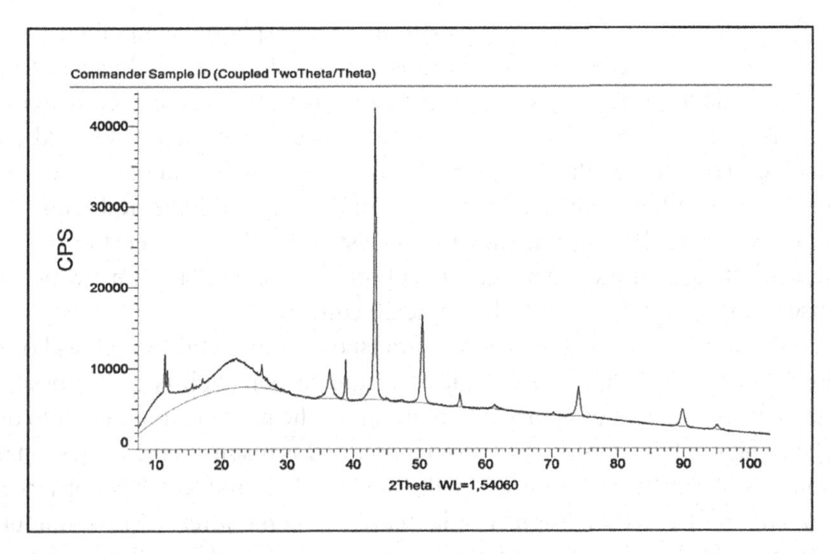

FIGURE 11.2 X-ray diffraction pattern of silicon-containing Cu/C NC (the reactants ratio 1:1).

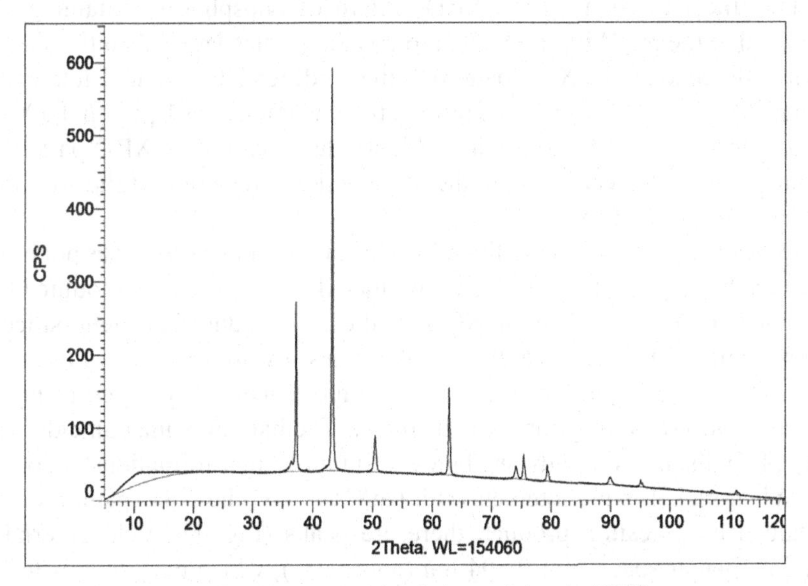

FIGURE 11.3 X-ray diffraction pattern of nickel-containing Cu/C NC.

In the presented XRD patterns there are lines which can be attributed to interplain distances corresponding to the phases containing reduced elements such as phosphorus, silicon, and nickel.

The comparison of the XRD patterns of phosphorus-containing Cu/C NCs obtained at the NC and APP ratios 2 and 1 shows the influence of the modifier thickness on the reactant interaction character. On the XRD patterns, the correlation of the lines were compared at angles 2θ equal to 43° and 16° which corresponds to the interplain distances d_{43} = 0.49 and d_{16} = 1.22 nm. On the basis of the EMF spectra of the excitation of the X-ray radiation, the predicted Cu–C–P–phase composition is established. However, the sum of the diameters of the copper and phosphorus atoms is 0.476 nm, which is close to interplain distance at the angle 2θ equal to 43°.

Taking into account the electrons transport from metal to a phosphorus atom through carbon, the deformation of the carbon shell is quite possible due to the movement of hydrocarbon groups of the polyene fragment into the surface layer of the modified NC Nan granule; this results in the formation of submolecular structures with size in the range of 1–10 nm which are observed as a halo on the XRD pattern. The increase of the modifier—APP—amount decreases the activity of its interaction with the NC and probability of such deformation, which in turn, decreases the electron flow through carbon. In fact, the halo area on the XRD pattern of phosphorus-containing NC obtained at the NC/APP ratio equal to 2 is 1.84 times larger than the similar area of the halo on the XRD pattern of the modified NC obtained at the ratio equal to 1. At the same time, the ratio of the line intensities I_{43}/I_{16} on the XRD pattern in Figure 11.1a is 2.5 times higher than that on the XRD pattern in Figure 11.1b. This can indicate the more active interaction of the APP thin layers with the Cu/C NC.

When the Cu/C NC is modified by silica, the reactant ratio does not influence the final result. Figure 11.2 shows the XRD pattern of the modified NC obtained at the ratio 1:1 of the NC and silica. The reduced element (silicon) is uniformly distributed over the modifier thickness independent of its thickness. On the XRD pattern (Fig. 11.2), the most intensive peak corresponds to the phase containing copper and silicon. The halo maxima coincide with interplain distances usually attributed to submolecular formations of silica.

When nickel oxide interacts with Cu/C NC, on the XRD pattern of the obtained modification product, there are peaks (Fig. 11.3) characterizing the interplanar spacing of 0.504 nm (about 43°), which corresponds to the Ni–Cu phase, and the interplanar spacing of 0.378 nm (about 63°), which corresponds to the $Ni_{0.95}Cu_{0.05}O$ phase. The reduced nickel is present in the product as an impurity.

It should be noted that the XRD pattern data are in good agreement with the data obtained by the XPS spectra. For studying the mechanism of the formation of the chemical bond between the atoms of metal, carbon, silicon, phosphorus, and nickel in the considered systems, the C1s, O1s, Cu3s, Ni3s, Si2p, and P2p core level spectra are investigated (Table 11.1; Figs. 11.4 and 11.5). The P2p spectra of the phosphorus-containing NCs obtained at the NC/APP ratio equal to 1 (Fig. 11.4a) and 2 (Fig. 11.4b) indicate that at the ratio of 2, the process of the phosphorus reduction proceeds by almost 50%.

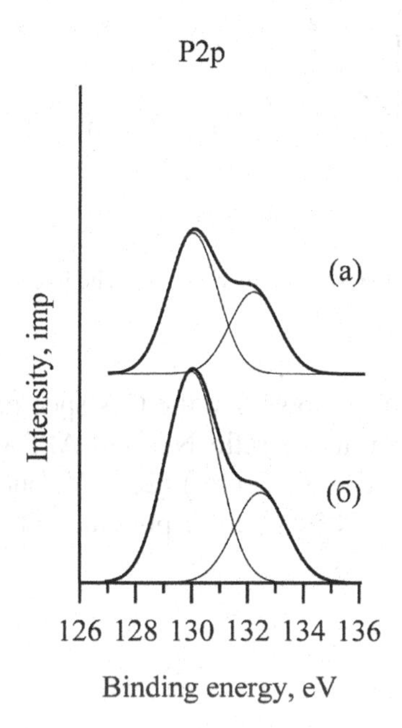

FIGURE 11.4 XPS P2p spectra: (a) Cu/C + APP, 1:1 and (b) Cu/C + APP 1:0.5.

The phosphorus peak corresponds to 130 eV, and the peak at 132.3 eV can possibly be ascribed to oxidized phosphorus or phosphorus attached to carbon by double bond (P=C).[6] In the C1s spectrum of phosphorus-containing NC obtained at the reactant ratio of 2, the increase in the C–H peak intensity is observed which is possibly due to the carbon shell deformation taking place in the redox process.

Figure 11.5 shows the Si2p spectrum of Cu/C NC modified with silica.

According to the intensity relation of the peaks at 99 eV (the peak corresponds to Si^0) and 101 eV (SiO_2), the silicon oxide reduction process proceeds by 50%. Such result is independent of the used amounts of the reactants (silica and Cu/C NC), which can be explained by the "screening" of the NC-active centers by the first layer.

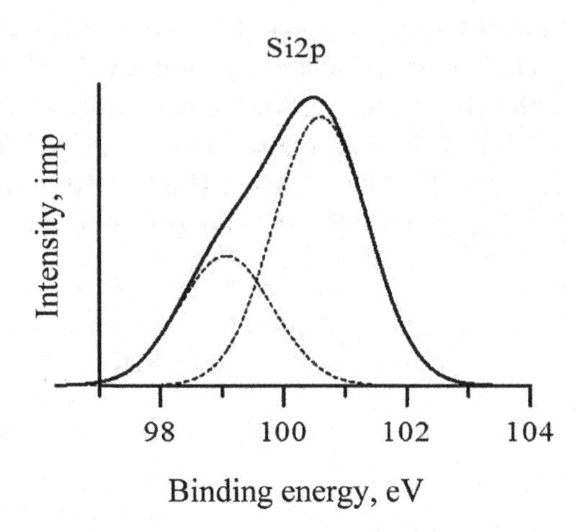

FIGURE 11.5 XPS spectrum Si2p of the product of the interaction of Cu/C NC + SiO_2 at the ratio 1:1.

The XPS investigation shows that the C1s spectrum of the phosphorus-containing NC obtained at the ratio NC and APP equal to 2 (Fig. 11.6) consists of two components C–C (sp^2) (284 eV) and C–C (sp^3) (286 eV), and the C–H component (285 eV) corresponding to the group of the polyene fragment.

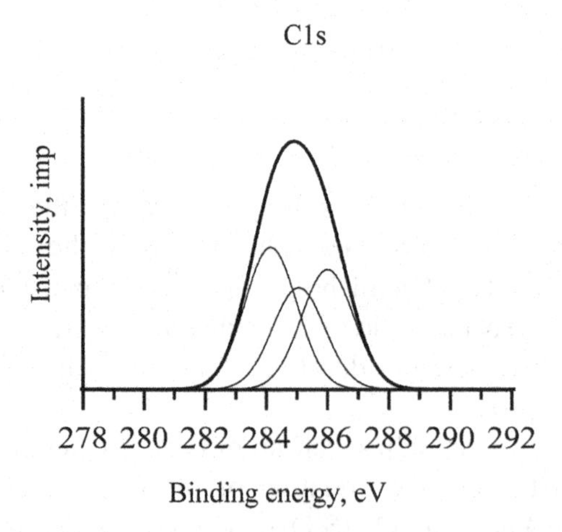

FIGURE 11.6 C1s spectrum of the phosphorus-containing NC obtained at the ratio of NC/APP = 2.

The intensities of the C–C (sp^2) and C–C (sp^3) components are similar, which corresponds to the Nan granule shape close to spherical. The intensity of the C–H component is almost three times larger than the intensities of the peaks in the spectra of the similar NCs obtained at smaller ratios of reactants. This indicates the deformation of the carbon shell accompanied by an increase in the polyene fragment amount in the surface layer.

As it follows from the C1s spectra, the similar deformation of the carbon shell takes place when Cu/C NC is modified with nickel oxide (Fig. 11.7).

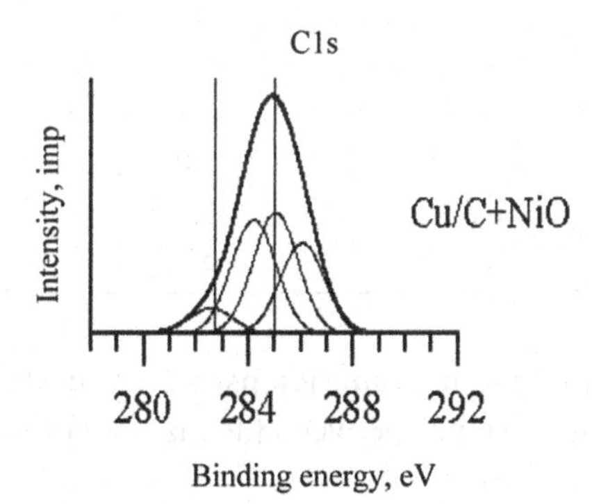

FIGURE 11.7 XPS C1s spectrum of nickel-containing Cu/C nanocomposite.

In this case, the polyene and carbide fragments move into the nanogranule surface layer. In the spectrum, the fourth component appears which can be referred to the sp hybridization. It is quite probable that the fourth component appears due to the Nan granule carbon shell deformation, which is also indicated by the growth of the intensity of the component which is referred to the electron energy of the C–H bond. The intensity of the C–H component in the C1s spectrum of the phosphorus-containing NC obtained at the NC/APP ratio equal to 2 (Fig. 11.6) is less than that of the same component in the spectrum of the initial Cu/C NC (Fig. 11.7).

From the comparison of the atomic magnetic moments of copper (Table 11.1), it follows that the largest increase of the copper atomic magnetic moment is reached for the phosphor-containing NC obtained at the ratio of Cu/C NC and APP equal to 2 (the magnetic moment is 4.2 μ_B). At the reactant ratio equal to 1, the magnetic moment value decreases more than by a factor of 2. This result can be explained by the decrease of the rate of the quantization of an electron at the growth of the thickness of the APP layer. In the case of obtaining the silicon-containing NC, a decrease in the thickness of the modifier layer does not lead to a change in the copper atomic magnetic moment. All the parameters remain the same (Table 11.1). When the Cu/C NC is modified with nickel oxide, a change in the copper and nickel atomic magnetic moments is observed.

TABLE 11.1 Parameters of the Multiple Splitting of the 3s Spectra of the Cu/C NC Modified with Phosphorus-, Silicon- and Nickel-containing Compounds.

Sample	I_2/I_1	Δ, eV	μ_{met}, μ_B
$Cu3s_{nano}$	0.2	3.5	1.3
$Cu3s_{nano}$ (P)	0.4	3.5	2.0
$Cu3s_{nano}$ (P × 2)	0.4	3.5	2.0
$Cu3s_{nano}$ (P × 1/2)	0.85	3.5	4.2
$Cu3s_{nano}$ (Si)	0.6	3.0	3.0
$Cu3s_{nano}$ (Si1/2)	0.6	3.0	3.0
$Cu3s_{nano}$ (Ni)	0.4_{Cu} (0.4_{Ni})	3_{Cu} (2_{Ni})	2_{Cu} (2.3_{Ni})

I_2/I_1, the relation of the maxima intensities of the lines of the multiple splitting; Δ, the energy distance between the maxima of the multiple splitting in Me3s-spectra.

The increase of the copper atomic magnetic moment due to the "unsparing" of d electrons and their movement to higher levels is to some extent confirmed by the growth of the spin number by two orders per gram on the Nan granule carbon shell of the phosphorus-containing Cu/C NC obtained at the NC/APP ratio = 2. The result has been obtained at the EPR investigation of the corresponding specimen.

Thus, the offered hypothesis on the unsparing and movement of copper d electrons to the Nan granule carbon shell for replenishing the consumption in the redox process is confirmed by the XRD patterns and ERP studies. It should be noted that the number of electrons participating in the process determines the growth of the copper atomic magnetic moment. It can be suggested that such mechanism of controlling the magnetic characteristics of metal atoms will take place for d and f elements of 5–7 periods to a larger degree.

11.4 CONCLUSION

In the present paper, a hypothesis is offered of the possibility of controlling the magnetic characteristics of d and f elements by redox processes with the participation of a certain number of electrons. When the number of electrons participating in the process is increased, the growth of the metals' atomic magnetic moments is expected due to the "pulling out" of the corresponding d(f) electrons to higher energy levels, and possibly, to carbon shells. The hypothesis is confirmed by the results of the investigation of Cu/C NC modified by phosphorus-, silicon-, and nickel-containing substances with the use of redox processes. The redox reactions lead to the reduction of the mentioned elements from the oxidized forms forming the bonds between the elements and the Nan granule metal.

KEYWORDS

- **copper/carbon nanocomposite**
- **mechanochemical modification**
- **X-ray photoelectron spectroscopy**
- **atomic magnetic moment**
- **electronic structure**

REFERENCES

1. The Method for Obtaining Carbon Metalcontaining Nanostructures. Applicant and Patent Holder—AO "IEMZ—KUPOL". Patent RF 2,393,110, 2010.
2. Shabanova, I. N.; Kodolov, V. I.; Terebova, N. S.; Trineeva, V. V. X ray Photoelectron Spectroscopy for the Investigation of Metal/Carbon Nanosystems and Nanostructured Materials.—Izd-vo "Udmurtskiy Universitet": M.-Izhevsk, 2012; p 252.
3. Kodolov, V. I.; Trineeva, V. V. New Scientific Trend—Chemical Mesoscopy. *Chem. Phys. Mesoscopy* **2017,** *19* (3), 454–465.
4. Ma'Mari, F. A.; Moorson, T.; Teobaldi, G.; et al. Beating the Stoner Criterion Using Molecular Interface. *Nature* **2015,** *524*, 69–73.
5. Kodolov, V. I.; Trineeva, V. V.; Kopylova, A. A. et al. Mechanochemical Modification of Metal/Carbon Nanocomposites. *Chem. Phys. Mesoscopy* **2017,** *19* (4), 230.
6. Wang, J. Q.; Wu, W. H.; Feng, D. M. The Introduction to Electron Spectroscopy (XPS/XAES/UPS)—National Defense Industry Press: Beijing, 1992; p 640.

CHAPTER 12

MECHANOCHEMICAL MODIFICATION OF METAL/CARBON NANOCOMPOSITES

VLADIMIR I. KODOLOV[1,2*], V. V. TRINEEVA[1,3], A. A. KOPYLOVA[1,2], R. V. MUSTAKIMOV[1,4], S. A. PIGALEV[1,2], N. S. TEREBOVA[1,5], T. M. MAKHNEVA[1,3], and I. N. SHABANOVA[1,5]

[1]*Basic Research—High Educational Center of Chemical Physics and Mesoscopy, Udmurt Scientific Centre, Ural Division, RAS, Izhevsk, Russia*

[2]*Kalashnikov Izhevsk State Technical University, Izhevsk, Russia*

[3]*Institute of Mechanics, Udmurt Scientific Centre, Ural Division, RAS, Izhevsk, Russia*

[4]*Scientific Innovation Centre, Izhevsk Electromechanical Plant—"KUPOL," Izhevsk, Russia*

[5]*Physicotechnical Institute, Udmurt Scientific Centre, Ural Division, RAS, Izhevsk, Russia*

[*]*Corresponding author. E-mail: kdol@istu.ru*

ABSTRACT

The conditions of copper (nickel)/carbon nanocomposites modification by such substances as silica (or silica gel), ammonium polyphosphate (or polyphosphoric acid), and ammonium thiosulfite (or sulfur) are substantiated in this chapter. The mechanochemical modification mechanism of correspondent nanocomposites is proposed. The results of modified nanocomposites production with the mechanism confirmation are discussed. The structures and properties of modified nanocomposites are investigated by the following methods: X-ray photoelectron spectroscopy, radiography, IR spectroscopy, transition electron microscopy. The structural changes of modified nanocomposites as well as increase in the metal atomic magnetic moments are presented. In the phosphorus containing Cu/C nanocomposites, the copper atomic magnetic moment is increased from 1.3 to 3 μB, and the oxidation state of phosphorus atom is decreased from 5 to 0, according to X-ray photoelectron spectroscopy.

12.1 INTRODUCTION

Some possible methods of nanostructures (nanosystems) modification are known.[1–3] The substances interaction processes with nanostructures occurred at the formation of covalent or coordinative bonds, and also possibly at dispersion interaction. These processes are realized when the gaseous or liquid substances act on nanostructures or, as in our case, when the interaction between solid nanocomposite (NC) and active components takes place. In other words, in our case, the mechanochemical modification is carried out, including reduction–oxidation process. For the investigation and the far production of NCs, following NCs and substances in pairs are used: cooper/carbon, nickel/carbon, iron/carbon NCs, and also silica, silica gel, ammonium polyphosphate, ammonium sulfite. The investigations of initial NCs show that metal/carbon NCs represent to the metal clusters associated with carbon fibers or the metal clusters within carbon shell consisting from three to four layers of carbon fibers. Each carbon fiber includes acetylene and carbine fragments with delocalized electrons on fragment joints. In other words, the metal/carbon NCs obtained by mechanochemical method have active surface, which contains double bonds and delocalized electrons.[4–7] These NCs are produced on the Izhevsk Electromechanical Factory "Kupol."

12.2 METHOD OF METAL/CARBON NCs MECHANOCHEMICAL MODIFICATION

The representation about metal/carbon NCs is based on the results of the following investigations: IR or Raman spectroscopy, X-ray photoelectron spectroscopy (XRPES), electron paramagnetic resonance (EPR), transition electron microscopy (TEM), and atomic force microscopy (AFM). The structure of copper/carbon NC is shown in Figure 12.1.

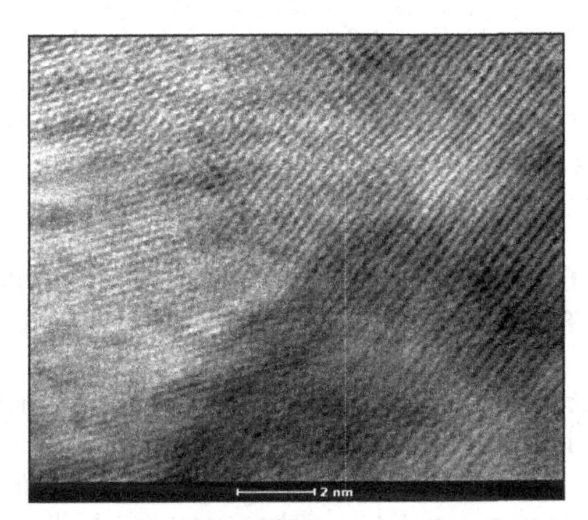

FIGURE 12.1 TEM microphotograph for copper/carbon nanocomposites.

The presence of active double bonds and delocalized electrons in carbon shell of metal/carbon NCs gives possibility for their modification by means of redox and addition processes. Therefore, the substances interaction, in which elements (Si and P) have the highest oxidation state (+4 or +5), with such electron-containing compounds as metal/carbon NCs, is evident. The plain scheme of reduction process is shown below.

$$\text{HK}\,(-\delta, \text{ or } \textstyle\sum e) + P^{+5} \rightarrow \text{HK} - P^{+3}\,(P^{+2}, P^{0})$$
$$\text{HK}\,(-\delta, \text{ or } \textstyle\sum e) + Si^{+4} \rightarrow \text{HK} - Si^{0}\,(Si^{+2})$$

Mechanism based on the chemical mesoscopics notions for electron transport across nanoparticle (metal cluster) from one to other bank (nanoreactor walls) is proposed in Figure 12.2.

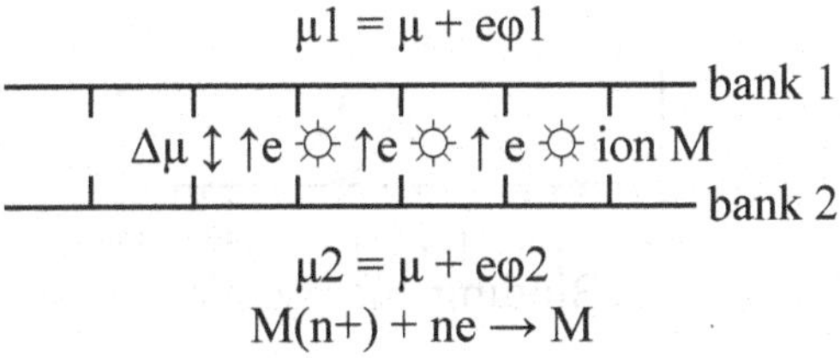

$$\mu 1 = \mu + e\varphi 1$$

bank 1

$\Delta\mu \updownarrow \uparrow e$ ☼ $\uparrow e$ ☼ $\uparrow e$ ☼ ion M

bank 2

$$\mu 2 = \mu + e\varphi 2$$
$$M(n+) + ne \rightarrow M$$

FIGURE 12.2 Scheme of electron transport and the redox reactions within nanoreactors.

Chemical activity of nanogranules obtained is determined by their interaction with polar substances containing p elements with the positive oxidation state, for example, Si^{+4}, P^{+5}, and S^{+6}.

In the metal/carbon NC-modification process, grinding of copper/carbon or nickel/carbon NCs with the corresponding substances containing P or Si is carried out. As the result of this mechanochemical process, the xerogel is formed, which is dried at 80°C.

12.3 INVESTIGATION OF METAL/CARBON NCs: PROPERTIES AND CHARACTERISTICS

The previously created method[5] is used for the copper/carbon NC modification mechanism investigation. The inner levels spectra for C1s, O1s, Si2p, P2p, S2p, and Cu3s are studied to investigate the mechanism of chemical bonds formation between metal or carbon atoms and silicon or phosphorus atoms. Cu3s spectra of nanosystems modified by Si-, P-, and S-containing compounds are presented in Figure 12.3.

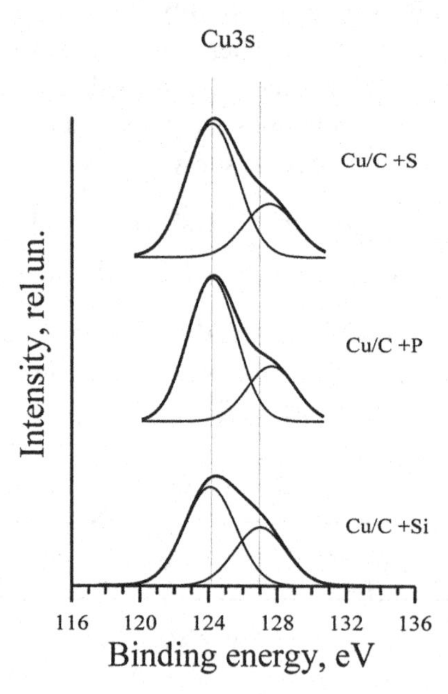

FIGURE 12.3 X-ray photoelectron spectra of copper/carbon nanocomposites modified by p element-containing substances.

In Ref. [1], it is shown that the copper atomic magnetic moment appears in copper/carbon NC. During modification, the atomic magnetic moments of metal/carbon NCs increased in comparison with nonmodified NCs (Table 12.1). The analysis of Table 12.1 data shows the influence of modifier nature on the magnetic characteristics as well as the influence of its layer thickness on the nanogranule surface. At the same time, it is shown that atomic magnetic moment grows in modifier (elements) row Si, P, and S at the identical conditions. It is established that during modification, all modifiers are changed because of elements reduction.

TABLE 12.1 The Parameters of the Multiple Splitting of the 3s Spectra of the Metal/Carbon Nanocomposites Modified with Phosphorus-, Silicon-, and Nickel-containing Compounds.

Sample	I_2/I_1	Δ, eV	μ_{Ni}, μ_B	μ_{Cu}, μ_B	μ_{Fe}, μ_B
Ni3s (crystal)	0.15	4.3	0.5		
Ni3s (nonmodified)	0.32	3.0	1.8		
Ni3s (NC/Si s – 1)	0.80	3.8	4.0		
Ni3s (NC/P s – 1)	0.60	3.6	3.0		
Ni3s (NC/S s – 1)	0.56	4.0	2.8		
Cu3s (nonmodified)	0.20	3.6		1.3	
Cu3s (NC/Si s – 1)	0.60	3.0		3.0	
Cu3s (NC/P s – 2)	0.85	3.5		4.2	
Cu3s (NC/P s – 1)	0.42	3.6		2.0	
Cu3s (NC/S s – 1)	0.40	3.6		1.8	
Cu3s (NC/Ni s – 1)	0.4Cu/ 0.4 Ni	3 Cu/ 2 Ni		2.0 Cu/2.3 Ni	
Fe3s (nonmodified)	0.42	3.9			2.2
Fe3s (NC/P s – 1)	0.50	4.0			2.5
Fe3s (NC/P s – 2)	0.60	4.0			3.2

I_2/I_1, the relation of the maxima intensities of the lines of the multiple splitting; Δ, the energy distance between the maxima of the multiple splitting in Me3s-spectra.

For instance, according to P2p spectrum (Fig. 12.4), phosphorus in phosphorus-containing copper/carbon NC changes the oxidation state from +5 to zero. The binding energy of P2p changes from 135 eV, corresponding to PO_4 group, to 129 eV for P^0. The process flows on 90%. The interaction of copper and phosphorus is possible in this case.

C1s spectrum of this NC (Fig. 12.5b) is distinguished from C1s spectrum of nonmodified NC (Fig. 12.5a); the corresponding form C–H on 15% smaller than in spectrum of nonmodified NC. In turn, the relation of intensities for sp^2 and sp^3 hybridization is increased that can be linked with the nanogranule increasing and approach its form to roundish.

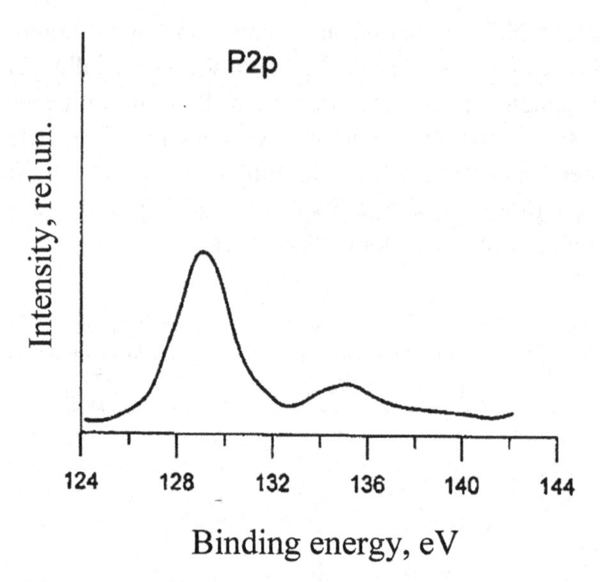

FIGURE 12.4 P2p spectrum of Cu/C nanocomposite modified by ammonium polyphosphate at the relation 1:1.

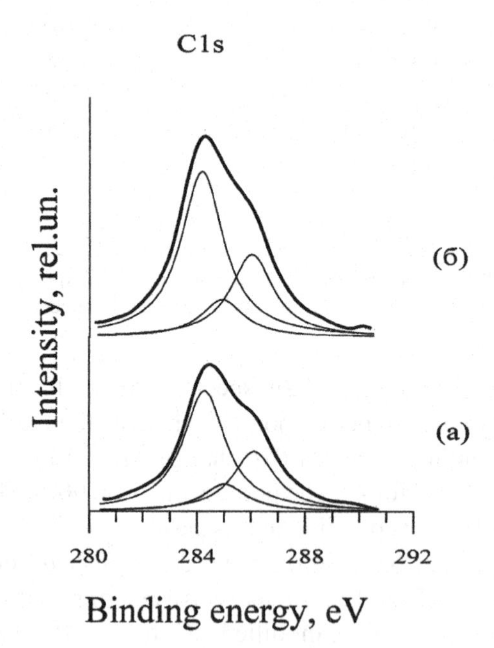

FIGURE 12.5 C1s spectra: (a) C1s spectrum of copper/carbon nanocomposite and (b) C1s spectrum of modified P-containing copper/carbon nanocomposite.

Analogous investigations are carried out[4,5] for copper/and nickel/carbon NCs modified by silicon-containing substances. Spectra Si2p and C1s for Cu/C NC modified by silica at the relation NC/silica = 1 leads to Figure 12.6.

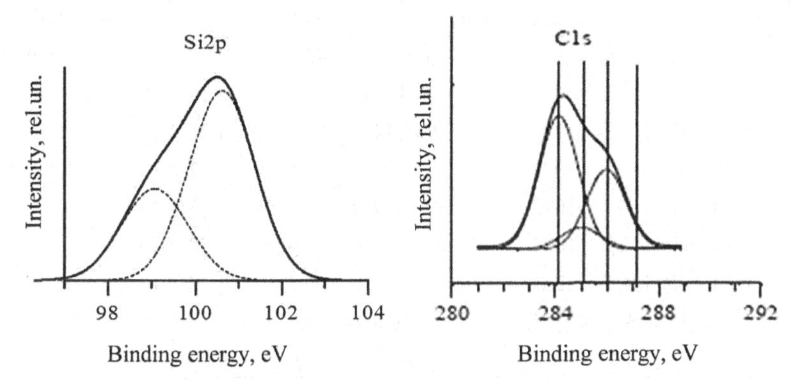

FIGURE 12.6 X-ray photoelectron spectra for modified Cu/C (Si) nanocomposite: (a) Si2p spectrum and (b) C1s spectrum.

According to Si2p spectrum, the relation of spectrum form intensities shows the redox process development of 51.4%. C–H intensity in spectrum of modified Si-containing NC was 65% smaller than the correspondent value in spectrum of initial NC. The thickness of Si-containing shell for Cu/C (Si) NC is four times higher in comparison with shell of modified P-containing NC.

For confirmation and development of notions about modification process and obtaining of modified metal/carbon NCs, the investigations are carried out with application of the following methods: radiography, IR spectroscopy, and TEM. It should be taken that in all the cases of metal/carbon NCs, modification by Si-, P-, S-containing substances results in reduction of these elements. For example, the X-ray pattern analysis of phosphorus-containing Cu/C NC shows the presence of peaks for groups Cu–C–P at θ equaled to 43°.

Modified NCs are studied by IR spectroscopy. For phosphorus-containing NCs, the highest intensity of spectra lines corresponds to P–O–C group vibration which is formed during the interaction of ammonium polyphosphate with the NC shell (Table 12.2).

The investigation of copper/carbon NC modified by silica using IR spectroscopy shows increase of medium self-organization index D (Fig. 12.7) which is calculated by formula:

$D = I/(a/2)$, where I—the line intensity and a/2—the half width of line.

TABLE 12.2 The Characteristics of Lines for IR Spectra of Samples Cu/C↔P (1:1), Cu/C↔P (1:0.5), and Cu/C↔P (1:1.5).

N lines	Spectrum of samples	Characteristics of spectrum lines				Corre-sponding groups
		Wave number, cm^{-1}	Intensity	Half width	D	
1	Cu/C↔P (1:0.5)	911	0.010519	42,324	2.49×10^{-4}	P–O–P
	Cu/C↔P (1:1)	904	0.011797	51,674	2.28×10^{-4}	P–O–P
	Cu/C↔P (1:1.5)	895	0.024485	57,447	4.26×10^{-4}	P–O–P
2	Cu/C↔P (1:0.5)	1073	0.054722	149,260	3.67×10^{-4}	P–O–C
	Cu/C↔P (1:1)	1072	0.039904	135,310	2.95×10^{-4}	P–O–C
	Cu/C↔P (1:1.5)	1073	0.019365	71,075	2.72×10^{-4}	P–O–C
3	Cu/C↔P (1:0.5)	1254	0.005739	50,479	1.14×10^{-4}	P=O
	Cu/C↔P (1:1)	1254	0.0090368	48,373	1.87×10^{-4}	P=O
	Cu/C↔P (1:1.5)	1253	0.022251	53,025	4.20×10^{-4}	P=O

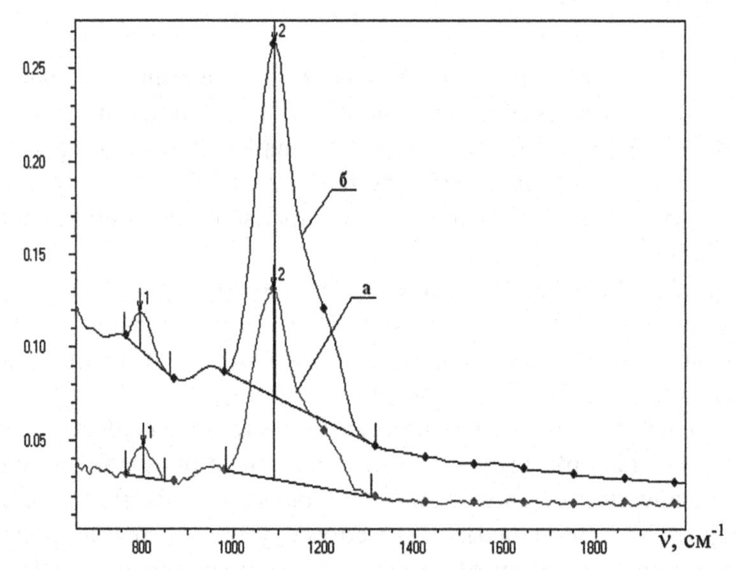

FIGURE 12.7 **(See color insert.)** IR spectra of samples: (a) silica and (b) silica + Cu/C nanocomposites.

According to data of Table 12.3, the self-organization is increased in field of 980–1300 cm^{-1} (52.11%). The line in this field corresponds to the vibration of Si–O and, probably, Si–C bonds.

TABLE 12.3 Characteristics of IR Spectra Lines for Samples Silica and System Silica with Cu/C Nanocomposites.

No. of lines	Sample	Characteristics of spectra lines					Corresponding to lines of spectrum
		Start, cm^{-1}	Finish, cm^{-1}	Intensity	Half width	D	
1	$(SiO_2)_n$	759.98	848.7	0.016287	53.711	3.032×10^{-4}	Si–O–H
	$(SiO_2)_n$ + Cu/C HK	763.83	854.49	0.022069	47.481	4.648×10^{-4}	Si–O–H Si–C
2	$(SiO_2)_n$	983.73	1303.9	0.10186	102.66	9.922×10^{-4}	Si–O Si–O–Si
	$(SiO_2)_n$ + Cu/C HK	983.73	1292.3	0.19596	102.2	19.174×10^{-4}	Si–O Si–O–Si Si–C

The investigation of modified metal/carbon NCs by means of TEM of energy resolution shows that the shell from carbon fibers on nanogranule surface is well preserved. For instance, the TEM image of phosphorus-containing Cu/C NC is presented in Figure 12.8.

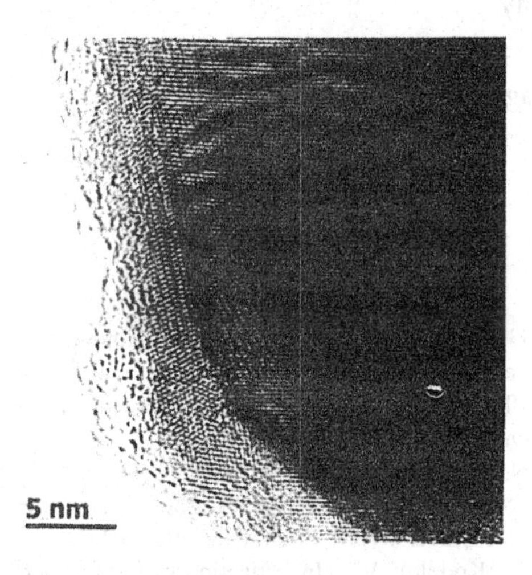

FIGURE 12.8 TEM image of phosphorus containing copper/carbon nanocomposite.

In contrast to Figure 12.1, in this case, there are bulges and strands of fibers. At the same time, the possible phases are found: carbon—70%; phases Cu–C–P–O and C–Cu–P–O in which the content of copper equals 15%; and phosphorus equals 7%.

12.4 CONCLUSION

For the first time, the metal/carbon NCs which contain silicon, phosphorus, or sulfur, are produced within active media by means of mechanochemical modification. In this case, the change of element oxidation states as well as the increasing of metal (copper, nickel, and iron) atomic magnetic moment takes place. At the same time, above elements and functional groups with them appear in carbon shell of NCs. These facts open a new era for further investigations and development of metal/carbon NCs application fields.

KEYWORDS

- **mechanochemical modification**
- **metal/carbon nanocomposites**
- **redox synthesis**
- **X-ray photoelectron spectroscopy**
- **radiography**
- **IR spectroscopy**
- **atomic magnetic moment**

REFERENCES

1. Patent RF 2,393,110, 2010. Applicant and Patent Holder "IEMZ—KUPOL." Method for Preparation of Carbon Metal-containing Nanostructures.
2. Kodolov, V. I.; Trineeva, V. V. New Scientific Trend—Chemical Mesoscopics. *Chem. Phys. Mesoscopy* **2017,** *19* (3), 454–465.
3. Shabanova, I. N.; Kodolov, V. I.; Terebova, N. S.; Trineeva, V. V. X ray Photoelectron Spectroscopy for Investigation of Metal/Carbon Nanosystems and Nanostructured Materials. Udmurt University: M.—Izhevsk, 2012; p 252.
4. Kopylova, A. A.; Kodolov, V. I. Investigation of the Copper/Carbon Nanocomposite Action with Silicon Atoms from Silica Compounds. *Chem. Phys. Mesoscopy* **2014,** *16* (4), 556–560.
5. Kopylova, A. A.; Zaitseva, E. A.; Kodolov, V. I. The Functionalization of Copper/Carbon Nanocomposite by Silicon Atoms. Abstracts of Fifth International Conference "From Nanostructures, Nanomaterials and Nanotechnologies to Nanoindustry," Izhevsk, April, 2015; p 93–95.

6. Mustakimov, R. V. The Elaboration of Intumescent Fireproof Coating Modified by Phosphorus Containing Copper/Carbon Nanocomposite. Magister Thesis, IzhSTU, Izhevsk, 2015.
7. Pigalev, S. A.; Kodolov, V. I. Quantum Chemical Modeling of Copper/Carbon Nanocomposite Functionalization by Sulphur Containing Compounds. Abstracts of Fifth International Conference "From Nanostructures, Nanomaterials and Nanotechnologies to Nanoindustry." Izhevsk, April, 2015; p 153.

CHAPTER 13

SELF-NANOEMULSIFYING DRUG DELIVERY SYSTEM: FORMULATION DEVELOPMENT AND QUALITY ATTRIBUTES

PANKAJ V. DANGRE* and SHAILESH S. CHALIKWAR

Department of Pharmaceutics, R.C. Patel Institute of Pharmaceutical Education and Research, Shirpur, Dhule, Maharashtra, India

**Corresponding author. E-mail: pankaj_dangre@rediffmail.com*

ABSTRACT

The discovered new chemical entities have high lipophilicity and poor aqueous solubility (Biopharmaceutics Classification System class II and IV) which result in poor oral bioavailability, lack of dose proportionality, and high intra- and intersubject variability. For better therapeutic regimen, solubility is the most imperative criterion to achieve desired concentration of drug in systemic circulation. Self-nanoemulsifying drug delivery system (SNEDDS) is an emerging novel technique which solves problems associated with the delivery of poorly soluble drugs. They are isotropic mixture of oil, surfactant, and cosurfactant and are vital tool in solving low bioavailability issue of poorly soluble drug. The technique involved encapsulation of drug in a lipid base with or without pharmaceutically acceptable surfactant. Very recently, the technique has attracted researchers and industries owing to their physical stability, easy manufacturing methods, and can be filled in soft gelatin capsules. On administration of SNEDDS, it generates a drug containing microemulsion with a large surface area upon dispersion in the gastrointestinal tract. The emulsions facilitate the absorption of the drug via intestinal lymphatic pathway and by partitioning of drug into the aqueous phase of intestinal fluids, which results in higher bioavailability of the drug. The present chapter deals with the formulation components, mechanism of self-emulsification, formulation methodology, evaluation parameters, and application of self-microemulsifying drug delivery system.

13.1 INTRODUCTION

The oral route is the most popular and commonly used route for drug therapy due to its convenience of administration, patient comforts, and economical and simple design of dosage form. However, oral delivery of 50% drug molecules is hindered due to their high lipophilicity.[1] Approximately 40% of newly developed new chemical entities (NCEs) show low water solubility which results into poor bioavailability.[2] The absorption of such molecules from the gastrointestinal tract (GIT) is typically controlled by rate of dissolution.[3] Various efforts have been put forth for developing suitable formulations to improve the solubility and bioavailability of Biopharmaceutical Classification System (BCS) class II or IV. The various approaches employed are, including the use of surfactant,[4] lipids-based formulation,[5] particle size reduction,[6] permeation enhancer,[7] cyclodextrin complexation,[8] nanoparticles,[9] and solid dispersions.[10] The main focus of these formulation approaches is to increase the bioavailability of class II or IV drugs.

Recently, oral delivery of lipid-based drug delivery systems is gaining more concern, especially self-nanoemulsifying drug delivery systems (SNEDDSs). Numerous literatures report that hydrophobic drugs are more readily absorbed from these self-emulsifying (SE) preparations and improve the oral bioavailability.[11] SEDDS improves the solubility of poorly water-soluble drug (PWSD) and maintains the drug in the dissolved state in a small droplet of oil which remain intact all over its transit through GIT.[12–14] Furthermore, the SNEDDS alters biopharmaceutical characteristics of drugs and enhances solubility and improves the lymphatic absorption of PWSDs.[15,16]

Lipid-based formulation system offers the advantage of improved absorption and consequently enhanced bioavailability of lipophilic drug molecules particularly those belonging to BCS class II and IV.[17] The lipids, typically triglycerides, alter the biopharmaceutical properties of drugs and thereby increase the rate of dissolution and solubility in the intestinal fluids. Furthermore, the drug remains in the dissolved state in the form of oil droplet which protects the drug from chemical as well as enzymatic degradation.

The administration of lipophilic drug along with lipid causes stimulation of lymphatic transport of drug and improves the drug absorption in to the portal circulation.

The Lipid Formulation Classification System (LFCS) was introduced by Pouton in 2000.[12] LFCS categories include four types of lipid formulations based on their composition, dilution effect, digestion, and the ability to prevent drug precipitation.

Type I: This system includes all the formulations in which drug present in triglycerides and/or mixed glycerides or in an emulsion (oil-in-water, o/w) which is stabilized by low amount of surfactants. The system shows poor aqueous dispersion and requires digestion by pancreatic lipase enzymes to form amphiphilic lipid which encourages drug transfer across colloidal aqueous phase. Type I is regarded as simple and biocompatible formulation.

Type II: This system is commonly regarded as SNEDDS. SNEDSS is an isotropic (single phase) mixture of oil, surfactant, and cosurfactant that form microemulsion after introduction into the aqueous media. Self-emulsification usually occurs at high surfactant concentration, that is, above 25% (w/w). However, at higher surfactant concentrations, more than 50–60% (w/w), the process of emulsification slows down with the formation of viscous liquid crystalline gels at the oil/water interface. The system has the potential ability of overcoming the slow dissolution rate which occurs mostly with water insoluble drugs due to effective partitioning of drugs between the oil and the aqueous phase which leads to faster absorption.

Type III: This system is typically regarded as SNEDDS. This system, besides lipids, includes the hydrophilic surfactants (hydrophilic and lipophilic balance, HLB > 12) and cosolvents such as ethanol, propylene glycol, and polyethylene glycol. This system is further divided into Type IIIA and Type IIIB formulations based on the hydrophilicity of the system. Type IIIB formulation is a more hydrophilic system where the content of hydrophilic surfactants and cosolvents is high to achieve greater dispersion rates when compared with Type IIIA. However, the chance of drug precipitation in formulation is high with Type IIIA as the content of lipid is low. SNEDDS formulations have wide contribution in the improvement of bioavailability of PWSDs.

Type IV: This system represents most hydrophilic formulations as these are devoid of oils. The system produces fine dispersion after introduction into the aqueous medium. A Type IV formulation is suitable for hydrophobic drug which has no lipophilicity.

13.2 MISCELLANEOUS TERMS

13.2.1 SELF-MICROEMULSIFYING DRUG DELIVERY SYSTEM

Self-microemulsifying drug delivery system (SMEDDS) is a single phase system consisting of oil, surfactant, and cosurfactant. The SMEDDS, after dilution in aqueous medium, forms microemulsion.[18] The peristaltic activity of the stomach and intestine resulted in the formation of microemulsion.[19] Generally, the size of the oil globules of the SMEDDS formulation after formation of microemulsion falls in the range between 100 and 250 nm.[20] The small size of oil droplets provides large surface area for the drug.[21] The numerous studies on SMEDDS formulation indicate its potential in enhancement of oral absorption as well as stability of the formulation.[22–25]

SMEDDS has shown to improve the incomplete dissolution of PWSD by enhancing its transportation via the lymphatic system and avoid the P-glycoprotein efflux, which results in increased drug absorption from the GIT.[26,27]

13.2.2 SELF-NANOEMULSIFYING DRUG DELIVERY SYSTEM

SNEDDSs are anhydrous isotropic liquid mixtures of oil, surfactant, and cosurfactant which simultaneously form o/w nanoemulsions upon dilution with the aqueous medium under gentle stirring.[28] The basic difference between SMEDDS and SNEDDS is between the sizes of the oil globules formed after the emulsification. The size of SMEDDS represents the globule size between the 100 and 250 nm, whereas in case with the SNEDDS, the size globule is considered less than 100 nm.[29,30]

Both SMEDDS and SNEDDS have showed to be the most booming approaches in the enhancement of solubility and bioavailability of the BCS class II and IV drugs.[31,32]

13.3 NOTEWORTHY FEATURES

13.3.1 BENEFITS FOR PATIENTS

The characteristic of SNEDDS depends on its preparation and formulation and the most essential property is rapid emulsification in the GIT. SNEDDS offers numerous advantages over the conventional emulsion. The advantages of SEDDS include:

- SNEDDS is physically more stable formulation as compare to the conventional emulsion.[31]
- SNEDDS produces fine oil droplets which promote rapid distribution of the drug throughout the GIT.[32]
- SNEDDS offers a large interfacial area for partitioning of the drug between oil and water.
- SNEDDS improves the oral bioavailability of PWSD by providing more consistent profiles of drug absorption and thereby enabling reduction in dose.[33]
- SNEDDS provides protection to the sensitive drugs from the hostile environment in the gut.
- SNEDDS offers protection to enzymatic degradation in the GIT.
- SNEDDS provides a selective drug targeting toward a specific absorption window in the GIT.[34,35]
- Control of delivery profiles.[36]
- SEDDS provides high loading capacity when compared to the conventional lipid solution.
- SNEDDS offers ease of manufacturing process and scale-up.
- The SNEDDS formulations can be filled into a capsule shell, either soft or hard, or after solidification it can be transferred into solid dosage form such as tablet.

13.3.2 BENEFITS OVER THE CONVENTIONAL DRUG DELIVERY SYSTEM

1. The fine droplets of the SNEDDS provide rapid distribution of the drug throughout GIT and thereby reduce the irritation associated with prolonged contact between bulk drug and the gut wall.[37]
2. The SNEDDS formulations are comparatively more physically stable than conventional emulsions. Furthermore, the physical stability of the liquid SNEDDDS can be improved by transforming it into the solid state.[38]
3. SNEDDS, after dilution, provides large interfacial area for partitioning of the drug between oily and aqueous phase which generally does not happen in case with the oily solutions of PWSDs.
4. As compared to the other lipid-based drug delivery systems, SNEDDS maintains the constant plasma drug concentration level and thus helps in better control of disease conditions.

5. Preparation of SNEDDS is relatively simple and requires simple mixing equipments; also the time needed for the formulation is less as compared to the emulsion.[39,40]

13.4 THE KEY COMPONENTS IN THE FORMULATION

13.4.1 OILS

The oil is considered as the most vital component in SEDDS. It has the ability to solubilize the required dose of the hydrophobic drug and facilitate self-emulsification as well as improve the lymphatic transportation of lipophilic drug which increases the absorption of the drug from GIT.[41,42] The saturated fatty oils (medium chain or long chain) and hydrolyzed vegetable oils have widely employed in the formulation of SNEDDS due to their formulation compliance and physiological advantages.[43,44] However, the SNEDDS comprises both saturated and unsaturated fatty oils. The selection of oil depends on their properties, composition, application, and HLB. The myristic, capric, caprylic, caproic, and lauric acid are the most widely employed oils in the formulation of SEDDS.[45] The semisynthetic oils are well reputed to produce good emulsification formulations; the efficiency of emulsification can be improved by the addition of suitable solubilizers in the system for oral administration.[46,47]

13.4.2 SURFACTANT

The surfactant is the most essential component of SNEDDS formulation. Surfactant imparts the essential emulsifying characteristics to the SNEDDS. Surfactants of natural origin are preferred because they are safer than the synthetic surfactants. The selection of a particular surfactant for formulation depends upon its HLB value and safety issue. The HLB of surfactant provides important information on potential application in the formulation of SNEDDS. The most widely used are the nonionic surfactants with a relatively high HLB. The nonionic surfactants are nontoxic as compared to the ionic ones.[48,49] The most commonly used nonionic surfactant includes Tween 80 and Pluronic F127.[49,50] Usually, a stable SNEDDS formulation requires high concentrations of surfactant (30–60%); however, high concentration of surfactant may cause irritation to the gastric mucosa. Furthermore, the droplet size of the emulsion is inversely affected by the concentration of the surfactant.[51,52] Surfactant has ability to dissolve relatively large amount of hydrophobic drug molecules and thus prevent the precipitation of the drug within the GIT.[53]

13.4.3 COSURFACTANT

The cosurfactant in SNEDDS formulation facilitates the dispersion process and improves the dispersion rate. Cosurfactant is more commonly employed to dissolve larger amounts of either the hydrophilic surfactant or the drug in the lipid base. The newer cosolvents such as Transcutol™ and Glycofurol™ have numerous advantages over the traditional ones, including better stability and less volatility.[54,55]

13.5 PREPARATIONS OF SEDDS

The preparation of SNEDDS is quite simple and involves mixing of oil, surfactant, and cosurfactant. The selection of the components depends upon the solubility of the drug molecule in the respective excipients. Thereafter, the pseudoternary diagram is constructed to select the optimum composition of the component. The schematic representation of steps involved in the preparation of the SNEDDS is depicted in the Figure 13.1.

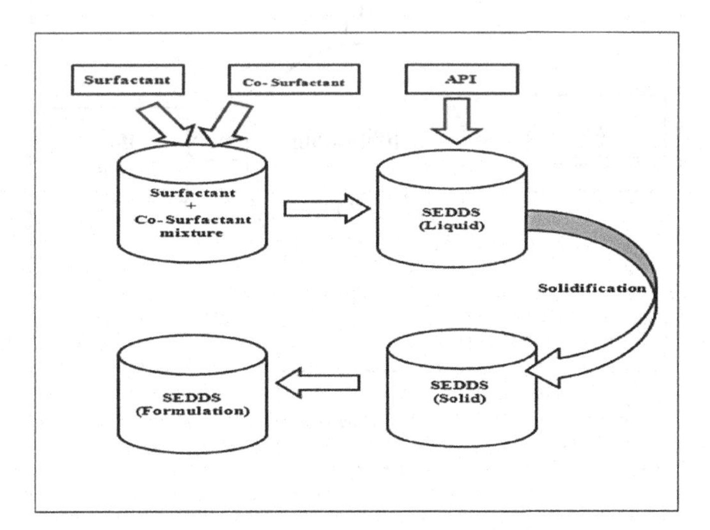

FIGURE 13.1 The schematic representation of steps involves in the preparation of SNEDDS.

- Construction of ternary phase diagram to find out optimum combinations of oil, surfactant, and cosurfactant.
- Solubilization of drug in mixture of surfactant and cosurfactant.
- The incorporation of oil phase into the above mixture.
- Transformation of liquid SNEDDS into solid by employing suitable solidification techniques.
- Finally, solid SNEDDS is developed into a suitable pharmaceutical dosage form.

13.6 SOLIDIFICATION TECHNIQUES

Generally, the formulated SNEDDS is liquid in state, but sometimes it could be in a semisolid state depending on the physical state of excipients used. Solid SNEDDS also offers added versatility in terms of possible dosage forms. The following description elaborates various liquid to solid SNEDDS conversion techniques (Fig. 13.2).

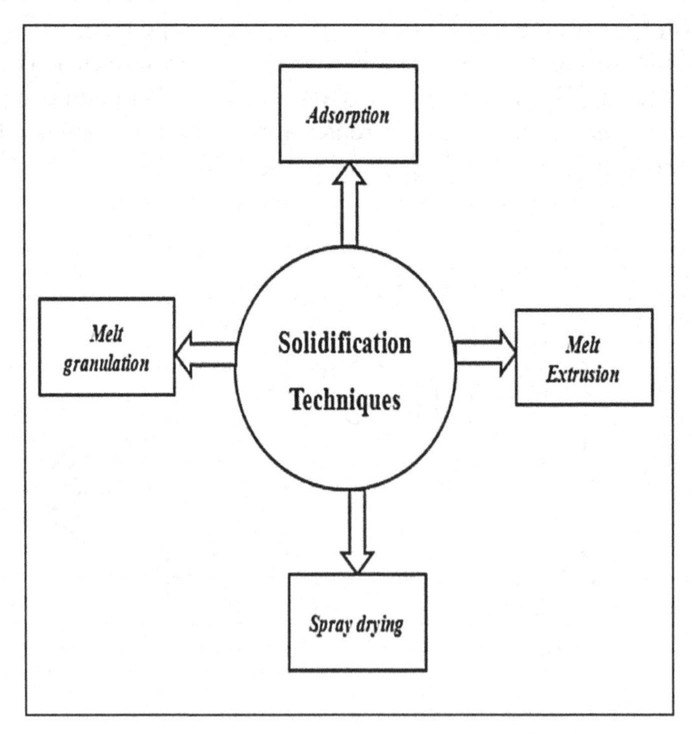

FIGURE 13.2 The liquid to solid conversion techniques (solidification) of SEDDS.

13.6.1 ADSORPTION TO THE SOLID SUPPORT

In this approach, the liquid SEDDS formulation is suitably adsorbed on inert solid support so as to obtain free-flowing powders. The process is quite simple and is accomplished by mixing liquid SNEDDS formulation with solid carrier in a suitable blender. The process is widely employed recently and offers numerous advantages, that is, good drug content uniformity and excellent drug-loading capacity (up to 70% w/w).[56] Solid carriers can be microporous[57] or/and cross-linked polymers.[58]

13.6.2 MELT GRANULATION

In this process, the agglomerate of the powder blend containing SNEDDS is formed. The process employed a meltable polymer and melting is performed at a relatively low temperature. A wide range of solid and semisolid lipids can be employed as meltable polymers. Among these, Gelucire is commonly employed. It also helps in the improvement in the dissolution rate.[59] The melt granulation process is commonly employed for adsorbing SNEDDS onto inert solid carriers.[60,61]

13.6.3 SPRAY DRYING

The process involves the solubilization of the drug in the liquid components so as to form a homogeneous mixture. The homogeneous liquid mixture is then atomized in the form of a spray of droplets. The droplets are brought into a drying chamber, so as to evaporate water contained in the liquid mixture resulting in the formation of dry particles. The careful evaluations are carried out to meet the desired specifications of the powder.

13.6.4 MELT EXTRUSION/EXTRUSION SPHERONIZATION

In this process, bulk material with plastic behavior is converted into a sphere particle (spheroids) of uniform size and density. This process has benefits of high drug-loading capacity (60%) and good content uniformity. In this process, the size of particles is controlled by the aperture size of an extruder. It is mostly employed in the pharmaceutical industry for the preparation of uniform sized pellets (spheroids).

The studies indicated that the maximum portion of this SNEDDS that can be solidified by extrusion spheronization takes 42% of the dry pellet weight.[62,63] The rheological characteristic of wet masses is measured by an extrusion capillary. SE pellets of diazepam, progesterone is prepared by the application of this process.[64]

13.7 CRITICAL QUALITY ATTRIBUTES

The relevant critical quality attributes (CQAs) of the SEDDS formulation can be justified by quality risk management system and experimentation. The general CQAs of the SNEDDS formulation should be considered during the product development. These properties are constitutional to the formulation and significantly affect the manufacturing process. Hereinafter, are described some of the critical common quality attributes that should be considered during the SNEDDS development.

13.7.1 PHYSICAL ATTRIBUTES

The liquid and solid SNEDDS formulation after preparation are generally packaged either in the capsule or transferred into another suitable dosage form. Therefore, the physical attributes, namely, color, odor, and appearance of the formulation are not considered as critical attributes. Furthermore, these attributes are not associated with efficacy and safety of the patient.

13.7.2 ASSAY AND DRUG CONTENT

The SNEDDS formulation being the homogeneous dispersion should contain the appropriate amount of the drug to ensure the desired therapeutic effect. The variability in the drug assay and content uniformity is likely to influence the drug safety and efficacy.[65] Therefore, these attributes are considered as moderately critical.

13.7.3 SELF-EMULSIFICATION TIME

It is the time taken by the SNEDDS to get emulsified in the aqueous medium. The disappearance of the SNEDDS in the medium gives an indication emulsification. The emulsification is prime essential for the SNEDDS and hence is considered as highly critical. The time of emulsification varies with the composition of the SNEDDS, as the more amount of oil in the formulation requires more time for emulsification.[66] However, the less time for emulsification indicates the maximum efficiency of the formulation.

13.7.4 GLOBULE SIZE

The globule size of SNEDDS is also considered as highly critical since the smaller globule size allows better penetration through GI epithelial lining and paracellular pathways.[67] The size of the globule depends upon the amount and chemical nature of the surfactant and cosurfactant in the SEDDS formulation.

13.7.5 THERMODYNAMIC STABILITY

The thermodynamic stability of SNEDDS is considered as critical attribute as the SEDDS should withstand the variation in temperature, mainly during the storage period. For this study, the SEDDS formulation is exposed to different stress conditions through freeze–thaw cycle (−21°C and +25°C) and heating–cooling cycle (4°C and 45°C) with the storage at each temperature of not less than 48 h.[68]

13.7.6 DISSOLUTION EFFICIENCY

Dissolution profile of the SEDDS formulation provides information about the drug release pattern as well as the solubilization of the drug in the dissolution medium. The performance of the SEDDS formulation completely depends upon its dissolution efficiency and hence is considered as critical attribute.

13.7.7 PERMEABILITY

The permeation of the drug molecule is highly desirable for achieving the therapeutically effective concentration in plasma blood concentration. SNEDDS is notably employed for enhancing the permeation characteristics of the drug, and therefore this parameter is considered as critical attribute.

13.7.8 BIOAVAILABILITY

The bioavailability is regarded as the amount of the drug that enters into the systemic circulation. The SEDDS is commonly employed to improve the bioavailability of the lipophilic drugs or those drugs which have less bioavailability. The efficiency of the SNEDDS formulation depends upon its ability to improve the bioabsorption of the drug, and therefore s considered as highly critical.

13.8 CHARACTERIZATION TECHNIQUES

13.8.1 SELF-EMULSIFICATION TIME

The determination of the self-emulsification time is carried out by using USP XXIV type II dissolution test apparatus. Approximately 1 g of SNEDDS sample is introduced into a basket holding 500 mL of distilled water maintained at 37°C with 50 rpm as speed of the paddle.[69] The assessment of the self-emulsification is carried out by visual observation.

13.8.2 GLOBULE SIZE DETERMINATION

In most instances, the measurement of the globule size is carried out by using Malvern Zetasizer Nano Series ZS90. For this, the SEDDS containing drug is diluted to 100-folds with purified water. Afterwards, the diluted sample is stirred for 1–2 min to ensure uniformity in the dispersion.[70]

13.8.3 DISSOLUTION EFFICIENCY

The in vitro dissolution efficiency of the SNEDDS is assessed by using USP dissolution test apparatus II. Particularly, in case with the liquid SNEDDS, the formulation is firmly sealed in the dialysis bag. The phosphate buffer (pH 6.8) is generally used as the dissolution medium.[71] The requisite amounts of the samples are withdrawn at specific time intervals, and the concentration of the drug is measured by using either ultraviolet spectrophotometer or high-performance liquid chromatography.

13.8.4 PERMEABILITY STUDY

In vitro or ex vivo permeation studies of the SNEDDS are performed on an isolated or perfused organ system. The most commonly employed methods are (1) in situ single-pass perfusion technique (SPIP) and (2) everted sac technique. In the first method, a perfusion solution containing drug is passed through the proximal part of the jejunum and all experimental conditions are maintained likely with the in vivo conditions. In the second method, a small part of intestine is laced at one end and is everted using glass rod or thread.[45] This method is commonly used to study the transport behavior and kinetic parameter of the drug.

13.8.5 BIOAVAILABILITY STUDY

In vivo bioavailability studies of the SNEDDS are mostly performed in Wistar rats or rabbits. The SEDDS containing drug is administered through oral gavage. The dose of the drug is adjusted based on the animal weight. After oral administration, approximately 0.5 mL of blood is collected at different time intervals and analyzed for drug concentration. The pharmacokinetic study is carried out using linear trapezoidal method and various pharmacokinetic parameters are estimated.

13.9 CONCLUSION

SNEDDS is basically mixture of oil, surfactant, and cosurfactant. The recent studies have shown its ability to improve the solubility and bioavailability of the PWSDs. The limitations of the liquid SNEDDS formulation can be easily overcome by just transferring it into a solid form. There are numerous promising techniques for the transformation of the liquid SNEDDS into a solid form. These techniques are now getting more concern to find their utility for commercial production.

There are some measures that should be accounted, particularly in the manufacturing and development of the SNEDDS. The CQAs should be carefully assigned to prevent batch failures or unaccepted situation. All the critical attributes should be properly organized and controlled in order to build the quality of the formulation. A sound knowledge of the system may be needed to control and surpass the unfortunate proceedings.

Lastly, it is essential that SNEDDS with suitable solidification should be exploited for the development of suitable pharmaceutical dosage form for the patient. Also, this technology paves the way for PWSDs to move to remarkably convenient pharmaceutical dosage form.

KEYWORDS

- **SNEDDS**
- **delivery system**
- **emulsion**
- **formulation**
- **solubility**
- **evaluation**

REFERENCES

1. Gursoy, R. N.; Benita, S. Self-emulsifying Drug Delivery Systems (SEDDS) for Improved Oral Delivery of Lipophilic Drugs. *Biomed. Pharmacother.* **2004,** *58*, 173–182.
2. Tang, B.; Cheng, G.; Gu, J. C.; Xu, C. H. Development of Solid Self-emulsifying Drug Delivery Systems: Preparation Techniques and Dosage Forms. *Drug Discov. Today* **2008,** *13* (13–14), 606–612.
3. Pouton, C. W. Formulation of Poorly Water Soluble Drugs for Oral Administration: Physiochemical and Physiological Issues and the Lipid Formulation Classification System. *Eur. J. Pharm. Sci.* **2006,** *29* (3–4), 278–287.
4. Torchilin, V. P. Structuer and Design of Polymeric Surfactant-based Drug Delivery Systems. *J. Control Release* **2001,** *73*, 137–172.
5. Odeberg, J. M.; Kaufmann, P.; Kroon, K. G.; Hoglund, P. Lipid Drug Delivery and Rational Formulation Design for Lipophilic Drugs with Low Oral Bioavailability, Applied to Cycloserine. *Eur. J. Pharm. Sci.* **2003,** *20*, 375–382.
6. Sekiguchi, K.; Obi, N. Studies on Absorption of Eutectic Mixture. I A Comparison of the Behavior of Eutectic Mixture of Sulfathiazole and That of Ordinary Sulfathiazole in Man. *Chem. Pharm. Bull.* **1961,** *9*, 866–872.
7. Aungst, B. J. Novel Formulation Strategies for Improving Oral Bioavilability of Drugs with Poor Membrane Permeation or Presystemic Metabolism. *J. Pharm. Sci.* **1993,** *82*, 979–987.
8. Ammar, H. O.; Salama, H. A.; Ghorab, M.; Mahmoud, A. A. Implication of Inclusion Complexation of Glimperide in Cyclodextrin–Polymer Systems on Its Dissolution, Stability and Therapeutic Efficacy. *Int. J. Pharm. Sci.* **2006,** *320*, 53–57.

9. Robinson, J. R. Introduction: Semi-solid Formulations for Oral Drug Delivery. *Bull. Tech. Gattefosse* **1996,** *89,* 11–13.
10. Serajuddin Abu, T. M. Solid Dispersion of Poorly Water-soluble Drugs: Early Promises, Subsequent Problems, and Recent Breakthrough. *J. Pharm. Sci.* **1999,** *88* (10), 1058–1066.
11. Jeevana, J. B.; Sreelakshami, K. Design and Evaluation of Self-nanoemulsifying Drug Delivery System of Flutamide. *J. Young Pharm.* **2011,** *3,* 4–8.
12. Pouton, C. W. Lipid Formulations for Oral Administration of Drugs: Nanoemulsifying, Self-emulsifying and Self-microemulsifying Drug Delivery Systems. *Eur. J. Pharm. Sci.* **2000,** *11,* S93–S98.
13. Prajapati, B. G.; Patel, M. M. Conventional and Alternative Pharmaceutical Methods to Improve Oral Bioavailability of Lipophilic Drugs. *Asian J. Pharm.* **2007,** *1,* 1–8.
14. Fatouros, D. G.; Karpf, D. M.; Nielsen, F. S.; Mullertz, A. Clinical Studies with Oral Lipid Based Formulations of Poorly Soluble Compounds. *Ther. Clin. Risk. Manag.* **2007,** *3,* 591–604.
15. Park, M. J.; Ren, S.; Lee, B. J. In Vitro and In Vivo Comparative Study of Itraconazole Bioavailability When Formulated in Highly Soluble Self-emulsifying Systems and in Solid Dispersions. *Biopharm. Drug Dispos.* **2007,** *28,* 199–07.
16. Amidon, G. L.; Lennernas, H.; Shah, V. P.; Crison, J. R. A Theoretical Basis for a Biopharmaceutic Drug Classification: The Correlation of In Vitro Drug Product Dissolution and In Vivo Bioavailability. *Pharm. Res.* **1995,** *12* (3), 413–419.
17. Bansal, A. K.; Pawar, Y. B.; Sarpal, K. Self-emulsifying Drug Delivery System: A Strategy to Improve the Oral Bioavailability. *Curr. Res. Inform. Pharm. Sci. (CRIPS)* **2010,** *11* (3), 43–49.
18. Ghosh, P. K.; Murthy, R. S. Microemulsion: A Potential Drug Delivery System. *Curr. Drug Deliv.* **2006,** *3,* 167–180.
19. Kommuru, T. R.; Gurley, B.; Khan, M. K.; Reddy, J. K. Self-emulsifying Drug Delivery System (SMEDDS) of Coenzyme Q_{10}: Formulation, Development and Bioavailability Assessment. *Int. J. Pharm.* **2001,** *212,* 233–246.
20. Kohli, K.; Chopra, S.; Dhar, D.; Arora, S.; Khar, R. K. Self-emulsifying Drug Delivery System: An Approach to Enhance Oral Bioavailability. *Drug Discov. Today* **2010,** *15,* 958–965.
21. Charman, W. N. Lipid, Lipophilic Drugs and Oral Drug Delivery—Some Emerging Concepts. *J. Pharm. Sci.* **2000,** *89* (8), 967–978.
22. Patel, A. R.; Vavia, P. R. Preparation and In Vivo Evaluation of SMEEDS (Self-microemulsifying Drug Delivery System) Containing Fenofibrate. *AAPS PharmSciTech* **2007,** *9* (3), E344–E352.
23. Dixit, A. R, Rajput, S. J, Patel, S. G. Preparation and Bioavailability Assessment of SMEEDS Containing Valsartan. *AAPS PharmSciTech* **2010,** *11* (1), 314–321.
24. Nekkanti, V.; Karatgi, P.; Prabhu, R.; Pillai, R. 2010. Solid Self-microemulsifying Formulation for Candesartan Cilexetil. *AAPS PharmSciTech* **2010,** *11* (1), 9–17.
25. Prajapati, S. T.; Joshi, H. A.; Patel, C. N. Preparation and Characterization of Self-microemulsifying Drug Delivery System of Olmesartan Medoxomil for Bioavailability Improvement. *J. Pharm.* **2013,** *1,* 1–9.
26. Humberstone, A. J.; Charman, W. N. Lipid-based Vehicles for the Oral Delivery of Poorly Water Soluble Drugs. *Adv. Drug Deliv. Rev.* **1997,** *25* (1), 103–128.

27. Porter, C. J.; Charman, W. N. Intestinal Lymphatic Drug Transport: An Update. *Adv. Drug Deliv. Rev.* **2001,** *50* (1–2), 61–80.
28. Date, A. A.; Deasai, N.; Dixit, R.; Nagarsenkar, M. Self-nanoemulsifying Drug Delivery System: Formulation Insights, Applications and Advances. *Nanomedicine 5* (10), 1595–1616.
29. Chakraborty, S.; Shukla, S.; Mishra, B.; Singh, S. Lipid an Emerging Platform for Oral Delivery of Drugs with Poor Bioavilability. *Eur. J. Pharm. Biopharm.* **2009,** *73* (1), 1–15.
30. Gershanik, T.; Benita, S. Self-dispersing Lipid Formulations for Improving Oral Absorption of Lipophilic Drugs. *Eur. J. Pharm. Biopharm.* **2000,** *50* (1), 179–188.
31. Katteboina, S.; Chandrasekhar, P.; Balaji, S. Approaches for the Development of Solid Self-emulsifying Drug Delivery Systems and Dosage Forms. *Asian J. Pharm. Sci.* **2009,** *4*, 240–253.
32. Jessy, S.; Diagambar, J. Newer Approaches to Self Emulsifying Drug Delivery System. *Int. J. Pharm. Pharm. Sci.* **2010,** *2*, 37–42.
33. Barakat, N. S. Enhanced Oral Bioavailability of Etodolac by SEDDS: In Vitro and In Vivo Evaluation. *J. Pharm. Pharmacol.* **2010,** *62* (2), 173–180.
34. Patel, P. A.; Chaulang, G. M.; Akolkotkar, A. Self Emulsifying Drug Delivery System: A Review. *Res. J. Pharm. Tech.* **2008,** *1*, 313–323.
35. Perlman, M. E.; Murdande, S. B.; Gumkowski, M. J. Development of a Self Emulsifying Formulation That Reduces the Food Effects for Torcetrapib. *Int. J. Pharm.* **2008,** *351*, 15–22.
36. Sapra, K.; Sapra, A.; Singh, S. K.; Kakkar, S. Self Emulsifying Drug Delivery System: A Tool in Solubility Enhancement of Poorly Soluble Drugs. *Indo. Global J. Pharm. Sci.* **2012,** *2* (3), 313–332.
37. Tang, J. L.; Sun, Z. G He. Self-emulsifying Drug Delivery System: Strategy for Improving Oral Delivery of Poorly Soluble Drugs. *Curr. Drug Ther.* **2007,** *2* (1), 85–93.
38. Bandari, S.; Kallakunta, V. R.; Jukanti, R.; Veerareddy, P. R. Oral Self Emulsifying Powder of Lercanidipine Hydrochloride: Formulation and Evaluation. *Powder Technol.* **2012,** *221*, 375–82.
39. Narang, A. S, Delmarre, D.; Gao, D. Stable Drug Encapsulation in Micelles and Microemulsions. *Int. J. Pharm.* **2007,** *345*, 9–25.
40. Constantinides, P. P.; Scalart, J. P.; Lancaster, C. et al. Formulation and Intestinal Absorption Enhancement Evaluation of Oil-in-water Micro-emulsions Incorporating Medium Chain-glycerides. *Pharm. Res.* **1994,** *11* (10), 1385–1390.
41. Khoo, S. M.; Humberstone, A. J.; Porter, C. J.; Edwards, G. A, Charman, W. N. Formulation Design and Bioavailability Assessment of Lipidic Self-emulsifying Formulation of Halofantrine. *Int. J. Pharm.* **1998,** *167* (1), 155–164.
42. Gershanik, T.; Benita, S. Self-dispersing Lipid Formulations for Improving Oral Absorption of Lipophilic Drugs. *Eur. J. Pharm. Biopharm.* **2000,** *50* (1), 179–188.
43. Constantinides, P. P. Lipid Microemulsion for Improving Drugs Dissolution and Oral Absorption: Physical and Biopharmaceutical Aspects. *Pharm. Res.* **1995,** *12* (11), 1561–1572.
44. Pouton, C. W. Formulation of Self-emulsifying Drug Delivery Systems. *Adv. Drug Deliv. Rev.* **1997,** *25* (1), 47–58.
45. Singh, B.; Bandopadhyay, S.; Kapil, R.; Singh, R.; Katare, O. P. Self-emulsifying Drug Delivery Systems (SEDDS): Formulation Development, Characterization, and Applications. *Crit. Rev. Ther. Drug Carr. Syst.* **2009,** *26* (5), 427–521.

46. Chambin, O.; Jannin, V. Interest of Multifunctional Lipid Excipients: Case of Gelucire 44/14. *Drug Dev. Ind. Pharm.* **2005,** *31* (6), 527–534.
47. Hauss, D. J.; Fogal, S. E.; Ficorilli, J. V.; Price, C. A.; Roy, T.; Jayaraj, A. A.; Keirns, J. J. Lipid-based Delivery Systems for Improving the Bioavailability and Lymphatic Transport of a Poorly Water-soluble LTB4 Inhibitor. *J. Pharm. Sci.* **1998,** *87* (2), 164–169.
48. Chistyakov, B. E.; Fainerman, V. B.; Möbius, D.; Miller, R. Theory and Practical Application Aspects of Surfactants (Chapter 6). In *Studies in Interface Science;* 2001; Vol. 13, pp 511–618.
49. Ofokansi, K. C.; Chukwu, K. I.; Ugwuanyi, S. I. The Use of Liquid Self-microemulsifying Drug Delivery Systems Based on Peanut Oil/Tween 80 in the Delivery of Griseofulvin. *Drug Dev. Ind. Pharm.* **2009,** *35* (2), 185–191.
50. Fernandez-Tarrio, M.; Yanez, F.; Immesoete, K.; Alvarez-Lorenzo, C.; Concheiro, A. Pluronic and Tetronic Copolymers with Polyglycolyzed Oils as Self-emulsifying Drug Delivery Systems. *AAPS PharmSciTech* **2008,** *9* (2), 471–479.
51. Nielsen, F. S.; Petersen, K. B.; Müllertz, A. Bioavailability of Probucol from Lipid and Surfactant Based Formulations in Minipigs: Influence of Droplet Size and Dietary State. *Eur. J. Pharm. Biopharm.* **2008,** *69* (2), 553–562.
52. Yap, S. P.; Yuen, K. H. Influence of Lipolysis and Droplet Size on Tocotrienol Absorption from Self-emulsifying Formulations. *Int. J. Pharm.* **2004,** *281* (1–2), 67–78.
53. Carvajal, M. T.; Patel, C. I, Infeld, M. H.; Malick, A. W. Selfemulsifying Drug Delivery Systems (SEDDS) with Polyglycolized Glycerides for Improving In Vitro Dissolution and Oral Absorption of Lipophilic Drugs. *Int. J. Pharm.* **1994,** *106*, 15–23.
54. Kale, A. A.; Patravale, V. B. Design and Evaluation of Self-emulsifying Drug Delivery Systems (SEDDS) of Nimodipine. *AAPS PharmSciTech* **2008,** *9* (1), 191–196.
55. Borhade, V. B.; Nair, H. A.; Hegde, D. D. Development and Characterization of Self-microemulsifying Drug Delivery System of Tacrolimus for Intravenous Administration. *Drug Dev. Ind. Pharm.* **2009,** *35* (5), 619–630.
56. Ito, Y.; Kusawake, T.; Ishida, M.; Tawa, R.; Shibata, N.; Takada, K. Oral Solid Gentamicin Preparation Using Emulsifier and Adsorbent. *J. Control. Release* **2005,** *105*, 23–31.
57. Fabio, C.; Elisabetta, C.; Remedia S. R. L. Pharmaceutical Composition Comprising a Water/Oil/Water Double Microemulsion Incorporated in a Solid Support. WO2003/013421, 2002 (Assignee).
58. Boltri, L.; Coceani, N.; De Curto, D.; Dobetti, L.; Esposito, P. Enhancement and Modification of Etoposide Release from Crospovidone Particles Loaded with oil–Surfactant Blends. *Pharm. Dev. Technol.* **1997,** *2*, 373–81.
59. Seo, A.; Holm, P.; Kristensen, H. G.; Schaefer, T. The Preparation of Agglomerates Containing Solid Dispersions of Diazepam by Melt Agglomeration in a High Shear Mixer. *Int. J. Pharm.* **2003,** *259*, 161–171.
60. Gupta, M. K.; Goldman, D.; Bogner, R. H.; Tseng, Y. C. Enhanced Drug Dissolution and Bulk Properties of Solid Dispersions Granulated with a Surface Adsorbent. *Pharm. Dev. Technol.* **2001,** *6*, 563–72.
61. Gupta, M. K. et al. Hydrogen Bonding with Adsorbent During Storage Governs Drug Dissolution from Solid-dispersion Granules. *Pharm. Res.* **2002,** *19*, 1663–72.
62. Newton, M.; Petersson, J.; Podczeck, F.; Clarke, A.; Booth, S. The Influence of Formulation Variables on the Properties of Pellets Containing a Self-emulsifying Mixture. *J. Pharm. Sci.* **2001,** *90*, 987–95.

63. Newton, J. M.; et al. Formulation Variables on Pellets Containing Selfemulsifying Systems. *Pharm. Tech. Eur.* **2005,** *17*, 29–33.
64. Tuleu, C.; Newton, M.; Rose, J.; Euler, D.; Saklatvala, R.; Clarke, A.; Booth, S. Comparative Bioavailability Study in dogs of a Self-emulsifying Formulation of Progesterone Presented in a Pellet and Liquid Form Compared with an Aqueous Suspension of Progesterone. *J. Pharm. Sci.* **2004,** *93*, 1495–502.
65. Beg, S.; Sandhu, P. S.; Batra, R. S.; Khurana, R. K.; Singh, B. QbD-based Systematic Development of Novel Optimized Solid Self-nanoemulsifying Drug Delivery Systems (SNEDDS) of Lovastatin with Enhanced Biopharmaceutical Performance. *Drug Deliv.* **2015,** *22* (6), 765–784.
66. Dangre, P.; Gilhotra, R.; Dhole, S. Formulation and Statistical Optimization of Self Micro-emulsifying Drug Delivery System of Eprosartan Mesylate for Improvement of Oral Bioavailability. *Drug Deliv. Transl. Res.* **2016,** *6* (5), 610–621.
67. Reiss, H. Entropy-induced Dispersion of Bulk Liquids. *J. Colloids Interface Sci.* **1975,** *53*, 61–70.
68. Ramasahayam, B.; Eedara, B. B.; Kandadi, P.; Jukanti, R.; Bandari, S. Development of Isradipine Loaded Self-nano Emulsifying Powders for Improved Oral Delivery: In Vitro and In Vivo Evaluation. *Drug Dev. Ind. Pharm.* **2015,** *41* (5), 753–763.
69. Bachynsky, M. O.; Shah, N. H.; Patel, C. I.; Malick, A. W. Factor Affecting the Efficiency of Self-emulsifying Oral Delivery System. *Drug Dev. Ind. Pharm.* **1997,** *23*, 809–816.
70. Singh B, Bandyopadhyay S, Katare OP. Optimized Self Nano-emulsifying Systems of Ezetimibe with Enhanced Bioavailability Potential Using Long Chain and Medium Chain Triglycerides. *Colloids Surf. B Biointerface* **2012,** *100*, 50–61.
71. Dangre, P. V.; Gilhotra, R. M.; Dhole, S. N. Formulation and Development of Solid Self Micro-emulsifying Drug Delivery System (S-SMEDDS) Containing Chlorthalidone for Improvement of Dissolution. *J. Pharm. Invest.* **2016,** *46* (7), 633–644.

CHAPTER 14

THE INVESTIGATION OF COPPER/ CARBON NANOCOMPOSITE AQUEOUS SOLS FOR APPLICATION AT THE CULTIVATION OF LILIES

A. A. LAPIN[1], V. M. MERZLJAKOVA[2], and VLADIMIR I. KODOLOV[3*]

[1]*Kazan State Energy University, Kazan, Russia*

[2]*Izhevsk State Agricultural Academy, Izhevsk, Russia*

[3]*Kalashnikov Izhevsk State Technical University, Izhevsk, Russia*

[*]*Corresponding author. E-mail: Vkodol@istu.ru*

ABSTRACT

To effectively grow flower crops in protected soil, it is necessary to introduce innovative technologies that ensure high yields, and also allow reduction in material costs and increase in profitability. In modern industrial floriculture, methods of nanotechnology (NT) make it possible to influence the process of growing flower crops, their productivity, and the quality of flowers. Scientific researches in the field of the use of NT in protected soil during the cultivation of flower crops are rare. In this connection, the aim of the work was to study the effect of metal/carbon nanocomposite (NC) based on copper on lily cultivation under protected soil conditions. For 2 years, the reaction of the lily to the treatment of copper/carbon NC was studied. The biometric indices of the lily were determined as the height of the flowering shoot in the budding phase, the number of buds, the diameter of the open flower, and the height of the stem when cutting plants. When treated with nanocomposites, there was a significant increase in the height of the flowering shoot of lilies from 4.9 to 8.4 cm. The highest plants were Santander lily plants that were treated with 0.01% NC. The productivity of the lily (cutting lilies) is determined by the number of buds in the inflorescence, the diameter of the flower, and the height of the plant. When cutting lilies, the number of buds on the plant remains the same as in the budding phase. There were also a greater number of buds in the Siberian lily variety than in Santander. Depending on the treatment with different concentrations, the maximum number of buds was noted at 0.01% concentration of copper/carbon NC. The diameter of the open flower in the Santander variety was significantly larger than that of the Siberian variety (21.3 cm). Treatment of bulbs with NC contributed to a significant increase in the diameter of the open flower. The largest diameter of the flower was noted when treated with 0.01% NC (22 cm). The height of the stem (at cutting lilies) was increased at treatment by copper/carbon NC from 131.0 to 141.0 cm in comparison with the control (99.0–102.3 cm).

Thus, the results of scientific and industrial experiments have shown that the treatment of lily bulbs with copper/carbon nanocomposites contributes to an increase in height during cutting of plants, the number of buds, and the diameter of the open flowers.

14.1 INTRODUCTION

To effectively grow flower crops in protected soil, it is necessary to introduce innovative technologies that ensure high yields, reduction in material costs, and increase in profitability. In modern industrial floriculture, methods of nanotechnology (NT) make it possible to influence the process of growing flower crops, their productivity, and the quality of flowers. A promising direction is the development and application of nanoelements (NE) with optimal particle sizes for maximum assimilation of macro- and microelements for improved plant growth. Changes in the main characteristics of substances and elements are due not only to small dimensions but also to the manifestation of quantum mechanical effects with the dominant role of interfaces. These effects occur at a critical size that is commensurable with the so called correlation radius of a physical phenomenon (e.g., with the mean free path of electrons, the size of the magnetic domain, etc.). An important feature of metal NE, which plays a key role in their use in agriculture, is their low toxicity. Technical specifications (TU U 24.6—35291116—001-2007) have been applied to the large group of NE based on Ag, Cu, Co, Mn, Mg, Zn, Mo, and Fe metals whose production was established by domestic producers.[1–7]

Studies in this direction are carried out on agricultural crops when cultivating in conditions of open ground. Thus, high efficiency as a means for presowing treatment of seeds of wheat, sunflower, rapeseed, maize, amaranth, and other crops has been revealed with respect to solutions of carbon nanotubes, ultradispersed suspensions, metal nanoparticles, and nonmetals. At the same time, the increase in yields, adaptation to unfavorable climatic conditions, and improvement of the quality of raw plant material was observed. The effect of NE on biosystems was of a prolonged nature. The stimulating effect of nanostructured suspensions, when soaking the bulbs is due to plants assimilating nutrient elements through the bulb, leaves, and stems. Possible mechanism of influence of copper/carbon nanocomposite on the lily is interaction at the cellular level. NEs are able to transmit their excess energy to molecules, increasing the speed of biochemical reactions taking place in plants. They actively participate in the processes of microelement balance, that is, they are bioactive.

14.2 EXPERIMENTAL

Scientific research was carried out in conditions of protected soil in industrial greenhouses. The experiment was laid in threefold replication, placing the variants by the method of randomized repetitions. The technology of forcing lilies in protected ground in lily containers is generally accepted. Copper-containing nanocomposite (NC) was used in the form of different concentrations. Experiment is two factored. Factor A—hybrids of lilies; factor B—concentration of nanocomposite. The subjects of the research were hybrids of the lily of the selection and seed-growing company "Van den Bose Flower bulbs" Siberia and Santander (The Netherlands). As a control, variants were used without bulb treatment and using water.

14.3 RESULTS AND DISCUSSION

The establishment received the bulbs of lilies in peat. Samples of this peat were selected and agrochemical analysis was performed using volumetric analysis (Table 14.1).

Agrochemical parameters of this peat contribute to the long storage and transportation of bulbs of long lilies and do not allow them to germinate.

TABLE 14.1 Agrochemical Characteristics of Peat During Storage of Lily Bulbs.

Index	Result	The value of the index of sphagnum peat
Moisture of peat, %	60.0	59.46
pH (suspension)	4.0	5.5–6.1
Specific conductivity of the suspension, mS/cm, 25°C	0.093	–
Nitrate nitrogen $N–NO_3$	*0.57*[a]	*26.3*
	0.1	29.0
Ammonia nitrogen $N–NH_4$	*7.36*	*135.6*
	1.5	149.0
P_2O_5	*1.64*	*291.0*
	0.3	320.0
K_2O	*34.56*	*353.0*
	7.0	388.0
$Ca^{2+} + Mg^{2+}$, mmol/g	*2.98*	–
	0.6	

[a]**Note**: numerator (for dry matter)—mg/100 g, denominator—mg/dm^3.

The lilies were cut out in peat mixture. A mixture of peat should have a good granulometric composition of the soil and have good water permeability at the entire depth of root growth (especially the top layer) during the entire vegetation period. Samples of peat mixtures were selected and agrochemical analysis was carried out (Table 14.2).

For growth of plants on peat, the optimum moisture content is within 78–85% of the mass. In this case, 35–50% of the pore volume is occupied by air; this is important for the normal functioning of the root system. But even with abundant watering, peat moss contains up to 20% air in the pores. The moisture content in the peat mix is normal.

The very important factor for the development of roots is the acidity of the soil. High acidity of soil leads to insufficient intake of phosphorus, magnesium, and iron. When growing Eastern hybrids of lilies, which include our varieties, the pH should be between 5.0 and 6.5.

Lilies are sensitive to soil salinity. Soil salinity is determined by the specific electrical conductivity. Lilies require low levels of electrical conductivity.

The content of mobile forms of nutrients is low; however, the level of plant nutrition is regulated by a computer program in accordance with the needs of lily bulbs. Due to the inherent peat strength of buffer and high sorption capacity, mineral fertilizers are not washed out and stored in an accessible form for plants; at the same time, the danger of creating a concentration of salts that is harmful to plants is lowered.

TABLE 14.2 Agrochemical Characteristics of the Peat Mixture at the Laying of the Experiment.

Index	Result
Moisture of peat, %	74.3
pH (suspension)	5.10
Specific conductivity of the suspension, mS/cm, 25°C	0.115
Nitrate nitrogen $N–NO_3$	*0.24*[a]
	0.2
Ammonia nitrogen $N–NH_4$	*6.01*[a]
	3.8
P_2O_5	*3.94*[a]
	2.5
K_2O	*32.65*[a]
	21.0
$Ca^{2+} + Mg^{2+}$, mmol/g	*3.82*[a]
	2.4

[a]**Note**: numerator (for dry matter)—mg/100 g, denominator—mg/dm³.

High productivity and quality of plants depend on the full provision of plants with microelements; therefore, when planting lily bulbs, an analysis was made of the content of micronutrients in the peat mix (Table 14.3).

TABLE 14.3 Chemical Analysis of Peat Mixture for the Content of Micronutrients.

Micronutrients (mg/kg dry weight)	The actual value of the indicators from the test results
Zinc	15.60
Copper	3.81
Iron	126.60
Manganese	57.30
Boron	8.27

By treating bulbs of lilies by metal/carbon copper-based NC, we contribute to replenishing the available forms of copper in lily plants. During the period of growth and development, the level of nutrition of lily plants is regulated by a computer program in accordance with the needs for macro- and microelements.

For 2 years, the reaction of the lily to the treatment by copper/carbon NC was studied. The biometric parameters of the lily were determined as the height of the flower bud in the budding phase (Table 14.4), the number of buds, the diameter of the open flower, and the height of the stem when cutting plants.

TABLE 14.4 Height of Flowering Shoot, cm.

Hybrids of lilies (factor A)	Concentration of nanocomposites (factor B)					Average by factor A HCP_{05} A = 1.7
	Without treatment (c)	Water (c)	0.01%	0.02%	0.05%	
Siberia (c)	79.6	71.9	87.7	87.9	87.2	82.8
Santander	77.9	81.6	89.1	82.2	82.3	82.6
Average by factor B HCP_{05} B = 2.7	79.2	76.9	88.5	85.2	84.6	–
HCP_{05} for individual differences	3.8					

The number of buds on the shoot was higher for the Siberian variety, than for the Santander variety by 1.1 pieces (for HCP = 0.06 pieces). During the processing of NC, a significant increase in the number of buds is observed (5.0–5.3 pieces) (Table 14.5).

TABLE 14.5 Number of Buds, Pieces.

Hybrids of lilies (factor A)	Concentration of nanocomposites (factor B)					Average by factor A HCP_{05} A = 0.06
	Without treatment (c)	Water (c)	0.01%	0.02%	0.05%	
Siberia (c)	4.7	4.7	5.8	5.6	5.4	5.2
Santander	3.5	3.5	4.7	4.5	4.6	4.1
Average by factor B HCP_{05} B = 0.1	4.1	4.1	5.3	5.1	5.0	–
HCP_{05} for individual differences				0.1		

The length of the bud in Siberian variety averaged 10.5 cm, while the Santander variety was 1.2 cm lower. While processing with nanocomposites, there was a significant increase in the length of the bud. The longest lengths were observed in lilies treated with a 0.01% copper/carbon nanocomposite suspension (Table 14.6).

The length of the bud and the number of buds depend on the photosynthetic activity, which is directly related to the width of the leaves; so the width of the lily leaf was measured during the budding phase (Table 14.7).

TABLE 14.6 Length of Bud, cm.

Hybrids of lilies (factor A)	Concentration of nanocomposites (factor B)					Average by factor A HCP_{05} A = 0.06
	Without treatment (c)	Water (c)	0.01%	0.02%	0.05%	
Siberia (c)	10.1	10.3	10.9	10.6	10.6	10.5
Santander	9.0	9.1	9.6	9.5	9.3	9.3
Average by factor B HCP_{05} B = 0.09	9.5	9.7	10.3	10.0	9.9	–
HCP_{05} for individual differences	0.1					

TABLE 14.7 Leaf Width, cm.

Hybrids of lilies (factor A)	Concentration of nanocomposites (factor B)					Average by factor A HCP_{05} A = 0.05
	Without treatment (c)	Water (c)	0.01%	0.02%	0.05%	
Siberia (c)	3.0	3.1	3.4	3.3	3.1	3.2
Santander	3.4	3.5	3.7	3.6	3.6	3.6
Average by factor B HCP_{05} B = 0.08	3.2	3.3	3.6	3.5	3.4	–
HCP_{05} for individual differences				0.1		

In the Santander variety, the leaves were wider than those of Siberia, which is one of the varietal features. The increase in the number and length of buds in both varieties when processing bulbs with copper/carbon nanocomposite was associated with an increase in leaf width.

The productivity of the lily (cutting lilies) is determined by the number of buds in the inflorescence, the diameter of the flower, and the height of the plant. In this regard, the number of buds on lily plants was counted (Table 14.8).

TABLE 14.8 Number of Buds, Pieces.

Hybrids of lilies (factor A)	Concentration of nanocomposites (factor B)					Average by factor A HCP_{05} A = 0.04
	Without treatment (c)	Water (c)	0.01%	0.02%	0.05%	
Siberia (c)	4.8	4.7	6.0	5.6	5.5	5.3
Santander	3.6	3.5	4.7	4.5	4.6	4.2
Average by factor B HCP_{05} B = 0.06	4.2	4.1	5.4	5.0	5.0	–
HCP_{05} for individual differences				0.09		

A: When cutting lilies, the number of buds on the plant remains the same as in the budding phase. There were also a greater number of buds in the Siberian lily variety than in Santander. B: Depending on the treatment with different concentrations, the maximum number of buds was noted at 0.01% concentration of copper/carbon NC.

The diameter of the open flower in the Santander variety (21.3 cm) was significantly larger than that of the Siberian variety (19.7 cm). Processing of bulbs with nanocomposites promoted a substantial increase in the diameter of the open flower. The largest diameter of the flower was noted when treated with a 0.01% nanocomposite (22.0 cm) (Table 14.9).

TABLE 14.9 Diameter of the Open Flower, cm.

Hybrids of lilies (factor A)	Concentration of nanocomposites (factor B)					Average by factor A HCP_{05} A = 0.03
	Without treatment (c)	Water (c)	0.01%	0.02%	0.05%	
Siberia (c)	18.7	18.3	21.3	20.0	20.0	19.7
Santander	20.0	20.0	22.7	22.0	21.7	21.3
Average by factor B HCP_{05} B = 0.5	19.3	19.2	22.0	21.0	20.9	–
HCP_{05} for individual differences				0.7		

The height of the lilies for cutting was increased by treatment with copper/carbon nanocomposite to 131.0–141.0 cm in comparison with the control (99.0–102.3 cm). The lily of Siberia was slightly taller than the Santander variety (Table 14.10).

TABLE 14.10 Height of Stem, cm.

Hybrids of lilies (factor A)	Concentration of nanocomposites (factor B)					Average by factor A HCP_{05} A = 0.96
	Without treatment (c)	Water (c)	0.01%	0.02%	0.05%	
Siberia (c)	102.3	120.3	138.7	131.0	141.0	126.5
Santander	99.0	116.3	139.0	134.0	135.7	124.8
Average by factor B HCP_{05} B = 1.5	100.7	118.3	138.8	132.5	137.8	–
HCP_{05} for individual differences				2.2		

Thus, processing of lily bulbs with copper/carbon nanocomposites promotes an increase in plant height, and the number and diameter of flowers.

Flowers—a product that does not carry any practical benefit in itself and its value is rather arbitrary, but nevertheless people are ready to go to great expenses for their purchase because any solemn event, congratulation, as well as simply expressing one's feelings and sympathies, cannot be imagined without flowers. Flowers are perishable goods. Not all flowers manage to go the way to their customers.

For a more objective evaluation of the results obtained, it is necessary to determine the economic efficiency. For comparison with the control variant (soaking the bulbs of lilies in water), a variant was chosen with treatment of lily bulbs with a 0.01% copper/carbon nanocomposite, under which the highest lily yield was obtained (Table 14.11).

TABLE 14.11 Economic Efficiency of Copper/Carbon Nanocomposite Application when Growing a Hybrid of Lily Siberia.

Index	Variant of experiment	
	Water (c)	0.01%
Cuts in total, pieces/m^2	45	45
Average number of buds on the stem, pieces	4.8	5.9
The average price of a cut plant, rub.	74.35	81.55
Cost of production, rub./m^2	3345.75	3669.75
Production costs, rub./m^2	2164.8	2283.4
Net income, rub./m^2	1180.95	1386.35
Level of profitability, %	54.5	60.7
Cost of production, rub./kg	48.11	50.74

The use of 0.01% copper/carbon NK in the cultivation of lilies affects the cost of production, only slightly increasing it, whereas the average selling price increased significantly (by 7 Rubles) per plant due to the increase in the number of buds, which led to an increase in the cost of sales and, accordingly, profit and profitability of production.

14.4 CONCLUSION

Treatment of lily bulbs with 0.01% copper/carbon nanocomposite allowed to obtain products of higher quality (the number of buds increased, the bud itself was larger than that of untreated plants, and the leaf was of higher quality in length, width, and color saturation).

KEYWORDS

- **metal/carbon nanocomposite**
- **copper**
- **suspension**
- **hybrids of lilies**
- **protected soil**
- **buds**
- **diameter of the branch height of the plant**

REFERENCES

1. Kaplunenko, V. G.; Kosinov, N. V.; Bovsunovsky, A. N.; et al. Nanotechnology in Agriculture. *Grain* **2008,** *4*, 47–55 (in Russian).
2. Lapin, A. A.; Zelenkov, V. N.; Ivanova, M. I.; Potapov, V. V. Antioxidant Properties of Samples Onions-Slizun (*Allium Nutans L.*), Treated Water Zols I Nanosilica. In *Non-traditional Natural Resources, Innovative Technologies and Products: Collection of Scientific Papers*, ed. 25; M.: Rans, 2017; p 110–115 (in Russian).
3. Lapin, A. A.; Zelenkov, V. N.; Merzlyakova, V. M. Influence of Infrared Radiation on Antioxidant Activity of Plant Raw Material and Structured Water Adsorbed Inside. Part 2. Features of Structured Water in Lily Samples. *Butlerov Commun.* **2016,** *47* (9), 73–78. ROI: jbc-02/16-47-9-73 (in Russian).

4. Potapov, V. V.; Sivashenko, V. A.; Zelenkov, V. N. The Use of Nanodispersed Silica in Agriculture. *Butlerov Commun.* **2015,** *43* (9), 40–48. ROI: jbc-02/15-43-9-40 (in Russian).
5. Sklyarevskaya, N. V.; Gladkaya, Y. A.; Tolstikov, S. V. The Content and Composition of Carbohydrates from Leaves of Some Species of the Genus *Filipendula. Butlerov Commun.* **2015,** *43* (9), 141–145. ROI: jbc-02/15-44-11-141 (in Russian).
6. Shakir, I. V.; Grosheva, V. D.; Karetkin, B. A.; Baurin, D. V.; Panfilov, V. I. Complex Processing of Renewable Plant Raw Materials for High Protein and Probiotic Fodder Products. *Butlerov Commun.* **2017,** *50* (5), 73–80. ROI: jbc-02/17-50-5-73 (in Russian).
7. Fedorenko, V. F.; et al. *Nanotechnologies and Nanomaterials in the Agro-industrial Complex.* (Under General Editing) Fedorenko, V. F.—M.: FGBSI "Rosinformmagrotech," 2011; p 312 (in Russian).

CHAPTER 15

METAL/CARBON NANOCOMPOSITES AND MODIFIED ANALOGOUS: POSSIBLE PARTICIPATION IN VITAL PROCESSES

VLADIMIR I. KODOLOV[1], V. V. TRINEEVA[2], A. A. LAPIN[3], and V. M. MERZLJAKOVA[4]

[1]*Kalashnikov Izhevsk State Technical University, Izhevsk, Russia*

[2]*Institute of Mechanics, Udmurt Scientific Centre, Ural Division, RAS, Izhevsk, Russia*

[3]*Kazan State Energy University, Kazan, Russia*

[4]*Izhevsk State Agricultural Academy, Izhevsk, Russia*

[*]*Corresponding author. E-mail: Kodol@istu.ru*

ABSTRACT

Untill now, the nanostructures and nanosystems are used with high effectiveness for the living tissues regeneration, and also for the organisms' protection from free active radicals. In this chapter, the metal/carbon nanocomposites and their analogous, containing phosphorus or silicon, are proposed as the active agents for recombination reactions as well as for reactions which promote mitosis. The activity of these substances stipulates by their content, which includes metal clusters and shell containing carbon fibers from carbine and polyene fragments. The delocalized electrons are formed on the joints of these fragments. In other words, the free stable radicals are created. The stabilization of nanogranules is ensured by the coordination of metals and shell fragments. The nanocomposites applications such as stimulators of needle, flower, and fruit plants' growth are given.

15.1 INTRODUCTION

The problems of tissues regeneration are usually linked[1] with the local growth of active free radicals' quantity within injure (sick) orgies. The effective method of organism defense from active free radicals is the carrying of recombination processes. In other words, it is necessary to realize the active free radicals' interactions with electron systems of antioxidants or organism which stimulate the free radicals reactions for the formation of stable substances without delocalized electrons.

The stable free radicals or the substances which contain double bonds are effective as the initiators of recombination processes. These substances interact with active radicals and "blow out" their activity. Above substances have negative charges. Therefore, these substances can react with positive ions which are present on membrane cells. Then, the start of mitosis mechanism is possible.

Metal/carbon nanocomposites consist from metal clusters formed within carbon shell from carbon fibers. Above fibers contain fragments from carbine and polyacetylene chains which have delocalized electrons. Consequently, the nanocomposites and their analogous can be active agents in recombination reactions and in the mitosis processes.

15.2 SIZE, MORPHOLOGY, AND CONTENT OF METAL/CARBON NANOCOMPOSITES AND THEIR MODIFIED ANALOGOUS

Metal/carbon nanocomposites and analogous, modified by silicon or phosphorus compounds, are obtained by the use of mechanochemical synthesis when the redox processes take place. Reduction of metals from their oxides as well as the reduction of phosphorus or silicon takes place from their corresponding oxidized forms. The results of mechanochemical process are estimated by means of radiography, X-ray photoelectron spectroscopy, transition electron microscopy, and electron paramagnetic resonance. The process of reduction of the above elements is accompanied by the increase in metal atomic magnetic moments.

The study of morphology and content for metal/carbon nanocomposites and modified analogous obtained within nanoreactors of polymeric matrices[2–5] was carried out. The average size of obtained nanoparticles is 25 nm.

The results of distribution on copper/carbon nanocomposite nanoparticles sizes in alcohol and water are shown in Figure 15.1.

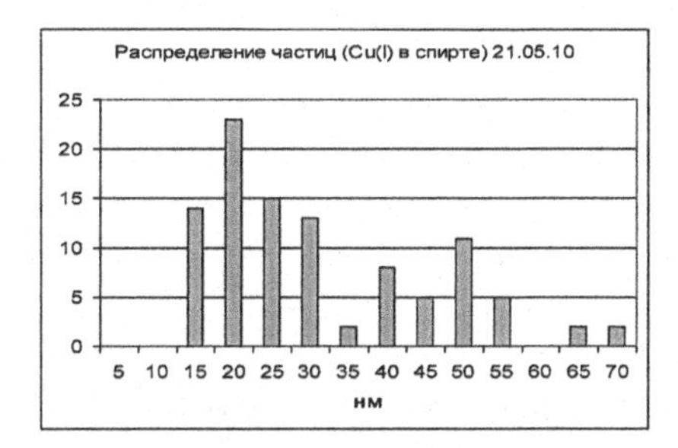

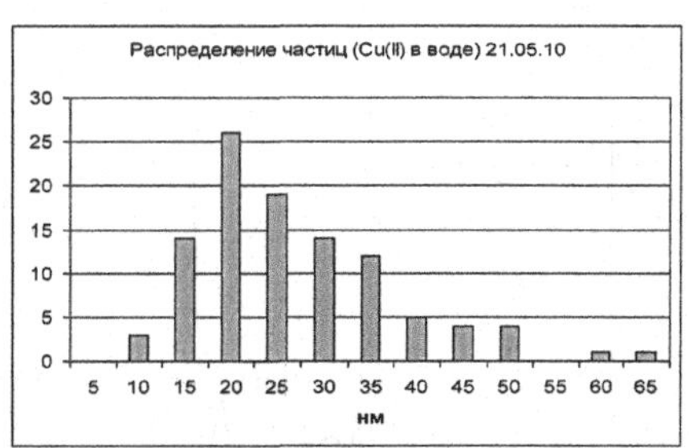

FIGURE 15.1 The distribution of copper/carbon nanocomposite nanoparticles in alcohol (a) and water (b).

The distribution character in these media is heterogeneous; in water (more polar medium), there are nanoparticles with size of 10 nm, and the quantity of nanoparticle with size range from 30 to 65 nm is less than 40%.

The energetic characteristics of copper/carbon nanocomposite are presented in Table 15.1.

TABLE 15.1 Energetic Characteristics of Copper/Carbon Nanocomposites.

Relation of copper and carbon content, %	50/50
Density of copper/carbon nanocomposite, g/cm^3	1.71
Summary mass of copper/carbon nanocomposite, au	36.75
Middle size of copper/carbon nanocomposite, d, nm	25
Specific surface of copper/carbon nanocomposite, m^2/g	160
Frequency of skeleton vibration of copper/carbon nanocomposite, s^{-1}	4×10^{11}
Middle vibration energy of copper/carbon nanocomposite, erg	1.6×10^{13}

It is necessary to note that the surface energy (vibration energy) of nanocomposites depends not only on the mass of nanoparticles but also on their specific surface. Therefore, the surface energy of iron/carbon nanocomposite is less than corresponding energy of copper/carbon nanocomposite.

According to diffractogram, the reduced forms of copper cluster prevail (Fig. 15.2).

Data of transition electron microscopy with high resolution testify that the carbon fiber consists of carbon atoms because of the correspondence of the fiber diameter and carbon atom diameter or C–H group diameter value (Fig. 15.3).

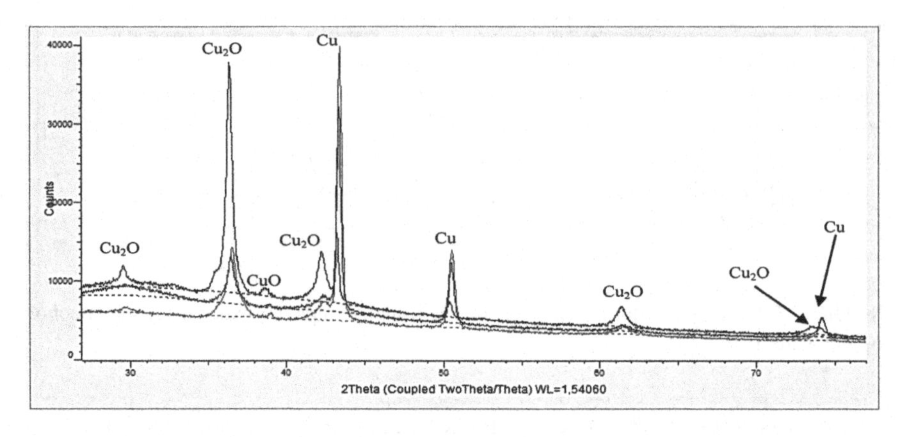

FIGURE 15.2 **(See color insert.)** Diffractogram of copper/carbon nanocomposites.

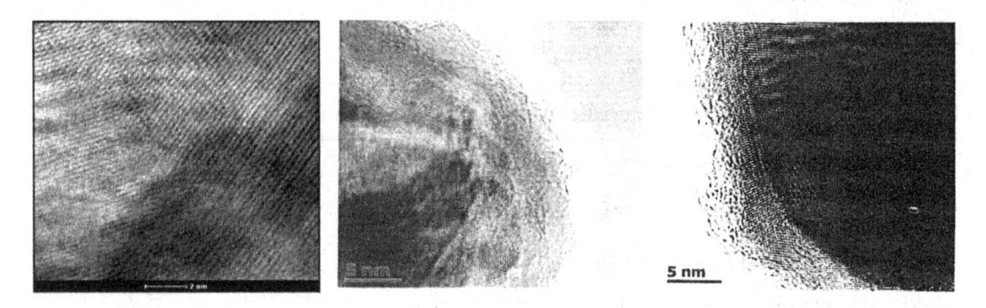

FIGURE 15.3 Microphotograph TEM of copper/carbon (a), nickel/carbon (b), and phosphorus-containing copper/carbon nanocomposites (c).

Thus, it is possible to suppose about the structure of carbon fiber from fragments of polyacetylene and carbine (Fig. 15.3a). Analogous structures are observed for nickel/carbon nanocomposite (Fig. 15.3b) and phosphorus-containing copper/carbon nanocomposite (Fig. 15.3c). The difference of two latest figures from Figure 15.3a concludes in the appearance of strands image on microphotographs Figure 15.3b and c. If the fragments alternate, then the delocalized electrons shall be on the joints. The increasing of joints number leads to the growth of delocalized electrons quantity, fiber flexibility, and the strand formation. This conclusion is confirmed by the investigations of nanocomposites with the use of X-ray photoelectron spectroscopy and electron paramagnetic resonance.

15.3 ELECTRON STRUCTURE AND PROPERTIES OF METAL/ CARBON NANOCOMPOSITES AND MODIFIED ANALOGOUS

The X-ray photoelectron spectroscopy investigations of copper/carbon nanocomposite and its analogous modified by Si, P, and Ni compounds are carried out. Results of these investigations are presented in the form of Cu3s, C1s, Si2p, and P2p spectra (Figs. 15.4–15.6). The model of metal atomic magnetic moments determination was elaborated on the basis of Van Flack theory. The comparative results on atomic magnetic moments for copper of Cu/C nanocomposite and its modified analogous are given. At the decipherment of C1s spectra, equally to components which ascribes C–H bond, there are components corresponding to sp, sp^2, and sp^3 hybridization of carbon.

The intensities relation for these components may be used for the determination of polyacetylene and carbine fragments content in fiber. At the same time, the intensities relation for components sp^2 and sp^3 gives the possibility of proposing the nanocomposite particle form. The form of nickel/ carbon nanocomposite particles is near to the cylinder. In this case, the equilibrium of sp^2 and sp^3 components intensities takes place. The form of copper/carbon nanocomposite particles is near to ellipsoid that corresponds to sp^2 component intensity growth in comparison with sp^3 component (Fig. 15.4).

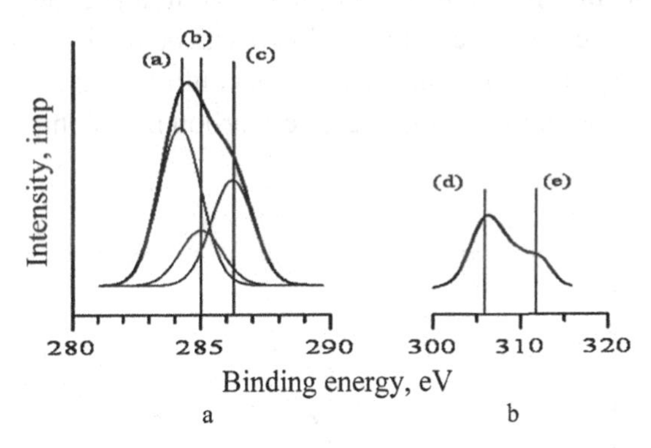

FIGURE 15.4 C1s copper/carbon nanocomposite X-ray photoelectron spectrum (a) and its satellites (b).

At the modification of metal/carbon nanocomposites by the compounds, which contain the elements with positive oxidation states, the redox processes arise because of electrons of carbon shell accompanied by elements' reduction. It can be seen from P2p and Si2p spectra of correspondent-modified copper/carbon nanocomposite (Fig. 15.5).

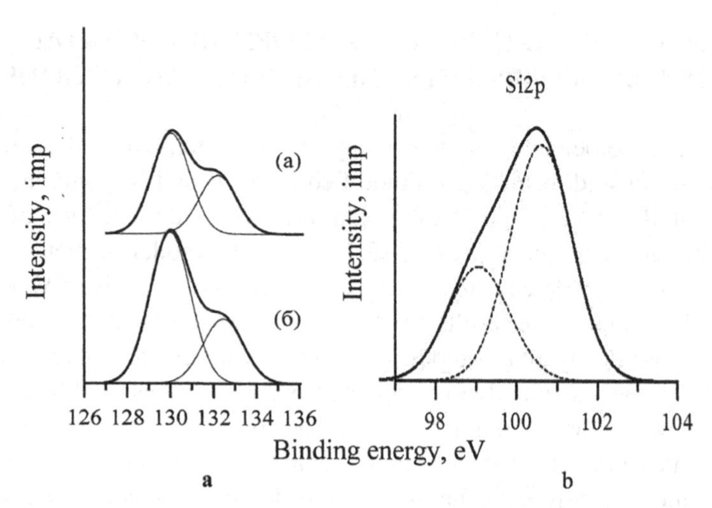

FIGURE 15.5 X-ray photoelectron spectra: P2p for phosphorus-containing (a) and Si2p for silicon-containing (b) Cu/C nanocomposites.

In Figure 15.5, the peaks corresponding to the reduced forms of phosphorus and silicon are present (for phosphorus, P^0 EP2p = 130 eV and for silicon, Si^0 ESi2p = 99 eV). It is necessary to note that the transformation degree, when the modification by phosphorus-containing substances occurs, is more, than in case, when the analogous process is carried out with silicon-containing participation. This fact is confirmed at the decipherment of phosphorus-containing modified copper/carbon nanocomposite diffractogram (Fig. 15.6).

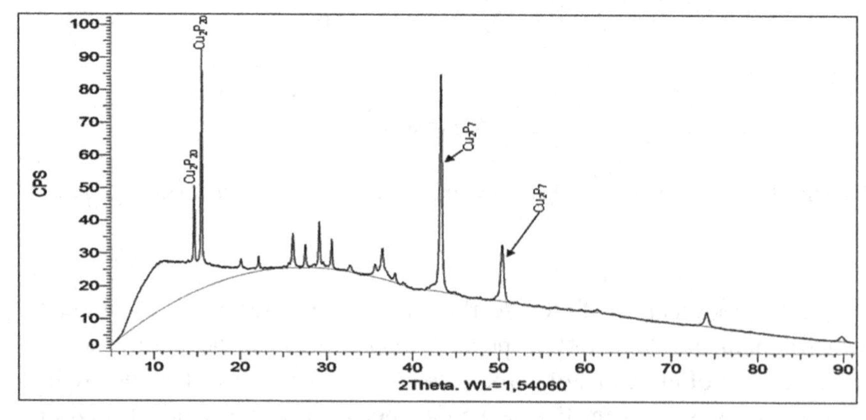

FIGURE 15.6 Diffractogram of phosphorus-containing Cu/C nanocomposites.

At the copper/carbon nanocomposite modification by means of nickel oxide, the redox process also takes place, only with distribution of participating electrons between copper and nickel (Fig. 15.7; Table 15.2).

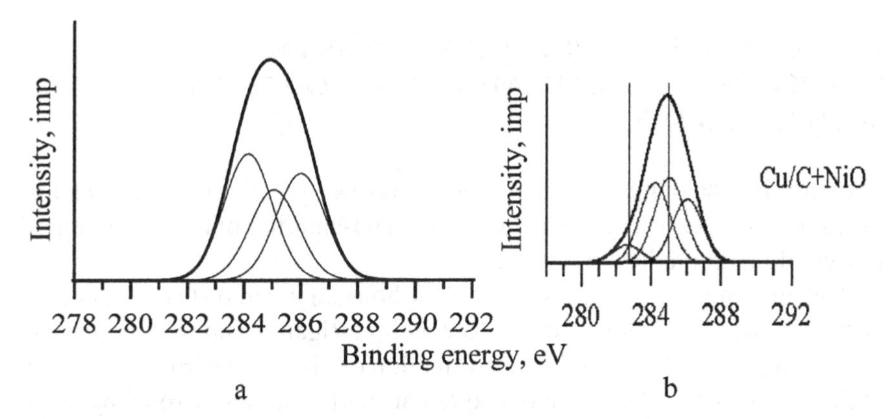

FIGURE 15.7 X-ray photoelectron C1s spectra of modified Cu/C nanocomposites: phosphorus containing (a) and nickel containing (b).

The comparison of copper atomic magnetic moments for different modified nanocomposites is presented in Table 15.2.

TABLE 15.2 Parameters of the Multiple Splitting of the 3s Spectra of the Cu/C NC Modified with Phosphorus-, Silicon- and Nickel-containing Compounds.

Sample	I_2/I_1	Δ, eV	μ_{met}, μ_B
$Cu3s_{нано}$	0.2	3.5	1.3
$Cu3s_{нано}$ (P)	0.4	3.5	2.0
$Cu3s_{нано}$ (P × 2)	0.4	3.5	2.0
$Cu3s_{нано}$ (P × 1/2)	0.85	3.5	4.2
$Cu3s_{нано}$ (Si)	0.6	3.0	3.0
$Cu3s_{нано}$ (Si1/2)	0.6	3.0	3.0
$Cu3s_{нано}$ (Ni)	0.4_{Cu} (0.4_{Ni})	3_{Cu} (2_{Ni})	2_{Cu} (2.3_{Ni})

I_2/I_1, the relation of the maxima intensities of the lines of the multiple splitting; Δ, the energy distance between the maxima of the multiple splitting in Me3s-spectra.

The great atomic magnetic moment for copper has arrived for phosphorus-containing Cu/C nanocomposite obtained at the relation of Cu/C nanocomposite and ammonium polyphosphate equaled to 2. The value of this achievement corresponds to 4.2 μ_B. Such results may be explained by the delocalized electron growth (in 10^2 spin/g) according to EPR investigations.

15.4 EXAMPLES AND PERSPECTIVES OF METAL/CARBON NANOCOMPOSITES APPLICATION AND MODIFIED ANALOGOUS IN VITAL PROCESSES

The participation of metal/carbon nanocomposites as active reduction agents in redox processes testify about the potential possibilities of these substances in recombination radical processes.

On our mind, the examples of metal/carbon nanocomposite participation in mitosis processes may be the following investigations on the stimulation of growth processes: (1) at the determination of pine seed germination,[6] (2) at the cultivation of lilies,[7] and also (3) at the formation of roots by slips of grapes.[8]

In the first case,[6] the application of copper/carbon nanocomposite (stabilized into sugar syrup) fine-dispersed suspension with the concentration of active substance equaled to 0.05% leads to the 70% increase in germination of pine seeds, the prophylaxis of soil organisms' negative influence, and the improvement of pine sowing indexes.

In the second case,[7] the results of scientific and industrial experiments have shown that the treatment of lily bulbs with copper/carbon nanocomposites contributes to an increase in height, when cutting plants; the number of buds; and the diameter of the open flowers. Besides, the effective concentration of active component (nanocomposite) is established as 0.01%, which leads to the improvement of following indexes:

Index	Number of buds of the open flower	Diameter	Height of stem	Net income	Level of profitability
% growth	25–30	14	36–40	17	11

The improvement in the leaf quality also takes place.

In Ref. [8] for the first time, the possibility of copper/carbon nanocomposite used as the stimulator of the roots formation by grapes slips is studied. When 0.01% of fine-dispersed suspension of copper/carbon nanocomposite is applied, the increase of roots quantity on one slip is observed as 2.5 times at the growth of shoot middle length of 17%.

15.5 CONCLUSIONS

The results of investigations of metal/carbon nanocomposites and their analogous as well as the examples of their application as the stimulators of cells growth and as the potential participants for free radicals reaction recombination testify about their perspective application in vital processes.

KEYWORDS

- **metal/carbon nanocomposite**
- **morphology**
- **analogous modified**
- **electron structure**
- **atomic magnetic moment**

REFERENCES

1. Shul'kin, A. V.; Kolesnikov, A. V.; Nikolaev, M. N.; Barenina, O. I. Role of Free Radicals in the Cornea Epithelium Regeneration. *Fundam. Investig.* **2013,** *7* (2), 451–455.
2. Vladimirov, Yu. A. Free Radicals in Biological Systems. *Soros Educ. J.* **2000,** *6* (12), 13–19.
3. Method of Carbon Metal-containing Nanostructures Production. Patent Holder Izhevsk EMP—KUPOL, Patent RF 2,393,110. 2010.
4. Kodolov, V. I.; Trineeva, V. V. New Scientific Trend—Chemical Mesoscopics. *Chem. Phys. Mesoscopy* **2017,** *19* (3), 454–465.
5. Shabanova, I. N.; Kodolov, V. I.; Terebova, N. S.; Trineeva, V. V. *X Ray Electron Spectroscopy for the Investigation of Metal/Carbon Nanosystems and Nanostructured Materials*; Udmurt University: M.—Izhevsk, 2012; p 252.
6. Patent RF 2,532,043, 2014, Bul. 30.
7. Lapin, A. A; Merzljakova, V. M.; Kodolov, V. I. Investigation of Copper/Carbon Nanocomposites Water Soles at the Lilies Cultivation. *Butlerovskie Commun.* **2017,** *52* (11), 131–137.
8. Fedorov, A. V.; Lekontseva, T. G.; Zoprin, D. A.; Hudjakova, A. V.; Trineeva, V. V. Effectiveness of Copper/Carbon Nanocomposite as Stimulator of the Shoots Formation by Cultural Grape Green Slips. *Mod. Sci. Investig. Cultiv.* **2017,** *7* (15), 345–347.

INDEX

M

N

P

S

T

U

V

Printed in the United States
by Baker & Taylor Publisher Services